Lecture Notes on Mathematical Modelling in the Life Sciences

D1628115

Series Editors:
Michael Mackey
Angela Stevens

More information about this series at http://www.springer.com/series/10049

Stephen P. Ellner • Dylan Z. Childs • Mark Rees

Data-driven Modelling of Structured Populations

A Practical Guide to the Integral Projection Model

 Springer

Stephen P. Ellner
Ecology and Evolutionary Biology
Cornell University, Corson Hall
Ithaca, NY, USA

Dylan Z. Childs
Animal and Plant Sciences
University of Sheffield
Yorkshire, UK

Mark Rees
Animal and Plant Sciences
University of Sheffield
Yorkshire, UK

ISSN 2193-4789   ISSN 2193-4797   (electronic)
Lecture Notes on Mathematical Modelling in the Life Sciences
ISBN 978-3-319-28891-8   ISBN 978-3-319-28893-2   (eBook)
DOI 10.1007/978-3-319-28893-2

Library of Congress Control Number: 2016933253

Mathematics Subject Classification (2010): 92D25; 92D40; 92C80; 92-02; 62P10; 97M60

Printed on acid-free paper

This Springer imprint is published by Springer Nature
The registered company is Springer International Publishing AG Switzerland

Preface

Over the past 30 years, one of the most rapidly expanding areas in ecology has been the development of structured population models, where individuals have attributes such as size, age, spatial location, sex, or disease status that affect their interactions and fate. These models can be roughly divided into conceptual models where the effects of biological structure are explored in the simplest possible setting to expose general principles and data-driven models in which the parameters and functions are estimated from observations of actual individuals, allowing inferences about population behavior. By far the most popular data-driven models are matrix projection models, which assume that the population can be divided into discrete classes. Much of the success of matrix models is a consequence of Hal Caswell writing a detailed "how to" book (Caswell 2001, 1st edition 1989) in which he described, with appropriate MATLAB code, how to build and use matrix models.

The Integral Projection Model (IPM), introduced by Easterling et al. (2000), is a generalization of matrix models to cover populations where individuals are classified by continuous traits. An IPM makes full use of hard-won data on the precise state of individuals, without an increase in the number of parameters to estimate; in fact it often needs fewer parameters, especially when the model includes environmental variability. The aim of this book is to do for IPMs what Caswell did for matrix models. We want this book to be useful and accessible to ecologists and evolutionary biologists interested in developing data-driven models for animal and plant populations. IPMs can seem hard as they involve integrals, and the link between data and model, although straightforward, can seem abstract. We hope to demystify IPMs so that they can become the model of choice for populations structured by continuously varying attributes. To do this, we structure the book around real examples of increasing complexity and show how the life cycle of the study organism naturally leads to the appropriate statistical analysis, which in turn leads to the IPM itself. We also provide complete, self-contained R code (R Core Team 2015) to replicate all of the figures and calculations in the book at https://github.com/ipmbook/first-edition.

We set out to summarize the state of the art, taking advantage of book format to start from the beginning, and to provide more details, more examples, more practical advice, and (we hope) clearer explanations. But we could not always resist the impulse to go beyond the current literature. Much of Chapters 3 and 9 is new. Chapter 4 takes a new approach to eigenvalue sensitivity analysis. Chapter 10 introduces a general way of putting demographic stochasticity into an IPM and some ideas about fitting IPMs to mark-recapture data that we hope might spur more applications to animal populations (or better yet, spur some statisticians to bemoan our naiveté and come up with something better). And while we were writing, the current literature often went beyond us, so we apologize to everyone whose new papers we ignored in the push to finish this book.

We also apologize to beginning IPM users and to advanced users. Advanced users will find that we start at the beginning, with things they already know. But we take a new approach to some of the first steps, so we hope that you will read the whole book. Beginners will find that we don't stop until we hit the limits of what we know, so we urge you to *not* read the whole book until you've gotten some hands-on experience with IPMs. And we hope that you won't give up completely when you hit something confusing. The start of the next section or chapter will probably be written for you, unless it starts with a warning to the contrary.

It has been just over three years since we started preparing this book. We have been fortunate to receive guidance and support from many friends and colleagues during this time. We have also been very fortunate to work with and learn from many students, postdocs, and other collaborators. The list starts with Philip Dixon (who got this started nearly 20 years ago by posing the question, histogram:kernel density estimate = matrix projection model:what?, and helping to answer it) – and it continues to expand. Those who have shaped our thinking over the years include Peter Adler, Ottar Bjørnstad, Ben Bolker, Yvonne Buckley, Hal Caswell, Peter Chesson, Margaret Cochran, Evan Cooch, David Coomes, Tim Coulson, Mick Crawley, Johan Dahlgren, Michael Easterling, Roberto Salguero-Gómez, Edgar González, Peter Grubb, Nelson Hairston Jr., Elze Hesse, Giles Hooker, Eelke Jongejans, Dave Kelly, Patrick Kuss, Jonathan Levine, Svata Louda, Marc Mangel, Jess Metcalf, Sean McMahon, Cory Merow, Heinz Müeller-Schärer, Arpat Ozgul, Satu Ramula, Richard Rebarber, Karen Rose, Sebastian Schreiber, Kat Shea, Andy Sheppard, Robin Snyder, Britta Teller, Shripad Tuljapurkar, Lindsay Turnbull, Larry Venable, and really each and every one of our students and postdocs.

Two anonymous reviewers of some early chapter drafts provided advice and much needed encouragement to carry on developing the text. A further three reviewers of the nearly complete manuscript offered yet more encouragement, as well as some valuable and constructive suggestions. Achi Dosanjh, our editor at Springer, has shown great patience throughout the process of taking the book from inception to publication. Parts of the book are based on our published papers, and we thank the reviewers and editors who read our manuscripts and helped us improve them.

A number of funding agencies provided direct and indirect support for our research on IPMs over the last 15 years, including the Natural Environment Research Council (DZC, MR), the Leverhulme Trust (DZC, MR), the US National Science Foundation (SPE), and the US National Institutes of Health through the program on Ecology and Evolution of Infectious Diseases (SPE).

Finally, we'd like to thank our families for their enduring support and patience.

Ithaca, NY, USA	Stephen P. Ellner
Sheffield, Yorkshire, UK	Dylan Z. Childs
Sheffield, Yorkshire, UK	Mark Rees

Contents

Chapter 1
Introduction

Abstract We provide a brief overview of how models can be used to scale up individual-level observations to the population level, focusing on the Integral Projection Model (IPM), and survey various extensions to the basic IPM. Advice on using the book and the things you need to know (maths, stats, and programming-wise) are then covered. A final section covers notation and highlights areas where confusion might occur as a result of symbols having different meanings depending on the context.

1.1 Linking individuals, traits, and population dynamics

It is often the case in population biology that things that are straightforward to measure are not the things you want to know. For example, in population biology individuals can often be marked and followed throughout their lives, yielding a mass of data on how survival, growth, and reproduction vary with, say, size and other aspects of the environment. This information, although of interest in its own right, is not all that we want to know; we really want to know how this individual-level data translates into population behavior. There are many population-level questions we might want to address. For example, one of the most basic things is whether the population will increase or decrease through time? If the population is harvested then the questions might be, how does harvesting large individuals affect population size and the evolution of, say, antler size? For a pest population we might want to explore different management strategies: culling young or mature, males or females? When you want to control the spatial spread of a pest, where in the life cycle do you get the biggest bang for the buck? Scaling up from individual-level data to population-level patterns is the key to answering all these questions. Much of the recent excitement with trait-based approaches in community ecology can be understood in this context. Traits are much easier to measure than population interactions, so if traits predict interactions, the world becomes simpler to study and understand.

Scaling up from complex state-dependent demography to population and community behavior is daunting. For example, individual size often influences

© Springer International Publishing Switzerland 2016
S.P. Ellner et al., *Data-driven Modelling of Structured Populations*,
Lecture Notes on Mathematical Modelling in the Life Sciences,
DOI 10.1007/978-3-319-28893-2_1

survival, growth, the chances of reproduction, how many offspring are produced, and how big they are. That's all in one year, and what an individual does in one year influences what happens in the next, and can, in some cases, have cumulative effects many years later. This is where models are essential. Integral Projection Models (IPMs) provide a way of synthesizing all this complex information and predicting population behavior. Remarkably all this complexity can often be summarized using regression models for the relationship between a continuous variable (such as size) and a continuous or discrete response (such as size next year or the probability of survival), and these lead directly to an IPM.

The "regression models" part of the previous sentence is important, as regression models are the foundation of modern statistics and now come in many flavors. GLMs are now fairly standard in population biology, allowing linear models to be fitted to data from a wide range of distributions. You want to model your fecundity data assuming that clutch sizes have a negative binomial distribution? No problem. Worried about assuming linearity or some other simple functional form? Then let the data decide by fitting a generalized additive model (GAM) which assumes the function is smooth but little else. Worried about using data where individual behaviors are correlated, say as a result of spatial structure, or genetics? Use a linear mixed model or generalized linear mixed model (GLMM). Thanks to R (R Core Team 2015, www.R-project.org) and the large community that contribute packages, the model fitting is now reasonably straightforward, and fitted models can be directly translated into an IPM. Mixed models, in particular, are enormously helpful for modeling demography in varying environments, because modeling variation with a few regression parameters is more parsimonious and much simpler than identifying and modeling the many possible correlations between matrix entries in a matrix projection model. Because of the close links between IPMs and regression models, a huge range of exciting biological questions can now be rigorously and efficiently addressed.

1.2 Survey of research applications

In the basic IPM, a single trait and nothing else influences demography. This is appropriate for some systems, but it will often be the case that the biology of the system is more complicated. To give you a feel for the range of the questions that can be addressed, let's have a look at some extensions to the basic model. Since the introduction of IPMs, numerous extensions have been developed (Table 1.1). These extensions can be divided into three categories: those dealing with aspects of the environment in which the population occurs, such as spatial or temporal variation in environmental quality; those dealing with more complex life histories, for example dormant states or stage structure where individuals are cross classified by stage (e.g., juveniles and adults) and a continuous state (e.g., size); and those dealing with interactions among individuals, such as size-dependent competition, within and between species. For all of these, however, the basic model considered in the next three chapters is the starting point.

Table 1.1 Applications extending the basic IPM to deal with additional biological properties.

Extension	Reference
Age structure	Childs et al. (2003), Ellner and Rees (2006)
Discrete & continuous states	Rees et al. (2006), Hesse et al. (2008), Williams (2009), Jacquemyn et al. (2010), Miller et al. (2009), Salguero-Gomez et al. (2012), Metcalf et al. (2013)
Stage structure	Ozgul et al. (2010), Childs et al. (2011), Ozgul et al. (2012)
Time lags	Kuss et al. (2008)
Multispecies interactions	Rose et al. (2005), Miller et al. (2009), Hegland et al. (2010), Adler et al. (2010), Bruno et al. (2011)
Costs of Reproduction	Rees and Rose (2002), Miller et al. (2009), Metcalf et al. (2009b)
Density dependence	Rose et al. (2005), Ellner and Rees (2006), Adler et al. (2010), Childs et al. (2011), Rebarber et al. (2012), Smith and Thieme (2013), Eager et al. (2014b)
Environmental covariates	Miller et al. (2009), Dahlgren and Ehrlén (2011), Metcalf et al. (2009b), Jacquemyn et al. (2010), Dahlgren et al. (2011), Nicole et al. (2011), Coulson et al. (2011), Simmonds and Coulson (2015)
Environmental stochasticity	Childs et al. (2004), Ellner and Rees (2007), Rees and Ellner (2009), Adler et al. (2010), Dalgleish et al. (2011), Eager et al. (2013), Eager et al. (2014a)
Demographic stochasticity	Vindenes et al. (2011, 2012); Schreiber and Ross (2016)
Spatial structure	Adler et al. (2010), Jongejans et al. (2011), Ellner and Schreiber (2012), Eager et al. (2013)
Eco-evolutionary dynamics	Coulson et al. (2010, 2011), Bruno et al. (2011), Smallegange and Coulson (2013),Vindenes and Langangen (2015), Rees and Ellner (2016), Childs et al. (2016)

1.3 About this book

Our main intended audience are population biologists interested in building data-driven models for their study systems. The more technical mathematical and programming material is presented in boxes or chapter appendices, so that you can focus on the key ideas. But we hope this book will also interest mathematicians and statisticians keen on developing the theory. There is much more to be done, as we discuss in Chapter 10 and elsewhere.

As we are great believers in learning by doing, we provide commented R code for each calculation that you can run yourself, and modify for your own applications. The script files that generated the figures and other results in this book are all available online, so that readers can implement an IPM for their study system by modifying our scripts. Go to https://github.com/ipmbook/first-edition and use the **Download ZIP** button to get a copy. Each figure caption in this book tells which scripts produced the figure. And when a calculation is done in the book, you should do the same one on your computer using your downloaded copy of our scripts.

Providing our scripts rather than an R package, or a GUI that lets you click a checkbox labeled *Eigenvalue Sensitivity Analysis*, means that you have to work a bit harder and you have to understand what the scripts are doing. There are two reasons why we do this. First, this is how we work ourselves, and we can only teach well what we know ourselves. Second, there already is an R package available, IPMpack (Metcalf et al. 2013), and we don't imagine that we could write a better one. R-literate ecologists (a rapidly growing population) will be able to learn from our R scripts exactly what different analyses involve and how they can be implemented. Then you have your choice of following our approach or using the package. Many examples of how to use IPMpack are available in the online Supporting Information for Merow et al. (2014).

In general when starting a new project we recommend that you start simple, and add complexity only when you have built and analyzed the simple model. For example, when building a stochastic environment IPM we would recommend building a constant environment model first, as we do in the next chapter, and then extending it so the kernel varies from year to year as we do in Chapter 7.

1.3.1 Mathematical prerequisites

Paraphrasing (or quoting) Roger Nisbet and Bill Gurney, you should remember that you have had a calculus class, but it's OK if you've forgotten most of it. For example, we never ask you to find the derivative or integral of a function. Don't remember what $\int \sin(1+x^3)dx$ is? Neither do we. But you do need to understand that the integral of a positive function represents "the area under the curve" in a plot of the function, because we usually evaluate integrals numerically using sums that approximate the area under the curve. You should also remember that the derivative of $f(x)$ is the slope of the curve in a plot of $y = f(x)$, and that a smooth function can be approximated by a Taylor series,

$$f(x + a) \approx f(x) + af'(x) + \frac{a^2}{2}f''(x) + \cdots \text{ for } a \approx 0. \qquad (1.3.1)$$

We won't expect you to find the derivative of $\exp(1+3x^2)$ or anything like that, but we sometimes use the basic rules for derivatives (product rule, quotient rule, chain rule) and integrals (e.g., the integral of $2f(z) + g(z)$ is twice the integral of $f(z)$ plus the integral of $g(z)$). You also need to understand what a partial derivative means (it's enough if you understand that when $z = 3x + 2y^2$, $\partial z/\partial x$ means the rate of change in z when x varies and y is held constant, so $\partial z/\partial x = 3$).

Some familiarity with matrix projection population models is needed, at least with Leslie matrix models for age-structured populations, and with the main concepts used to derive their properties: matrix multiplication, eigenvalues, and eigenvectors. If you know that the dominant eigenvalue is the long-term population growth rate, that the dominant right eigenvector is the stable age- or stage-distribution, and the dominant left eigenvector is relative reproductive value, you are probably ready to go. If you can also enter a Leslie matrix into R and compute all of those things, you're definitely ready. And if you're not ready, there are many short reviews of the matrix algebra you need. Three of our favorites are Caswell (2001, Appendix A), Ellner and Guckenheimer (2006, Chapter 2), and Stevens (2009, Chapter 2).

1.3.2 Statistical prerequisites and data requirements

For almost the entire book we assume that you have data on individuals tracked over time, so that for an individual observed in one census you know its fate (survived or died, and how many offspring it produced) and state (e.g., how big it is, or where it is) at the next census. Other types of data are briefly covered in Section 10.5, where we describe methods for mark-recapture and count data. In mark-recapture studies individuals are followed over time, but an individual may be unobserved or partially observed at some censuses and some deaths may be unobserved. With count data individuals are not followed through time, so we need to estimate the parameters of an IPM from changes in the state distribution from one year to the next. Methods for count and mark-recapture data are less well developed (see Chapter 10), so most IPMs have been built from data on completely tracked individuals, but that situation is starting to change.

Our main statistical tools are regression models. If you are happy with linear and generalized linear models (i.e., factors and continuous variables, contrasts and link functions, and working with the lm and glm functions in R) then you should be fine, although in Chapter 7 we do use mixed models. There are thousands of books covering regression models, but Fox and Weisberg (2011) is hard to beat and also provides an excellent introduction to R. We occasionally use some basic properties of standard probability distributions, such as the fact that a Poisson with mean μ also has variance μ, and that a Binomial(N, p) random variable has mean Np and variance $Np(1 - p)$. One introductory statistics course will have covered everything you need to know, except for a few technical sections where we will warn you of the additional prerequisites.

1.3.3 Programming prerequisites

We assume that you know the basics of using R for data analysis, including

- Creating and working with vectors, matrices, and data frames.
- Basic functions for summary statistics (mean,var,sum,sd, etc.) and data manipulation (extracting rows or columns, using apply, etc.).
- Creating simple plots of data points and curves.
- The functions for fitting linear and generalized linear models (lm,glm) and for later chapters linear mixed models using the lme4 package and spline models using the mgcv package.

One R-based statistics course would have taught you all that you need. If you've ever used R to load data from a text or CSV file, fit a linear regression model, and plot the regression line overlaid on the data points, you can pick up the rest as you go along. The most important R skill is using the Help system, any time we use a function that you don't recognize, or use one in a way that you're not familiar with.

We also assume that you know the basics of using R for general-purpose scientific programming,

- Basic mathematical functions like `log`,`exp`,`sqrt`, etc.
- For-loops and (occasionally) if/else constructs.
- The `r`,`d`,`p` functions for probability distributions such as `rnorm` to generate Gaussian (normal) random numbers, `dnorm` for the Gaussian probability density, and `pnorm` for Gaussian tail probabilities.
- Matrix operations: elementwise `(+,-,*,/,^)`, and matrix-vector/ matrix-matrix multiplication using `%*%` .

Some short introductions to these skills are available at `cran.r-project.org` (click on <u>Contributed</u> under *Documentation*) – we recommend *The R Guide* by Owen, and one available through publisher's web page for Ellner and Guckenheimer (2006).

1.4 Notation and nomenclature

We have tried to follow a few general rules:

1. z is initial state of an individual, z' is subsequent state, z_0 is state at birth.
2. When size is the only individual-level state variable we use z for size; when size is one of several state variables we use z for the vector of state variables, and x for size.
3. Functions of one variable, such as z or z', are denoted using lowercase Latin or Greek, e.g., $s(z), \sigma(z)$.
4. Any function of (z', z) is denoted using uppercase Latin or Greek, such as $G(z', z)$ and $\Omega(z', z)$.

Table 1.2 summarizes some of the most important notation used in the book; it's for you to refer to as needed, not to read now and memorize. Other notation is summarized in Table 3.2 for life cycle analysis and in Table 7.2 for sensitivity analysis of stochastic growth rate.

It is often helpful to use notation that makes an IPM look like a matrix projection model. Both kinds of model project forward in time by transforming one distribution function (population state now) to another (population state at the next census); mathematicians call this an *operator* on distribution functions. For matrix models, we write

$$\mathbf{n}(t + 1) = \mathbf{A}\mathbf{n}(t) \tag{1.4.1}$$

as a shorthand for the calculations in matrix multiplication, because equation (1.4.1) is much easier to read and understand than $n_i(t+1) = \sum_{j=1}^{m} A_{ij} n_j(t)$. We can do the same for IPMs. Instead of writing

$$n(z', t + 1) = \int_L^U K(z', z) n(z, t) \, dz \tag{1.4.2}$$

for an IPM,[1] we can write

$$n(t + 1) = Kn(t). \tag{1.4.3}$$

This looks a lot like (1.4.1), because it is a lot like (1.4.1). The notation also helps with writing computer code. IPMs are usually implemented on the computer with the state distribution n approximated by a big state vector, and the kernel K represented by a big iteration matrix. The notation $Kn(t)$ then translates to a matrix-vector multiplication, K%*%nt in R, and K*nt in MATLAB. The same is true for more complicated operations, as we explain later. We think that operator notation is helpful enough that you should get comfortable with it and use it yourself. We will explain operator notation in detail when we start to use it seriously in Chapters 3 and 4.

As a result of trying to be consistent within the book and also with past usage, there are a few places where you need to be careful. We have an affection for c's it seems. In Chapter 8 c is wave speed, elsewhere it's an arbitrary initial state distribution for a population or cohort, with total population size 1, c_0 is the offspring size distribution when this is independent of parental size, and of course in R c is concatenate. There are also several C's; C is a perturbation kernel, whereas C_0 and C_1 are offspring size kernels - in Chapters 3 and 6 C is also used as an arbitrary constant and in Chapter 6 a set of individual states - the context should make the meaning clear. R_0 is the net reproductive rate, R is the next-generation operator in Chapter 3, and $\mathcal{R}$ is the total number of recruits per year in Chapter 5. In Chapter 4 $\mathbf{s}$ is sensitivity, not to be confused with s which is survival; in Chapter 8 s is also the wave shape parameter, and again context should make the meaning clear. In Chapter 4 ν is the linear predictor for a regression, whereas v is reproductive value. $\mathbf{e}(\mathbf{z}', \mathbf{z})$ is the kernel elasticity function in Chapter 4, but in Chapter 3 it is the constant function $\mathbf{e}(z) \equiv 1$ to keep with established notation.

---

[1] If you haven't seen this equation before, don't worry: it's explained in Chapter 2.

Table 1.2 Some of the general notation used in this book.

Notation	Meaning and/or formula
	Projection kernel and its components
$\mathbf{Z}$	Set of all possible individual-level states.
$[L, U]$	Closed interval from L to U, the set of numbers z such $L \leq z \leq U$. For an IPM with a single individual-level state variable such as size, $\mathbf{Z} = [L, U]$ for some L and U.
$[L, U)$	Half-open interval that excludes the right endpoint U; $(L, U]$ excludes the left endpoint L.
h	Width of a size bin in a basic IPM implemented using midpoint rule.
P	Survival kernel $P(z', z)$.
F	Reproduction kernel $F(z', z)$.
K	Complete IPM projection kernel $K = P + F$.
G	Growth kernel $G(z', z)$ giving the frequency distribution of state z' for survivors at time $t + 1$ conditional on their state z at time t.
s	Survival probability function $s(z)$.
p_b	Probability of reproducing function $p_b(z)$.
b	Fecundity function $b(z)$, expected number of offspring regardless of their state, for individual who reproduce. Depending on the timing of census and breeding, the number of recruits may be lower due to mortality before the first census.
C_0	Offspring state kernel $C_0(z', z)$.
c_0	Frequency distribution of offspring state $c_0(z')$ that is not a function of parent size z.
	Stable population theory: constant kernel
λ	Stable population growth rate (dominant eigenvalue).
w	Stable state distribution (dominant right eigenvector).
v	Reproductive value (dominant left eigenvector).
$\mathbf{s}, \mathbf{e}$	Eigenvalue sensitivity and elasticity functions.
g_k	Number of individuals in generation k descended from some initial cohort.
R_0	Stable generation-to-generation population growth rate.
	Additional notation
λ_S	Stochastic population growth rate of a randomly time-varying kernel
c	Wave speed for a population expanding as a traveling wave (Chapter 8 only).
c	Initial state distribution for a population or cohort, with total population size 1, $\int_{\mathbf{Z}} c(z)\, dz = 1$
n_0	Arbitrary initial state distribution for a population or cohort, with total population size not necessarily equal to 1.
$M(s)$	Moment generating function of dispersal kernel in a spatial IPM.

Chapter 2
Simple Deterministic IPM

Abstract Easterling et al. (2000) originally proposed a size-structured IPM as an alternative to matrix projection models for populations in which demographic rates are primarily influenced by a continuously varying measure of individual size. That model was deterministic and density-independent, analogous to a matrix projection model with a constant matrix. Nothing could be simpler within the realm of IPMs. In this chapter we use that case to introduce the basic concepts underlying IPMs, and to step through the complete process of building and then using an IPM based on your population census data. To illustrate the process of fitting an IPM to population data, we generate artificial data using individual-based models of the monocarpic perennial *Oenothera glazioviana*, and of the Soay sheep (*Ovis aries*) population on St. Kilda (Clutton-Brock and Pemberton 2004).

2.1 The individual-level state variable

In this chapter, we assume that differences between individuals are completely summarized by a single attribute z, which is a continuous variable (e.g., length) rather than discrete (e.g., juvenile versus adult). Ideally z will be the character most strongly linked to individual survival, growth, and reproduction, though z cannot be (for example) an individual's realized growth rate or fecundity. The premise of an IPM is that individuals with the same current state z have the same odds of different future fates and states, but what actually happens involves an element of chance. Individuals within a population typically vary in many different ways. But as a starting point, we assume that to predict an individual's chance of living to next year, how many offspring it will have, etc., it helps to know z but knowing anything more doesn't lead to better predictions. We often refer to z as "size," meaning some continuous measure of body size such as total mass, volume, or log of longest leaf length. But z can be unrelated to size – for example, it could be the individual's spatial location in a linear habitat such as a riverbank, or a bird's first egg-laying date. However, z must

© Springer International Publishing Switzerland 2016

9

S.P. Ellner et al., *Data-driven Modelling of Structured Populations*,
Lecture Notes on Mathematical Modelling in the Life Sciences,
DOI 10.1007/978-3-319-28893-2_2

have finite limits: a minimum possible value L and a maximum value U. In Chapter 8 we consider models on infinite spatial domains, but those behave very differently from models with finite limits to the value of a trait.

In practice, to predict an individual's fate many possible measures of individual size or state could be used (e.g., the mass or snout-vent length of a crocodile; a plant's total leaf area or the length of its longest leaf; and so on), these may be transformed (log, square root, etc.), and in some cases additional covariates might be used. We cannot emphasize strongly enough the importance of finding the best models to forecast survival, fecundity, and changes in size. This step is like any other statistical analysis of demographic data, so there is no all-purpose recipe, but there are standard tools readily available. For now we assume that you've made the right choices, but we come back to this issue in Section 2.7.

2.2 Key assumptions and model structure

The state of the population at time t is described by the size distribution $n(z,t)$. This is a smooth function of z such that:

> The number of individuals with size z in the interval $[a,b]$
>
> at time t is $\displaystyle\int_a^b n(z,t)\,dz.$

$\qquad(2.2.1)$

A more intuitive description is that

> The number of individuals in the size interval $[z, z+h]$
>
> at time t is approximately $n(z,t)h$ when h is small;

$\qquad(2.2.2)$

as $h \to 0$, this approximation becomes exact (the relative error goes to zero). Note that $n(z,t)$ is not a probability distribution: its integral over the range of all possible sizes is the total population size, not 1. We will use *probability distribution* or *frequency distribution* to denote a distribution whose integral or sum is necessarily 1.

The model operates in discrete time, going from times t to $t+1$ (the unit of time is often a year, but it can be any duration such as one decade, one day, etc.). Between times t and $t+1$, individuals can potentially die or change in size, and they can produce offspring that vary in size. To describe the net result of these processes we define two functions $P(z',z)$ representing survival followed by possible growth or shrinkage, and $F(z',z)$ representing per-capita production of new recruits. In both of these, z is size at time t and z' is size at time $t+1$. For an individual of size z at time t, $P(z',z)h$ is the probability that the individual is alive at time $t+1$ and its size is in the interval $[z', z'+h]$ (as with $n(z,t)$ this is an approximation that is valid for small h, and the exact probability is given by an integral like equation 2.2.1). Similarly, $F(z',z)h$ is the number of new offspring in the interval $[z', z'+h]$ present at time $t+1$, per size-z individual at time t.

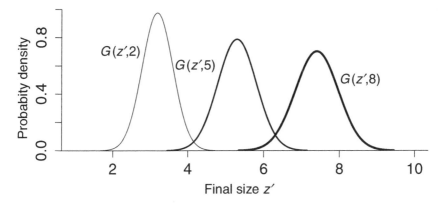

Fig. 2.1 Probability densities $G(z', z)$ of subsequent size z' conditional on survival, for prior sizes $z = 2, 5$, and 8. In this example, z' size has a normal distribution with mean $\bar{y} = 0.8 + 0.7z$ and variance that increases with prior size. Source file: `GrowthDensity.R`

It is often convenient to break P up into two processes: survival or death, and size transitions from $z \to z'$:

$$P(z', z) = s(z)G(z', z) \qquad (2.2.3)$$

where s is the survival probability, and $G(z', z)$ describes size transitions. In this case we have assumed that both survival and growth depend on initial size, z, as this is often the case, but there are other possibilities, as described in the next section.

With P separated into survival and growth, the kernel $G(z', z)$ can be thought of as a family of univariate probability densities for subsequent size z', that depend on initial size z (see Figure 2.1). In particular, it's always the case that $\int_L^U G(z', z)\, dz' = 1$ for all z; making sure that this is actually true is an important check on how an IPM has been constructed and implemented. If you aren't familiar with probability densities, please read Appendix 2.9. But we recommend that you do this after reading the rest of this chapter, so that you'll have seen some examples of how probabilities densities are used to model population processes.

The net result of survival and reproduction is summarized by the function

$$K(z', z) = P(z', z) + F(z', z) = s(z)G(z', z) + F(z', z) \qquad (2.2.4)$$

called the *kernel* - we will refer to $P(z', z)$ and $F(z', z)$ as the survival and reproduction components of the kernel. The population at time $t + 1$ is just the sum of the contributions from each individual alive at time t,

$$n(z', t+1) = \int_L^U K(z', z)n(z, t)\, dz. \qquad (2.2.5)$$

The kernel K plays the role of the projection matrix in a matrix projection model (MPM), and equation 2.2.5 is the analog for the matrix multiplication that projects the population forward in time.

P and F have to be somewhat smooth functions in order for equation (2.2.5) and the theory developed for it to be valid. We have previously assumed that they are continuous (Ellner and Rees 2006), but it's also sufficient if P and F have some jumps.[1] Consequently an IPM can include piecewise regression models, such as a fecundity model that jumps from zero to positive fecundity once individuals reach a critical "size at maturity" (e.g., Bruno et al. 2011), or a growth model in which individuals cannot shrink.

Looking at equation (2.2.5), it can be tempting to think of $n(z,t)$ as the *number* of size-z individuals at time t, but this is incorrect and will sometimes lead to confusion. It is important to remember the actual meaning of $n(z,t)$ as explained in equations (2.2.1) and (2.2.2). One time when it's essential to keep that in mind is when you want to move from one scale of measurement to another. For example, suppose that you have built an IPM in which z is log-transformed size, and you iterate it to project a future population $n(z, 100)$. Now you want to plot the distribution of actual size $u = e^z$; let's call this $\tilde{n}(u, 100)$. It's tempting to think that $\tilde{n}(u, 100) = n(\log(u), 100)$ because having size u is the same as having log-transformed size $z = \log(u)$. But this is wrong, because $n(z,t)$ really isn't the number of size-z individuals. We explain the right way in Appendix 2.9. You should read that section eventually, but that should wait until you've read the rest of this chapter and spent a while working with IPMs on the computer.

2.3 From life cycle to model: specifying a simple IPM

The kernel functions s, G, and F describe all the possible state-transitions in the population, and all possible births of new recruits. But where do these functions come from? Our goal in this section is to answer that question by showing how to translate population census data into a simple deterministic IPM for a size-structured population. We will describe how information on growth, survival, and reproduction is combined to make a kernel. As IPMs are data-driven, our aim is to show how to arrive at a model which is both *consistent* and *feasible*. By consistent, we mean a model that accurately reflects the life cycle and census regime. By feasible, we mean that the model can be parameterized from the available data. In the case studies later in this chapter, we take the next step of fitting specific models to data.

The key idea is that the kernel is built up from functions that describe a step in the life cycle of the species, based on the data about each step. Throughout, we assume that the data were obtained by marking individuals, and following them over their life cycle with evenly spaced censuses at times $t = 0, 1, 2, \cdots$.

---

[1] Technically, it's sufficient if there is a finite number of curves that divide the square $\mathbf{Z}^2 = \{L \leq z', z \leq U\}$ into a set of closed regions, and P and F are defined as continuous functions on the interior of each region ("closed" means that each region includes its boundary as well as the interior of the region). The value of K on the dividing curves can be assigned arbitrarily, since this has no effect on the value of (2.2.5).

(A) Pre-reproductive census

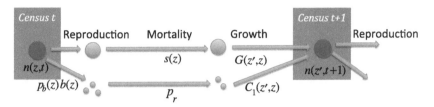

(B) Post-reproductive census

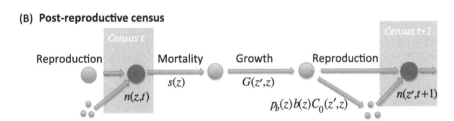

Fig. 2.2 Life cycle diagram and census points for (A) pre-reproductive and (B) post-reproductive census. The sequence of events in the life cycle is the same in both cases. However, the diagrams are different because reproduction splits the population into two groups (those who were present at the last census [large circles], and new recruits who were not present at the last census [small circles]), while at a census time the two groups merge. Reproduction is described by a two stage process with $p_b(z)$ being the probability of reproduction and $b(z)$ being the size-specific fecundity. Each new offspring has a probability p_r of successful recruitment, and its size at the next census is given by $C_1(z', z)$. The pre-reproductive census leads to IPM kernel $K(z', z) = s(z)G(z', z) + p_b(z)b(z)p_r C_1(z', z)$ where $C_1(z', z)$ is the size distribution of new recruits at age 1 (when they are first censused). The post-reproductive census leads to the IPM kernel $K(z', z) = s(z)G(z', z) + s(z)p_b(z)b(z)C_0(z', z)$ where $C_0(z', z)$ is the size distribution of new recruits at age 0 (when they are first censused). The term p_r is absent in the post-reproductive census because new recruits are censused "immediately" after birth, before any mortality occurs.

We strongly recommend that you begin by drawing a life cycle diagram indicating when demographic processes and census points occur, see Figure 2.2. For this first example (Figure 2.2A), we assume that at each time step there is a single census point immediately prior to the next occurrence of reproduction (i.e., there is a *pre-reproductive* census). At that time you record the size of each individual, and we assume that you can distinguish between breeders (those that attempt reproduction) and nonbreeders. At time $t + 1$, the population will include survivors from time t, and new recruits. We assume, for now, that you can assign offspring to parents, and therefore can record how many offspring each individual has.

Combining these data with the size measurements taken at time $t + 1$ gives you a data table like this, suitable for statistical analysis:

Size t	Offspring	Survive	Reproduced	Size $t+1$
3	NA	0	0	NA
7	5	1	1	8
8	4	1	1	10
5	NA	1	0	4

$$(2.3.1)$$

Reading across the top row we have the size of an individual in a census, how many offspring it produced (newborns at the next census), whether or not the individual survived to the next census, whether or not it reproduced (this is necessary because an individual might attempt breeding but have no offspring that survive to the next census), and the size of the individual at the next census. Following rows are for other individuals. An NA (indicating "missing data") appears in the Offspring column for any individual that did not attempt reproduction. This is important because it guarantees that only data on breeders will be used to estimate the relationship between size and fecundity. If you cannot distinguish breeders from nonbreeders at census time, the Reproduced column would be absent, the Offspring entry would be zero for all nonbreeders, and fecundity would then be the average over breeders and nonbreeders.

To define the structure of the IPM, we can begin by ignoring individual size and constructing a model for the dynamics of $N(t)$, the total number of individuals at census t, that reflects the life cycle and data. Each individual at time t can contribute to $N(t+1)$ in two ways: survival and reproduction.

- *Survival:* Having observed how many individuals survive from each census to the next, you can estimate an annual survival probability s. At time $t+1$ the population then includes $sN(t)$ survivors from time t.
- *Reproduction:* We could similarly define a per-capita fecundity b, and let $p_r bN(t)$ be the number successful recruits at time $t+1$, with p_r being the probability of successful recruitment. But the data distinguish between "breeders" and "nonbreeders," so we can be more mechanistic. Let p_b denote the probability that an individual reproduces, and b the mean number of offspring produced among individuals that reproduce. Then the number of new recruits at time $t+1$ is $p_b p_r bN(t)$.

Combining survivors and recruits, we have the unstructured model

$$N(t+1) = sN(t) + p_b p_r bN(t) = (s + p_b p_r b)N(t). \qquad (2.3.2)$$

The "kernel" for this model is $K = s + p_b p_r b$, which is just a single number because at this point the model ignores the size structure of the population.

The next step is to incorporate how the size at time t affects these rates: the probability of surviving, the probability of reproducing, and the number of offspring are all potentially functions of an individual's current size z:

$$s = s(z), \quad p_b = p_b(z), \quad b = b(z).$$

Our prediction of $N(t)$ can now take account of the current size distribution, $n(z,t)$,

$$N(t+1) = \int_L^U (s(z) + p_b(z)p_r b(z)) \, n(z,t) \, dz \ . \tag{2.3.3}$$

At this next level of detail, the kernel is a function of current size z, $K(z) = s(z)+p_b(z)p_r b(z)$. What's missing still from model (2.3.3) is the size-distribution at time $t+1$. To forecast that, we need to specify the size-distributions of survivors and recruits. These are given by the growth kernel for survivors, $G(z',z)$, and the size-distribution of recruits, $C_1(z',z)$, as described in Section 2.2, giving us the complete kernel

$$K(z',z) = s(z)G(z',z) + p_b(z)p_r b(z)C_1(z',z). \tag{2.3.4}$$

for the general IPM (equation 2.2.5).

2.3.1 Changes

We have seen that the structure of the kernel is jointly determined by the life cycle and when the population is censused. To emphasize this essential point, we will now consider how changes in these can lead to changes in the kernel.

Going back to the life cycle diagram (Figure 2.2), what would happen to the structure of the kernel if you had conducted a post-reproductive census? The first thing to notice is that order of events has changed: mortality now occurs *before* reproduction. This has important implications for the structure of the kernel and for the statistical analysis of the data. With a post-reproductive census, the data file will now look like this:

Size t	Offspring	Survive	Reproduced	Size $t+1$	
3	NA	0	NA	NA	
7	5	1	1	8	(2.3.5)
8	4	1	1	10	
5	NA	1	0	4	

A crucial difference here is that there are now NA's in both the Offspring and Reproduced column for all individuals that die before the next census. This is because there is no information about what the reproductive success would have been, for individuals who died prior to the next breeding period. As a result, the structure of kernel in this case is

$$K(z',z) = s(z)G(z',z) + s(z)p_b(z)b(z)C_1(z',z). \tag{2.3.6}$$

In order to reproduce, individuals now have to survive, hence $s(z)$ is a factor in both the survival and reproduction components of the kernel. The absence of the p_r term is a consequence of censusing the population immediately after reproduction: there is no mortality between birth and first census. Newly produced individuals do suffer mortality before their next census (at age 1), but this is included in the $s(z)$ term because already at age 0 they are part of $n(z,t)$ (Figure 2.2B). The functions $p_b(z)$ and $b(z)$ are now the probability of reproducing, and mean number of offspring produced, for individuals that survive the time step. The NAs in the data table make sure that you "remember" this, because they ensure that only data on survivors will be used to fit p_b and b.

When reproduction occurs just before the next census (as in Figure 2.2B), p_b and b could be fitted instead as functions of z', the size at the post-breeding census which is also the size when reproduction occurs. In that approach, the steps to producing size-z' offspring are: survive and grow to some size z^*, breed or not (depending on z^*) and if so have $b(z^*)$ offspring, some of which are size z'. The fecundity kernel needs to total this up over all possible sizes z^*, so we have

$$F(z', z) = s(z) \int_L^U G(z^*, z) p_b(z^*) b(z^*) C_0(z', z^*) dz^*. \qquad (2.3.7)$$

Alternatively, the interval between censuses can be broken up into survival and growth in t to $t + \tau$ followed by reproduction in $t + \tau$ to $t + 1$, with separate kernels:

$$n(z^*, t + \tau) = \int_L^U s(z) G(z^*, z) n(z, t) dz$$

$$\qquad (2.3.8)$$

$$n(z', t + 1) = n(z^*, t + \tau) + \int_L^U p_b(z^*) b(z^*) C_0(z', z^*) n(z^*, t + \tau) dz^*$$

We think that equation (2.3.6) is simpler (though you are free to disagree and use equations 2.3.8). When you fit p_b and b as functions of z, you are in effect letting the data do the integrals with respect to z^* for you, because the fitted models will represent the average breeding probability and fecundity with respect to the distribution of possible sizes when reproduction occurs.

The size distribution of new recruits will also vary depending on the timing of the census. In the post-reproductive census, $C_0(z', z)$ is the size distribution of recruits at age 0 immediately after their birth (or so we assume). In the pre-reproductive census, recruits were born immediately after the previous census, so they have already undergone a period of growth before they are first observed, and so $C_1(z', z)$ is the distribution of new recruits aged 1.

Next, let's change the order of events in the life cycle and run through this procedure again, as it's important and potentially confusing. Assume now that death occurs before reproduction – see Figure 2.3. This might describe a temperate deciduous tree or shrub population censused in early spring just before leafout, with growth occurring over the spring and summer and virtually all mortality in winter. Going through the same procedure to construct the kernel, hopefully you can convince yourself that for both the pre- and post-reproductive censuses the structure of the data files and kernels are exactly the same as they were if reproduction occurs before death. The reason for this is that in both cases there is a single census each year, so all demographic rates are based on the size z measured at the start of the time interval. For example in Figure 2.3A,

(A) Pre-reproductive census

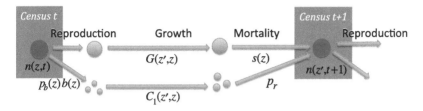

(B) Post-reproductive census

Fig. 2.3 As in Figure 2.2 but with the order of growth and mortality reversed.

the adult survival probability $s(z)$ is based on initial size even though individuals have already grown over the summer, and $s(z)$ then represents the average overwinter survival of trees that were size z at the start of the previous spring.

If there are multiple censuses per year, the additional measurements can be used in constructing the kernel, and the order of events in the life cycle diagram can reflect the timing of events relative to each census. Multiple censuses also mean that we have to decide when we are going to project the population's state. This will typically be one of the census times, but we could also use the average or maximum size over several censuses as our measure of individual size. Alternatively, the annual projection could be broken up into several substeps as in equation 2.3.8, like a seasonal matrix projection model. In Appendix 2.10 we use the Soay sheep case study to present one approach to constructing an IPM when there are several censuses within an annual cycle.

2.4 Numerical implementation

The one-variable IPM (2.2.5) is easy to implement numerically with a numerical integration method called *midpoint rule* (Ellner and Rees 2006). Define *mesh points* z_i by dividing the size range $[L, U]$ into m artificial size classes ("bins") of width $h = (U - L)/m$, and let z_i be the midpoint of the ith size class:

$$z_i = L + (i - 0.5)h, \quad i = 1, 2, \ldots, m. \tag{2.4.1}$$

The midpoint rule approximation to equation (2.2.5) is then

$$n(z_j, t + 1) = h \sum_{i=1}^{m} K(z_j, z_i) n(z_i, t). \tag{2.4.2}$$

A one-variable IPM can always be summarized by the functions $P(z', z)$ and $F(z', z)$. So we assume that you have a script that defines functions to calculate their values:

```
P_z1z <- function(z1,z,m.pars) {
    # your code, for example
    # return( s(z)*G(z1,z) )
    # using functions s and G that you have defined
}

F_z1z<- function(z1,z,m.pars) {
    # your code
}
```

Here m.pars is a vector of model parameters. The next step is to compute the mesh points:

```
L <- 1
U <- 10      # size range for the model
m <- 100     # size of the iteration matrix
h <- (U-L)/m
meshpts <- L + (1:m)*h - h/2
```

The R function outer makes it easy to compute the iteration matrix:

```
P <- h * (outer(meshpts, meshpts, P_z1z))
F <- h * (outer(meshpts, meshpts, F_z1z))
K <- P + F
```

If you are (like us) always worried that outer has not put values in the right places, the next step is to visualize the matrix:

```
matrixImage(meshpts, meshpts, K)
```

The function matrixImage which we have written displays a matrix the way they are usually printed (first row at the top, first column at the left). Source matrixImage.R and then try it out with:

```
A <- diag(1:10)
A <- matrixImage(A)
```

Box 2.1: Implementing midpoint rule in R

In R (see Box 2.1) we arrange the $hK(z_j, z_i)$ terms in a matrix called the *iteration matrix*. This allows us to iterate the model by matrix multiplication, and use the wide range of numerical tools available for matrices.

The only drawback to midpoint rule is that it's sometimes not very efficient relative to other methods. That becomes an issue when it takes a very large value iteration matrix to get accurate results. In Section 2.7 we give some pointers on choosing the size of the iteration matrix, and the size limits L, U. In most of this book we'll assume that midpoint rule is efficient enough, but we also discuss alternatives and how to implement them in Section 6.8.

2.5 Case study 1A: A monocarpic perennial

So far we've looked at translating your study system into an IPM, and how to solve the model numerically. In this section we will put all this together to show you that building a basic IPM is really pretty straightforward: there is no black magic or black boxes. To do this we will develop case studies for idealized plant and animal systems, based on published empirical studies. For each, we will simulate data from a individual-based model (IBM) - a simulation that tracks individuals - and analyze the resulting data to build an IPM. To mimic real life, our IBM treats every demographic event as stochastic. Survival versus death is a "coin-toss" (using random numbers) based on a size-dependent survival probability; size at next census for survivors is a random draw from a probability distribution that depends on initial size; and similarly for breeding or not, number of offspring, and so on. All of the individuals in the model follow the same "rules of the game," but apart from that, the model makes them as unpredictable as real plants or animals. The advantage of using simulated data instead of real data is that we know the "truth" and so we can see how various modeling assumptions influence our results, for example what happens if we incorrectly translate the life cycle into an IPM.

2.5.1 Summary of the demography

Monocarpic plants have been extensively modeled using IPMs because of the simplicity of their life cycles, and ease with which the cost of reproduction can be quantified - reproduction is fatal. Our IBM is based on *Oenothera glazioviana*, a monocarpic plant that often occurs in sand dune areas. Its demography has been extensively studied by Kachi and Hirose (Kachi 1983; Kachi and Hirose 1983, 1985). Plant size was measured using rosette diameter - the plants are rosettes of leaves and so can be thought of as disks - and the population censused every month or bimonthly. Plant size is measured in cm and was log-transformed before analysis (see Box 2.2 for an explanation of why we do this). Following Kachi and Hirose (1985) we will use the May census to construct an IBM. This is a pre-reproductive census, so the life cycle is as shown in Figure 2.2A with the additional feature that breeders cannot survive to the next census.

Plants present at the May census flower in July/August (and then die) with probability $\text{logit}(p_b(z)) = -18 + 6.9z$. Those that don't flower survive with probability $\text{logit}(s(z)) = -0.65 + 0.75z$. For those that don't flower and survive, their size next May is given by $z' = 0.96 + 0.59z + \epsilon$, where ϵ has a Gaussian distribution with mean 0 and standard deviation 0.67. Seed production of the flowering plants was estimated in the field, and for a size z plant $b(z) = \exp(1 + 2.2z)$ seeds are produced. The probability a seedling recruits, $p_r=0.007$, was estimated by dividing the number of seedlings by seed production. Recruits' size at the next census had a Gaussian distribution with mean -0.08 and standard

We are sometimes asked to explain why we often adopt a log transformation of size when building a new IPM. The short answer is that it often works, in that log transformation results in a linear growth model in which the error variance (i.e., the variation in growth) is independent of size. This means that we can model growth using just linear regression rather than more sophisticated methods that use additional parameters to describe the size-variance relationship. And when growth variance still depends on size after log transformation, the dependence is often weak, so that a simple linear or exponential model with just one more parameter is adequate. Another practical advantage of log transformation is that the IPM cannot generate individuals with negative sizes.

The slightly longer answer is that the log transformation makes biological sense when using a linear model to describe growth. For the moment, let x denote some absolute measure of size and $z = \log x$, and assume that growth is completely deterministic. Fitting a linear model using absolute size, $x' = A + Bx$, the expected growth increment $\Delta x = x' - x$ is a strictly decreasing or increasing function of size. That is, $\Delta x = A + (B - 1)x$. This is a decreasing function of size if individuals exhibit determinate growth ($B < 1$). However, in many species (e.g., trees) we observe a hump-shaped relationship between size and the absolute growth increment. This is precisely the relationship that emerges if we instead assume that the expected change in log size is a linear function of log size, and therefore fit a linear regression to successive values of log size, $z' = a + bz$. For species with determinate growth ($b < 1$), this implies that the relative growth rate $\log(x') - \log(x) = z' - z = a + (b - 1)z$ is a decreasing function of size. Under this model the relationship between the absolute growth increment and size is $\Delta x = e^a x^b - x$. This is hump-shaped when $b < 1$.

Growth is a complex phenomenon, reflecting resource availability, competition, and resource allocation. However, one fairly general explanation for the hump-shaped pattern arises from a consideration of energy acquisition and maintenance costs. All else being equal, larger individuals typically acquire more resources than smaller conspecifics, which means they have more energy available to spend on growth, reproduction, and maintenance. When individuals are small, maintenance costs increase slowly with size relative to acquisition, resulting in a positive relationship between size and absolute growth rate. Later in life when individuals are large, maintenance costs increase more rapidly with size relative to acquisition, leading to a negative relationship between size and growth. Such ideas can be formalized using dynamic energy budget theory.

Ultimately, the growth model should be guided by the data, and should produce good model diagnostics without over-fitting the data. It is also worth keeping in mind that you can fit a demographic model on one scale (e.g., log-transformed size) and building the IPM to work on another. Appendix 2.9 at the end of this chapter explains how to do this.

Box 2.2: On log-transforming individual size.

deviation 0.76. As we don't know which seedling was produced by which parent, the offspring size distribution is a probability density $c_0(z)$ that is independent of parental size z.

2.5.2 Individual-based model (IBM)

With information about the life cycle and the associated census regime in hand, we are now in a position to build an IBM and generate some artificial data for use in constructing our first "data-driven" integral projection model. The IBM is parameterized from the fitted linear and generalized linear models describing survival, growth, and recruitment as functions of individual plant size described in the previous section (Kachi and Hirose 1985; Rees et al. 1999). This ensures that the IBM is realistic, in the sense that the demography is roughly comparable to that of the original *Oenothera* population, but that we still know the true parameters underlying the simulated data.

To make the code simpler to follow we have divided the code into 3 sections, 1) Monocarp Demog Funs.R contains the demographic function that describe the how size influences fate, and also the functions used to define and make the iteration kernel; 2) Monocarp Simulate IBM.R simulates the individual-based model, and 3) Monocarp Calculations.R implements various calculations. If you're not interested in how the data was generated then you can ignore Monocarp Simulate IBM.R for now.

In the Monocarp Demog Funs.R script we first store the parameters estimated from the field in m.par.true, and each is named to make the subsequent formulae easier to interpret.

```
m.par.true <- c(surv.int =  -0.65,
                surv.z   =   0.75,
                flow.int = -18.00,
                flow.z   =   6.9,
                grow.int =   0.96,
                grow.z   =   0.59,
                grow.sd  =   0.67,
                rcsz.int =  -0.08,
                rcsz.sd  =   0.76,
                seed.int =   1.00,
                seed.z   =   2.20,
                p.r      =   0.007)
```

Here .int indicates the intercept, .z the size slope, and .sd the standard deviation, so to access the survival regression intercept we use m.par.true ["surv.int"]. The various demographic functions are then defined, these are described in detail in Section 2.5.4. Finally we wrapped up the code to make the iteration matrix (Box 2.1) in a function mk_K which we pass the number of mesh points m, parameter vector m.par, and the integration limits L,U.

The Monocarp Simulate IBM.R script runs the IBM, and for those interested in this we provide in the rest of this subsection a brief overview of what the

code does - this is not essential. We simulate the population as follows. Starting
with reproduction, we generate a Bernoulli (0 or 1) random variable, $Repr \sim Bern(p_b(z))$, for each plant to determine which ones reproduce. The function
$p_b(z)$ is a size-dependent probability of reproduction estimated from a logistic
regression with a logit link function. This means that by applying the inverse of
the logit transformation we can write this probability as $p_b(z) = (1 + e^{-\nu_f})^{-1}$,
where the estimated linear predictor is $\nu_f = -18 + 6.9z$. Each plant that flowers
then produces a Poisson number of seeds, $Seeds \sim Pois(b(z))$, where $b(z)$ is
the expected number of seeds predicted from a Poisson regression with a log
link function, such that $b(z) = \exp(1 + 2.2z)$. To go from seeds to new recruits,
we then simulate *a single binomial random variable* for the whole population,
$Recr \sim Bin(S_T, p_r)$, where p_r is the fixed establishment probability ($=0.007$)
and S_T is the total seed production by all flowering plants. To complete the
recruitment process we assign a size to the recruited individuals by simulating
a Gaussian random variable, $Rcsz \sim Norm(\mu_c, \sigma_c)$, where μ_c ($= -0.08$) and σ_c
($= 0.16$) are the mean and standard deviation of new recruit sizes.

Having dealt with flowering and recruitment, we now need to simulate
survival and death of established individuals. For each nonflowering plant
($Repr = 0$) we first simulate a Bernoulli random variable, $Surv \sim Bern(s(z))$,
where $s(z)$ is the size-dependent probability of survival. This was estimated
from a logistic regression, and therefore has the form $s(z) = (1 + e^{-\nu_S})^{-1}$,
where $\nu_S = -0.65 + 0.75z$. Following the survival phase we allow the surviving ($Surv = 1$) individuals to grow by simulating a Gaussian random variable,
$z1 \sim Norm(\mu_G(z), \sigma_G)$, where $\mu_G(z) = 0.96 + 0.59z$ is the expected size of an
individual at the next census given their current size and σ_G is the standard
deviation ($= 0.67$) of the conditional size distribution, estimated from a linear
regression. Note, the IBM is slightly more complicated than it needs to be as
we also track the ages of each individual, as this is needed in the next chapter.

2.5.3 Demographic analysis using lm and glm

The results of the simulation are stored in an R data frame (sim.data) and to
keep life simple the columns are named in such a way that they correspond to
the random variables we used in the IBM. Using the IBM we can generate a
data set of any size we want, but to make the example realistic we sampled
1000 individuals from the simulated data. Before we do any analysis we should
always check the data file has the right structure, for example flowering plants
(Repr= 1) need NA's in the Surv (survival) and z1 (next year's size) columns. This
is because flowering is the first event in the life cycle and flowering plants die
and so they have already been removed from the population before we assume
mortality and growth occur. Table 2.1 shows a snippet from the data frame,
which has the NAs in the right places.

In order to begin building an IPM we need to fit a series of models capturing the size-dependent rates of survival and reproduction, along with a pair of
functions capturing the growth of established individuals and size distribution
of recruits. The R code for fitting the required models is fairly simple. Survival
is modeled by a logistic regression, so we fit it by

Table 2.1 A few lines from the data frame produced by the monocarp IBM.

z	Repr	Seeds	Surv	z1	age
1.80	0	NA	1	0.92	5
-0.62	0	NA	0	NA	0
0.54	0	NA	1	0.42	1
1.45	0	NA	0	NA	1
-0.12	0	NA	0	NA	0
1.90	0	NA	0	NA	2
3.03	1	2161	NA	NA	1
0.45	0	NA	1	-0.01	1
1.67	0	NA	0	NA	3
-0.37	0	NA	1	0.34	0
0.53	0	NA	1	0.51	0

```
sim.data.noRepr <- subset(sim.data, Repr==0)
mod.Surv <- glm(Surv ~ z, family = binomial, data = sim.data.noRepr)
```

The first line selects the nonreproductive individuals ($Repr = 0$), and then we fit the glm to this subsetted data frame. We remove the flowering individuals because flowering is fatal and occurs before other events in the life cycle. We could have just fitted the glm using sim.data as the NA's would have been removed in the analysis, however by explicitly subsetting the data frame it is easier to see what the analysis is doing. The probability of flowering analysis is essentially the same,

```
mod.Repr <- glm(Repr ~ z, family = binomial, data = sim.data)
```

For these plants that flower ($Repr = 1$), we fit the fecundity function as

```
sim.data.Repr <- subset(sim.data, Repr==1)
mod.Seeds <- glm(Seeds ~ z, family=poisson, data=sim.data.Repr)
```

For growth and recruit size we fit the models using linear regression,

```
mod.Grow <- lm(z1 ~ z, data = sim.data)
sim.data.Rec <- subset(sim.data, age==0)
mod.Rcsz <- lm(z ~ 1 , sim.data.Rec)
```

where we use the NAs to remove the flowering or dead individuals from the growth analysis, and for recruit size fit a model with just an intercept as parental size doesn't influence recruit size, so recruit size has a Gaussian distribution with constant mean. Finally we estimate the probability of recruitment by dividing the number of recruits by total seed production. The statistical analysis presented here (R code in Monocarp Calculations.R) is simplified as we know the form of the functions that generated the data; in a real example we would plot residuals, test the regression assumptions, and explore a range of alternative models (see Section 2.7), but for now we can cheat a bit. Having fitted the various models we then store the parameter estimates in m.par.est, using the same order and names a m.par.true.

```
m.par.est <- c(surv     = coef(mod.Surv),
               flow     = coef(mod.Repr),
               grow     = coef(mod.Grow),
               grow.sd  = summary(mod.Grow)$sigma,
               rcsz     = coef(mod.Rcsz),
               rcsz.sd  = summary(mod.Rcsz)$sigma,
               seed     = coef(mod.Seeds),
               p.r      = p.r.est)

names(m.par.est) <- names(m.par.true)
```

2.5.4 Implementing the IPM

We will use the midpoint rule to implement our IPM and so we need to define the functions $P(z', z)$ and $F(z', z)$, see Box 2.1. The first step is to write down the form of the kernel,

$$K(z', z) = (1 - p_b(z))s(z)G(z', z) + p_b(z)b(z)p_r c_0(z'). \qquad (2.5.1)$$

As reproduction occurs first there is no $s(z)$ in the reproduction component of the kernel, and because reproduction is fatal only nonflowering plants survive to next year, hence the initial $(1 - p_b(z))$ in the survival component. The survival component of the kernel, defined in Monocarp Demog Funs.R, is given by

```
P_z1z <- function (z1, z, m.par) {
    return((1 - p_bz(z, m.par)) * s_z(z, m.par) * G_z1z(z1, z, m.par))
}
```

For clarity we have written the various functions in the order they occur in the kernel, and the naming of the R functions follows those used in the kernel, equation (2.5.1). For example, survival $s(z)$ is called s_z. $s(z)$ was fitted by logistic regression, so the logit of the survival probability is a linear function of the covariates. To calculate $s(z)$ we first calculate the linear predictor using the coefficients of the fitted model stored in m.par, and then transform to the probability scale with the inverse logit (i.e., logistic) transformation:

```
s_z <- function(z, m.par) {
    # linear predictor:
    linear.p <- m.par["surv.int"] + m.par["surv.z"] * z
    # inverse logit transformation to probability:
    p <- 1/(1+exp(-linear.p))
    return(p)
}
```

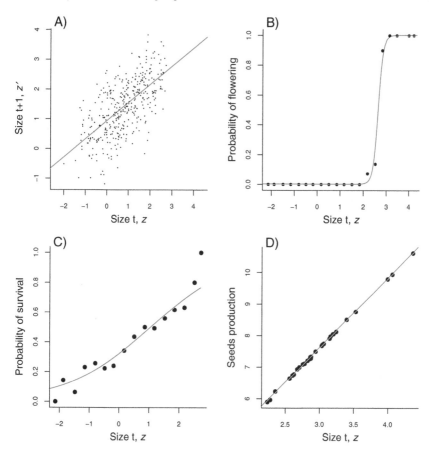

Fig. 2.4 The main demographic processes in the *Oenothera* life cycle. A) Plant size at the next census, B) the probability of flowering, C) the probability of survival, and D) seed production as functions of size at the current census. Source file: `Monocarp Calculations.R`

The $G(z', z)$ function is slightly more complicated as we are going to calculate the probability density function for size at the next census, z' conditional on current size z. To do this we use the regression of z' on z to calculate a plant's expected size at the next census given its current size, which in this case is `mu <- m.par["grow.int"] + m.par["grow.z"] * z`. The standard deviation of z' is determined by the scatter about the regression line in Figure 2.4A, which we also estimate from the fitted regression. Finally because the deviations about the fitted line are assumed to follow a Gaussian distribution, we calculate the probability density function (pdf) of subsequent plant size z' given its current size z using `dnorm`.

```
G_z1z <- function(z1, z, m.par) {
    mu <- m.par["grow.int"] + m.par["grow.z"] * z # mean size next year
    sig <- m.par["grow.sd"]                       # sd about mean
```

```
    p.den.grow <- dnorm(z1, mean = mu, sd = sig) # pdf of new size z1,
                                                 # for current size = z
    return(p.den.grow)
  }
```

For the reproduction component of the kernel we have

```
F_z1z <- function (z1, z, m.par) {
    return(p_bz(z,m.par) * b_z(z,m.par) * m.par["p.r"] * c_0z1(z1,m.par))
  }
```

As before we use the same naming convention as in equation 2.5.1 for the R
functions. With these functions defined and the limits of integration specified we
can compute the iteration matrix (see Box 2.1) using the function mk_K defined
in Monocarp Demog Funs.R, as explained below.

2.5.5 Basic analysis: projection and asymptotic behavior

For the *Oenothera* IBM there is no density dependence and so we might expect the population to grow or decline exponentially and the size distribution
to converge to some stable distribution, independent of the initial population
structure. Let's simulate and see. The result, Figure 2.5, certainly seems to
suggest that the total population size grows exponentially and the size structure
settles down to a stable distribution. For the true, m.par.true, and estimated,
m.par.est, parameters we can calculate the iteration matrices using mk_K and then
calculate their dominant eigenvalue using eigen, an industrial strength numerical
method for calculating eigenvalues and vectors, and compare this with the IBM.
The dominant eigenvalue describes the asymptotic rate of population increase,
see Section 3.1.1 for more detail. Here's the R-code:

```
> IPM.true <- mk_K(nBigMatrix, m.par.true, -2.65, 4.5)
> IPM.est  <- mk_K(nBigMatrix, m.par.est, -2.65, 4.5)

> Re(eigen(IPM.true$K)$values[1])
[1] 1.059307
> Re(eigen(IPM.est$K)$values[1])
[1] 1.040795

> fit.pop.growth <- lm(log(pop.size.t) ~ c(1:yr))
> exp(coef(fit.pop.growth)[2])
 c(1:yr)
1.072316
```

The function mk_K has four arguments: the number of mesh points, the parameter
vector, and the two integration limits, which in this case have been set slightly
outside the observed size range from the simulation. Using 250 mesh points
(nBigMatrix<-250) and true parameters we first calculate the iteration matrix and
store the result in IPM.true, we then do the same with the estimated parameters

and store the result in `IPM.est`. Using `eigen` we then calculate the real part `Re` of the dominant eigenvalue - note `eigen` calculates all the eigenvalues and vectors of the matrix and stores then in decreasing order. For the IBM we estimate the finite rate of increase by regressing `log(pop.size)` against time. There is good agreement between the IBM and the true and estimated IPMs, although even in this ideal case where we know the correct model and have a reasonable sample size (1000 observations) discrepancies of a couple of percent can occur, particularly in high fecundity systems - a single *Oenothera* plant can produce 10,000s of seeds. For the IBM there's not much else we can do except have multiple runs and hope they all show the same pattern. For the IPM the situation is different, as the conditions for stable population growth are well understood (Ellner and Rees 2006). The simplest sufficient condition is that some iterate of the kernel is a strictly positive function; this can be checked in practice by making sure that in some sufficiently high power of the iteration matrix, all entries are > 0 ("high power" here refers to matrix multiplication, e.g., K^3 means `K%*%K%*%K` in R). Other sufficient conditions, which are less easy to check, will be discussed in Chapter 6.

Mean plant size and the mean flowering plant size seem to settle down to some value (Figure 2.5B and C). How can we calculate these from the IPM? We can use `eigen` to calculate the dominant eigenvector, $w(z)$, for the iteration matrix.[2] For IPMs, like matrix models, the dominant right eigenvector is the stable distribution of the individual-level state variable: age, stages, or (in this case) sizes. To calculate average values we need the probability density function (pdf) of size - these have the property that they integrate to 1 which $w(z)$ doesn't. To convert our $w(z)$ to a pdf we just divide by $\int w(z)\, dz$. Average size is then given by

$$\bar{z} = \frac{\int w(z)\, z\, dz}{\int w(z)\, dz}. \tag{2.5.2}$$

In R we can do this as follows:

```
> meshpts <- IPM.true$meshpts
> w.est <- Re(eigen(IPM.est$K)$vectors[,1])
> stable.z.dist.est <- w.est/sum(w.est)
> mean.z.est <- sum(stable.z.dist.est*meshpts)
> mean.z.est
[1] 0.4556423
```

Let's unpack that a bit. First we store the mesh points defined in Box 2.1 in `meshpts`, so the subsequent formulae look a bit neater. Then we use `eigen` to find the dominant eigenvector w of the iteration matrix, and scale w so that it becomes a discrete probability density. The entries in `stable.z.dist.est`

---

[2] An eigenvector of an IPM is a function of z so some authors call it an eigenfunction; we don't.

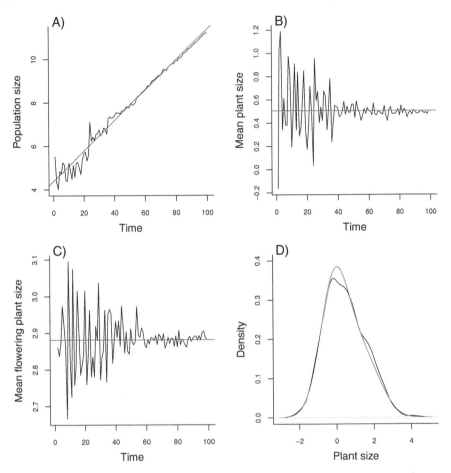

Fig. 2.5 Simulation of the *Oenothera* IBM showing A) log population size, B) mean plant size and C) mean flowering plant size plotted against time. In D) we have plotted the density estimates for size at the end of the simulation. The red lines are calculated quantities from the estimated IPM. Source file: `Monocarp Calculations.R`

then represent the proportion of individuals in each of the size categories $[z_i - h/2, z_i + h/2]$ centered on the mesh points z_i. Finally we compute the mean size by multiplying each z_i by the fraction of individuals whose size is z_i, and summing.

To compute the mean size of flowering plants, we need to take the stable size distribution and weight it by the probability of flowering. The timing of the *Oenothera* census implies that flowering occurs before death, so the stable distribution of flowering plants is $w_b(z) = p_b(z)w(z)$. Then we have to normalize w_b to a probability density, so that it represents the frequency distribution of sizes amongst individuals that flower. The calculations are very similar to those for computing mean size:

```
> wb.est <- p_bz(meshpts,m.par.est)*w.est
> stable.flowering.dist.est <- wb.est/ sum(wb.est)
> mean.flowering.z.est <- sum(stable.flowering.dist.est*meshpts)
> mean.flowering.z.est
[1] 2.985299
```

In each case the calculated means from the fitted IPM provide an excellent description of the simulation data from the IBM (Figure 2.5B and C). If the life cycle and census were such that flowering occurs after mortality we would have to first generate the density of survivors, $s(z)w(z)$, and then multiply by the probability of flowering, so the distribution of flowering plants would be $p_b(z)s(z)w(z)$.

What else might you want to know? The variance of plant size, or of flowering plant size? The fraction of the population that flowers? The effect on population growth of decreasing the flowering probability by 10%? The advantage of using midpoint rule is that intuitive calculations, like those above for mean size, actually work. We *could have* computed means size by approximating the integrals in (2.5.2) using midpoint rule; that would be

```
> stable.flowering.dist.est <- wb.est/ sum(h*wb.est)
> mean.flowering.z.est <- sum(h*stable.flowering.dist.est*meshpts)
```

which gives exactly the same result, as the h's cancel. But we don't have to do that. Instead, we think of the model's state vector $n(z_i, t)$ as a histogram summarizing a size-classified sample from the population, and compute things the way you would compute them from a histogram of data. The results are just as valid as they would be on real data. In this way, any quantity of interest can be computed by iterating the model or by finding its stable distribution, and treating the output as if it was data.

What the IPM leaves out is variation that goes away when you look at the aggregate dynamics of a very large population. Regardless of how large the population is, some kinds of variation don't average out. Consider, for example, a cohort of newborn individuals at time t who are all the same size z. If some individuals of a given size grow, and others shrink, then regardless of how many individuals there are in the cohort, after one time step there will be variation in size. The IPM includes this kind of variation, modeled by the growth kernel $G(z', z)$. The situation is different for *demographic stochasticity*, the random differences in fate among individuals who are separately "playing the same game": two individuals may have (nearly) the same survival probability, but one lives and the other dies: Section 10.3 describes methods for exploring demographic stochasticity in IPMs. In a small population, these chance outcomes can have a big impact. In a large one, however, the the *fraction* of individuals that survives will be less variable, and (in the limit of infinite population size) survival rates equal exactly the size-specific survival probability $s(z)$. You can see this in Figure 2.5A where the deviations from the trend line decrease as the population size increases. This is because the IPM averages out the demographic stochasticity and assumes that exactly $s(z)$ survivors are present next year for each size-z individual present now. So an IPM like the ones in this chapter can't tell us how the mean flowering size will vary between years as a result of the inherent randomness of mortality. For that you need an individual-based model that really "tosses coins" to decide who lives or dies. Similarly, the IPM does account for variance in recruit size, because with infinitely many recruits, there would still be variance in recruit size. But it does not account for variance in seed production among flowering plants, because with infinitely many parents, the per-capita average fecundity would exactly equal the expected per-capita fecundity.

2.5.6 Always quantify your uncertainty!

Finally, a fitted IPM leaves out the uncertainty that always comes with having a finite sample. If you estimate that $\lambda = 1.1$, what does this mean? You may have been told: it means the population will increase by 10% each year in the long run. But really *it doesn't mean anything*, until you determine how much the estimate of λ is likely to vary from one replicate data set to another.

The simplest and most general way of doing this is by bootstrapping. We will be brief, but only because the topic is covered well by Caswell (2001) and the methods and practical issues are exactly the same for IPMs. It's important, and you should learn how to do it. Quantification of the uncertainty in polar bear population projections (Hunter et al. 2010) was a key ingredient in the decision to list the species as Threatened under the US Endangered Species Act; previous analyses that did not include an uncertainty analysis had resulted in no decision (Hal Caswell, *personal communication*).

The idea behind bootstrapping is to mimic the process by which the sample of observations was drawn from the population, using your data as if they were the full population. Samples generated in this way are used to construct confidence intervals or test hypotheses (Efron and Tibshirani 1993; Davison and Hinkley 1997).

As an illustration, consider a data set of N marked individuals monitored for two years, with information on initial size and on subsequent fate (survival, reproduction, growth, etc.). We can construct an IPM and use this to estimate λ. To quantify the uncertainty in the estimate of λ, we repeatedly draw a sample of size N from the data with replacement. If X is a data matrix with row i containing data on the ith individuals, then Xboot=X[sample(1:nrow(X),replace=TRUE),] generates a bootstrap dataset. In any such dataset, some individuals occur multiple times and others are left out. Then using each bootstrap dataset in turn, we estimate the IPM parameters and calculate a λ value. The distribution of the estimated λs can then be used to construct confidence intervals (as explained by Efron and Tibshirani 1993; Davison and Hinkley 1997, and implemented in the boot package for R, which we recommend).

As an example, the script file Monocarp Lambda Bootstrap CI.R uses our monocarp IBM to generate a dataset with 1000 observations, and then bootstrap λ. To generate a bootstrapped distribution of λ the script uses the function boot, which has arguments specifying the data and the function to calculate λ; the bootstrapped estimates of λ are stored in the object boot.out. There are several

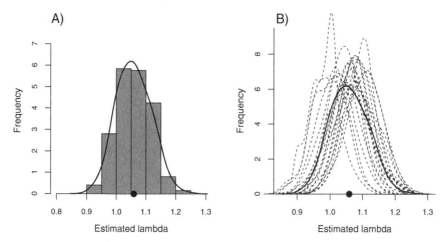

Fig. 2.6 Quantifying the uncertainty in estimates of λ based on a sample of size 1000 (here, coming from simulations of the monocarp IBM). Panel A) shows the distribution of λ estimates across replicate independent samples of size 1000, the histogram and a smooth estimate of the probability density using density with bw.SJ for the bandwidth. The solid circle at the bottom shows the value of λ for the parameters used to simulate the IPM. Panel B), dashed curves show the probability density of λ estimates from bootstrapping, repeated for 20 different samples. The solid curve is the density from Panel A. Source file Monocarp Lambda Bootstrap.R, see also Monocarp Lambda Bootstrap CI.R.

ways of calculating bootstrap confidence intervals, and these are implemented in the boot.ci function.

```
> boot.ci(boot.out, type = c("norm", "basic","perc"))
BOOTSTRAP CONFIDENCE INTERVAL CALCULATIONS
Based on 999 bootstrap replicates

CALL :
boot.ci(boot.out = boot.out, type = c("norm", "basic", "perc"))

Intervals :
Level     Normal            Basic            Percentile
95%   ( 0.8817, 1.0679 )  ( 0.8739, 1.0677 )  ( 0.8802, 1.0741 )
Calculations and Intervals on Original Scale
```

Reassuringly the confidence intervals are all similar. Rather less reassuring is the fact that despite having 1000 observations and knowing the correct demographic models the 95% confidence intervals range from ≈ 0.88 to 1.07, so we can't tell if the population is increasing or decreasing; the "true" value of λ was 1.06.

Let's step back a bit and, instead of resampling the data, we use the IBM to generate many replicate datasets. We can then use those to explore how well bootstrapping approximates the variability of estimates across replicate datasets. In Figure 2.6A we've plotted the distribution obtained by repeatedly sampling from the IBM. This defines the ideal, which we are going to approximate by bootstrapping. In Panel B we generated one sample of 1000 plants from the IBM and then bootstrapped λ values from the sample; we repeated this 20 times. From the plot it seems the bootstrapped samples have slightly less variance than the "truth," but in all cases the true value of λ is within the estimated distribution, so the bootstrap distributions have reasonable coverage.

For more complex demographic datasets the bootstrap can be adapted to mimic how the data were sampled (Kalisz and McPeek 1992). For example, if individuals are followed across multiple years, a bootstrapped sample can be constructed by sampling from the list of individuals. Likewise, if the population is stratified by location then bootstrap samples can be drawn from each location and combined. The same is true for sampling designs stratified by time, life stage, genotype, etc.

Confidence intervals on λ can also be constructed by sampling from the posterior distribution of parameter estimates if you've fitted your demographic models in a Bayesian framework; see Metcalf et al. (2009c) for an ecological example involving missing data and density dependence. Within the frequentist setting, we can sample from the estimated distributions of the regression parameters, incorporating the covariance when appropriate, and use these to quantify the uncertainty in the model and its predictions (a good example with explanation of the methods is Pacala et al. 1996). However, because IPMs are built from several separate demographic models, there can be correlations among parameters from different models that use the same data. This is most likely when marked individuals are followed over multiple censuses, so data on one individual may be used in the survival, growth, and fecundity models. An appropriately structured bootstrap accounts for this automatically, but these correlations are omitted when parameters are sampled independently from the estimated parameter distributions of two demographic models.

2.6 Case study 2A: Ungulate

In plant populations, some continuous measure of individual size is often a reliable and easily measured predictor of demographic performance, and key life history transitions such as flowering often depend more on size than on age. An IPM accommodates such demography in a straightforward manner. In contrast, an individual's performance in an animal population is often well predicted by their age or life stage (e.g., mature versus immature), and matrix models yield a good description of the population level processes. Nonetheless, body size or mass is still an key determinant of performance in many animal populations (Sauer and Slade 1987). All else being equal, larger individuals tend to exhibit greater survival and fecundity, so including this state variable can improve projections. In other cases the whole purpose of developing the model is to understand how a continuous state variable impacts demographic processes. For example, IPMs were used to understand how changes in body mass associated with environmental change mediated a major shift in the population dynamics of yellow-bellied marmots, *Marmota flaviventris* (Ozgul et al. 2010). In such cases an IPM is the best tool for the job.

In this section we develop a second case study to illustrate how an IPM can be used to explore body mass-structured dynamics of an archetypal ungulate population. The example is motivated by the well-studied feral Soay sheep (*Ovis aries*) population from the island of Hirta in the St. Kilda archipelago, off the northwest coast of Scotland. We have chosen to base our example on this system because it is one of the richest animal demographic datasets and has consequently been a major target of research into the dynamics and evolution of wild populations. Sheep do not stand still and wait to be counted, but if they're on a small island it's still possible to find them all and weigh them.

In this section our analysis will again be based on a dataset constructed from an individual-based simulation (IBM) parameterized using the Soay dataset. This allows us to simplify some of the details of the life history and keep the case study relatively simple, while again permitting straightforward comparison of the model predictions with "truth." Our aim is to demonstrate that the approach we have sketched out for moving from an individually structured dataset to a fully parameterized IPM applies to almost any life history that can be approximated as a sequence of transitions in discrete time.

2.6.1 Summary of the demography

We will assume that our simulated population is similar to the real Soay sheep population, but with a few important simplifications discussed below. The St. Kilda population has been studied in detail since 1985. Each year newborn individuals are caught, weighed, and tagged shortly after birth in the spring, and body mass measurements are taken from approximately half the population each summer during a catch in August. Maternity is inferred from detailed field observations, while periodic censuses and mortality searches ensure that individual mortality status and population density are very well characterized. Since body mass data on both established individuals and new recruits is only available during the August catch, it makes sense to choose this date as our census point to project the dynamics from. Almost all of the mortality in the system occurs during the winter months when forage availability is low and climate conditions are harsh.

These features of the life history and census regime mean that the life cycle diagram for the Soay system corresponds to the post-reproductive census case in Figure 2.2B, i.e., each annual census occurs after the year's new offspring have been produced but prior to the key mortality period. A potential problem with assuming this sequence of events is that only adults that survive from one summer census until the next should contribute new recruits to the population, despite the fact that lambing occurs several months earlier in the intervening spring. This is demographically equivalent to assuming that any reproducing individual that dies between giving birth in the spring and the summer catch will fail to raise viable offspring. With a handful of exceptions, this is precisely what is observed in the Soay system, so we can consider this to be a reasonable assumption.

To keep our example tractable we make a number of simplifications: (1) we only consider the dynamics of females, that is, we assume the population is demographically controlled by females; (2) we ignore the impact of age-structure; (3) we assume that the environment does not vary among years, either as a result of density-dependence (e.g., resource limitation) or abiotic factors (e.g., winter weather); (4) we assume that Soay females only bear singletons, though in reality they produce twins at a rate 10–15% in any given year. All of these assumptions can be relaxed, although the resulting model is rather more complicated (Childs et al. 2011). Many of these interesting features of the system will be examined in later chapters. Finally, we work with natural log of body mass as the size measure z.

2.6.2 Individual-based model

With the life cycle in mind we can construct an IBM in which individual performance is determined by (log) body mass, z. The IBM was parameterized from a series of linear and generalized linear models fitted to the field data that describe survival, growth, and components of recruitment as functions of individual body mass (DZC, *unpublished analysis*).[3] This ensures that as in the monocarpic plant the demography of the simulated population is similar to the real Soay sheep population, yet we know the true parameters underpinning the data. We adjusted the intercept of survival function so that the population growth rate $\lambda \approx 1.01$.

The IBM is implemented as follows. First, for each individual in the current summer population we simulate a Bernoulli (0-1) random variable, $Surv \sim Bern\,(s(z))$, where $s(z)$ is the size-dependent probability of survival estimated from a logistic regression, and therefore has the form $s(z) = (1 + e^{-\nu_S})^{-1}$, where $\nu_S = -9.65 + 3.77z$. Following the survival phase we allow the surviving individuals to grow by simulating a Gaussian random variable, $z' \sim Norm(\mu_G(z), \sigma_G)$, where $\mu_G(z) = 1.41 + 0.56z$ is the expected mass of an individual next summer given their current size and σ_G is the standard deviation ($= 0.08$) of the conditional size distribution, estimated from a linear regression.

Following survival and growth of established individuals, we simulate a sequence of three processes to add new recruits to the population. For each surviving individual we simulate a Bernoulli random variable which captures reproduction, $Repr \sim Bern\,(p_b(z))$, where $p_b(z)$ is the size-dependent probability of reproduction estimated from a logistic regression. This function is $p_b(z) = (1 + e^{-\nu_b})^{-1}$, where $\nu_b = -7.23 + 2.60z$. Since we assume that a single lamb is born at each reproductive event, the next step is to simulate a Bernoulli random variable, $Recr \sim Bern\,(p_r)$, for each of the reproducing individuals that describes the recruitment of their offspring to the established population next summer. Note, that in this model, the probability of recruitment p_r is independent of parent size. Finally, we assign a mass to the rec-

---

[3] The data analysis included year-to-year parameter variation; the parameter values used here describe an average year. Likelihood ratio tests were used to assess whether or not keep size in a particular model

ruited individuals ($Recr = 1$) by simulating a Gaussian random variable, $Rcsz \sim Norm(\mu_c(z), \sigma_c)$, where $\mu_c(z) = 0.36 + 0.71z$ is the expected mass of a recruit given their mother's mass, estimated from a linear regression of offspring summer mass against maternal mass *in the previous summer*, and σ_c is the standard deviation ($= 0.16$) of the conditional size distribution.

Starting with an initial population density of 500, we simulated the population until the density reached 5000 individuals and then selected a random sample of 3000 individuals from the simulated dataset to be used in the following analysis. The R code for the demographic functions and the IBM simulation can be found in Ungulate Demog Funs.R and Ungulate Simulate IBM.R, respectively. The code for running everything and carrying out the analysis we discuss next is in Ungulate Calculations.R.

2.6.3 Demographic analysis

The results of the IBM simulation are stored in an R data frame sim.data that is essentially identical in structure to the one we used to estimate the parameters of the IBM in the first place. We have named each column so that it matches the definition of the random variables for each life cycle transition described above. We should check this carefully to make sure it has the structure we are expecting. Here is snippet of the data frame:

```
z    Surv  z1    Repr  Recr Rcsz
3.07  0    NA    NA    NA   NA
3.17  1    3.24  1      1   2.47
3.02  1    3.17  0     NA   NA
2.92  0    NA    NA    NA   NA
3.20  1    3.19  1     NA   NA
2.97  1    3.09  1     NA   NA
3.14  1    3.25  0     NA   NA
3.06  1    3.00  0     NA   NA
```

Casual inspection suggests this dataset is in good shape. For example, the one individual that survives ($Surv = 1$) but fails to reproduce ($Repr = 0$) has a sequence of missing values (NA) for the three remaining variables describing offspring recruitment.

Since we know how the data was generated we'll just fit the "right" models to the data and skip the model criticism step. Don't be distracted by the different data frames used to fit each model (e.g., surf.plot.data and repr.plot.data). We created these from sim.data so we use the correct data in each analysis and to help us plot the data and produce summary figures that follow. Survival is modeled by a logistic regression, so we fit it by

```
mod.Surv <- glm(Surv ~ z , family = binomial, data = surf.plot.data)
```

The same is true for whether or not a female reproduced ($Repr$), and whether or not that lamb survived to recruit into the population their first summer ($Recr$),

```
mod.Repr <- glm(Repr ~ z, family = binomial, data = repr.plot.data)
mod.Recr <- glm(Recr ~ 1, family = binomial, data = sim.data)
```

Note that mod.Recr does not have any dependence on mother's size z, so it is estimating a single number: the recruitment probability. The repr.plot.data data frame was constructed by subsetting the full dataset on the condition that the *Surv* variable equals one, because individuals that do not survive cannot reproduce (survival precedes reproduction). You do not have to do this if you are careful about putting NAs in the right place, but it is good practice as it may help prevent errors cropping up. The subsequent sizes of new recruits and of surviving adults are fitted by linear regression,

```
mod.Grow <- lm(z1 ~ z, data = grow.plot.data)
mod.Rcsz <- lm(Rcsz ~ z, data = rcsz.plot.data)
```

The fitted models are summarized in Figure 2.7. All of the models look fairly reasonable (as they should!). Finally, as in the monocarp example, we extract the parameter values from each of the fitted models and store them in a named parameter vector m.par.est that will be used in the IPM script.

2.6.4 Implementing the IPM

The next step is to write down the kernel and check that its formulation matches our knowledge of the life cycle and the data collection protocols,

$$K(z', z) = s(z)G(z', z) + s(z)p_b(z)p_r C_0(z', z)/2 \qquad (2.6.1)$$

In this instance both the survival and reproduction kernels contain the $s(z)$ term, because the main period of mortality occurs prior to reproduction (Figure 2.2B). The rest of the survival kernel is just $G(z', z)$ because reproduction is usually not fatal in Soay sheep. The rest of the reproduction kernel is simply the product of the probability of having an offspring, $p_b(z)$, the probability of an offspring surviving to its first census, p_r, and the offspring size distribution, $C_0(z', z)$. The factor of $1/2$ appears because we are tracking only females and assume an equal sex ratio.

This looks like the correct kernel, so we now need to implement the model. We use the approach given in Box 2.1 again so we need to specify the $P(z', z)$ and $F(z', z)$ functions:

```
## Define the survival-growth kernel
P_z1z <- function (z1, z, m.par) {
    return( s_z(z, m.par) * G_z1z(z1, z, m.par) )
}
```

```
## Define the reproduction kernel
F_z1z <- function (z1, z, m.par) {
    return( s_z(z, m.par) * pb_z(z, m.par) *
            (1/2) * pr_z(m.par) * C_0z1(z1, z, m.par) )
}
```

There is nothing new here. These are just R translations of the two kernel components, and as before, each function is passed a numeric vector, m.par,

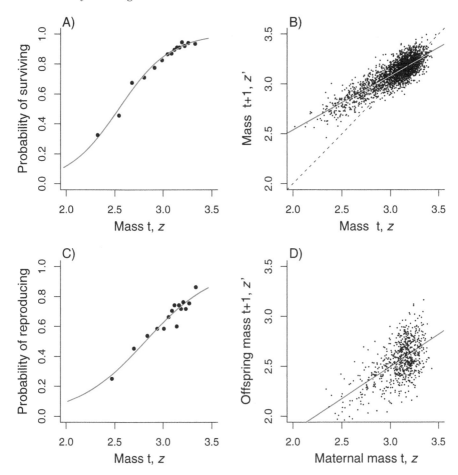

Fig. 2.7 The main mass-dependent demographic processes in the Soay sheep life cycle. A) the probability of survival, B) female mass in the next summer census, C) the probability of reproduction, and D) offspring mass. Source file: Ungulate Calculations.R

that holds the parameter values for the underlying demographic regressions. In order to complete the implementation of our model we next need to define the various functions called within P_z1z and F_z1z. Again, these correspond exactly to the statistical models we fitted to data. For example, the three functions describing the probabilities of survival, reproduction, and recruitment were fitted by logistic regression, as in the *Oenothera* model. So we again obtain the IPM functions by using the estimated coefficients to construct the linear predictor, and the inverse-logit transformation to get the probability:

```
s_z <- function(z, m.par) {
    # linear predictor:
    linear.p <- m.par["surv.int"] + m.par["surv.z"] * z
    # inverse-logit transformation:
```

```
    p <- 1/(1+exp(-linear.p))
    return(p)
}


pb_z <- function(z, m.par) {
    linear.p <- m.par["repr.int"] + m.par["repr.z"] * z
    p <- 1/(1+exp(-linear.p))
    return(p)
}


pr_z <- function(m.par) {
    linear.p <- m.par["recr.int"]
    p <- 1/(1+exp(-linear.p))
    return(p)
}
```

As in the previous example we then need to calculate the probability density function (pdf) for the size at the next census, z' given current size, z, which is done in exactly the same way.

```
G_z1z <- function(z1, z, m.par) {
    mu <- m.par["grow.int"] + m.par["grow.z"] * z # mean size next year
    sig <- m.par["grow.sd"]                       # sd about mean
    p.den.grow <- dnorm(z1, mean = mu, sd = sig)  # pdf for size z1
    return(p.den.grow)
}
```

Finally we calculate the probability density function for recruit size at the next census, given parental size in the current census, using the same approach.

```
> C_0z1z <- function(z1, z, m.par) {
    mu <- m.par["rcsz.int"] + m.par["rcsz.z"] * z # mean size next year
    sig <- m.par["rcsz.sd"]                       # sd about mean
    p.den.rcsz <- dnorm(z1, mean = mu, sd = sig)  # pdf offspring size z1
    return(p.den.rcsz)
}
```

The final step in implementation is choosing the size range and number of mesh points. Because large individuals tend to shrink and their offspring are much smaller than themselves (Figure 2.7), the upper limit just needs to be slightly larger than the largest observed size, and we set $U = 3.55$. The smallest individuals tend to grow, but they have a small chance of breeding and having offspring who are even smaller than themselves. To make sure that the IPM includes those individuals, we can compute the mean offspring size for the smallest observed size ($z \approx 2$) and subtract off two standard deviations of offspring size:

```
> m.par.est["rcsz.int"]+m.par.est["rcsz.z"]*2 - 2*m.par.est["rcsz.sd"];
> rcsz.int
[1] 1.523043
```

So we take $L = 1.5$. The total size range is $U - L \approx 2$ units on log scale. We will use 100 mesh points so that the increment between mesh points is ≈ 0.02 units on log scale, which is about a 2% difference in body mass.

2.6.5 Basic analysis

There is no density dependence in the Soay IBM so we expect the population to grow or shrink exponentially. This is indeed what we find. The finite growth rate of the population (λ) estimated from the simulation is approximately 1.023, so the population is growing by about 2% each year in the IBM. As with the *Oenothera* example (Section 2.5.5), we can estimate the growth rate using the IPM as follows: 1) use the mk_K function with the estimated model parameters m.par.est to make an iteration matrix; 2) use the eigen function to compute the dominant eigenvalue of this matrix, which is our estimate of the population growth rate. When we do this we find that the estimated IPM predicts a lambda of 1.022, which is very close to the value we calculated directly from the IBM output.

In Section 2.5.5 we also showed how to calculate the stable size distribution, $w(z)$, and mean size, $\bar{z}$, using the eigen function. This is just a matter of extracting the dominant eigenvector w, normalizing it to a discrete probability density function (stable.z.dist.est <- w/sum(w)), and using it to compute the mean. What if we want to calculate other central moments of the stable size distribution, such as the variance, σ_z^2? The same logic applies, and with $w(z)$ and $\bar{z}$ in hand such calculations are straightforward. For example, the variance can be written as

$$\sigma_z^2 = E(z^2) - \bar{z}^2 = \frac{\int w(z)\, z^2\, dz}{\int w(z)\, dz} - \bar{z}^2 \qquad (2.6.2)$$

where $E(z^2)$ is the expected value of z^2 with respect to the normalized stable size distribution. Assuming that we have stored the stable size distribution (stable.z.dist.est) and mean (mean.z.est) for the estimated model, the R code for implementing this calculation is

```
> var.z.est <- sum(stable.z.dist.est * meshpts^2) - mean.z.est^2
> var.z.est
[1] 0.07855819
```

That is, we first calculate $E(z^2)$ by multiplying each z_i^2 by the proportion of individuals in the $[z_i - h/2, z_i + h/2]$ size category, and sum these. We then subtract the square of the mean, $\bar{z}^2$, to arrive at the variance. Not surprisingly, this is very close to the size variance estimated directly from the IBM data (=0.078).

More generally, the expected value of any smooth function of size with respect to the stable size distribution can be approximated in this way: first evaluate the function at the meshpoints, then multiply each of these by the corresponding value of the normalized stable size distribution, and sum. For example, if you want to know the mean size of female sheep on the untransformed size scale -

remember, the Soay model works with log body mass - you just need to apply
the exponential function to the mesh points first.

```
> mean.z.ari.est <- sum(stable.z.dist.est*exp(meshpts))
> mean.z.ari.est
[1] 20.57083
```

What else might we like to do with the model? In some settings it can be
useful to know something about the stable size-age structure implied by the
model. For example, knowledge of the implied age structure can provide a useful
sanity check; you might worry if the oldest individual ever observed in your study
population was 9 years old but your model predicts that 20% of individuals
should be older than this. There may be good reasons for such discrepancies
(e.g., if the population was recently perturbed) but it they may also indicate
that it is time to pause and revisit the model.

The stable size-age distribution can be calculated by using the reproduction
and survival kernels to project forward one time-step a cohort derived from a
population at the stable size distribution, and separating out the number of
newborns, one-year olds, and so on. There is one important snag though: in a
growing or shrinking population the relative number of individuals in a given age
class are influenced by both the demographic rates and the population growth
rate. To begin, we calculate the size distribution of new recruits, $w_0(z)$, using
the fecundity component of the kernel and the stable size distribution

$$w_0(z') = \frac{1}{\lambda} \int F(z', z) \, w(z) \, dz \tag{2.6.3}$$

scaling by $1/\lambda$ to compensate for the change in total population. To see why we
do this scaling, recall that $\lambda w = (P + F)w$ so that

$$w = \frac{1}{\lambda}(P + F)w.$$

Equation (2.6.3) calculates the part of the w on the left-hand side in the last
equation that are newborns, namely $\frac{1}{\lambda}Fw$. Knowing the part of w that are
newborns, we can use that in the right-hand side w to calculate the one-year
olds as

$$w_1(z') = \frac{1}{\lambda} \int P(z', z) \, w_0(z) \, dz. \tag{2.6.4}$$

The size distribution associated with older individuals can be calculated itera-
tively in the same way,

$$w_a(z') = \frac{1}{\lambda} \int P(z', z) \, w_{a-1}(z) \, dz. \tag{2.6.5}$$

You can probably guess what the R code associated with these calculations (for
the first 4 age classes) looks like.

```
> a0.z.dist.est <- IPM.est$F %*% stable.z.dist.est / lam.est
> a1.z.dist.est <- IPM.est$P %*% a0.z.dist.est / lam.est
```

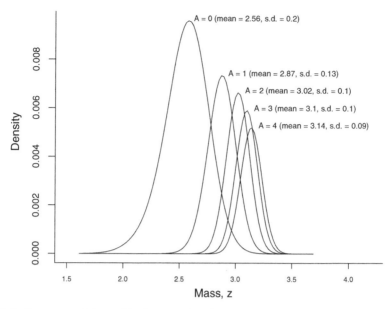

Fig. 2.8 Summary of the stable size-age distribution of the estimated Soay model. Only the first 5 age classes are shown. Source file: Ungulate Calculations.R

```
> a2.z.dist.est <- IPM.est$P %*% a1.z.dist.est / lam.est
> a3.z.dist.est <- IPM.est$P %*% a2.z.dist.est / lam.est
> ## and so on...
```

Figure 2.8 plots these distributions to summarize the estimated stable size-age structure of the Soay population.

As in the monocarp case study, the calculations in this section are essentially simulation-based: use the fitted IPM to simulate a population, and then compute statistics from model output in exactly the way you would compute them from data. The calculation of var.z.est, for example, is exactly what you would do with a table giving the numbers of individuals in a set of size categories. The calculation of a3.z.dist.est and so on corresponds to the experiment: mark a set of individuals at birth, and come back in each successive year to observe how many survived and what their sizes are. Anything you could do with those data can be done in exactly the same way with output from the IPM.

However, many age-related properties (e.g., mean age at first reproduction) and other demographic parameters (e.g., R_0) can be calculated analytically from the IPM kernels and functions. Those methods are the focus of the following chapter.

2.7 Model diagnostics

Each population is a unique situation, so developing a good model is an iterative process of developing candidate models based on the biology and life history of the species, probing those models for faults, and then trying to resolve them.

It's important to double-check a model at all steps, from diagramming the life cycle through implementing the model on the computer.

2.7.1 Model structure

A model such as equation (2.3.4) is a sentence that you can "read out loud" to see if it matches what you believe about the population. The term $s(z)G(z', z)$ says:

> An individual of size z at time t will be size z' at time $t + 1$ if it survives, and then grows (or shrinks) to size z'.

The next term $p_b(z)p_r b(z)C_1(z', z)$ says:

> Production of new offspring is a multistep process. Starting from the current census: an individual has a size-dependent probability of breeding $(p_b(z))$. If the individual breeds, it produces a size-dependent number of offspring ($b(z)$ on average), each of which has probability p_r of surviving to the next census. The size distribution of new recruits that survive $(C_1(z', z))$ is dependent on parent size.

When you read your kernel aloud in this way, it should match your understanding of the species' life cycle.

2.7.2 Demographic rate models

The functions that make up the kernel are statistical models that can be interrogated with standard model diagnostics. Statistical models are often chosen based on tradition, such as logistic regression for survival probability. But tradition is often a reflection of what was computationally feasible half a century ago. Modern computers and R allow us to let the data "speak for themselves" about what models we should fit and let us carefully test the adequacy of simple models.

Figure 2.9 shows a few simple diagnostics for the *Oenothera* growth model $G(z', z)$, again using the artificial data from the IBM held in the data frame sim.data. The growth model is a linear regression,

```
mod.grow <- lm(z1 ~ z, data = sim.data)
```

here using a subset of the simulated data so that growth is estimated from observations on about 120 individuals. The first 3 panels are similar to what you would get from R's built-in diagnostics for a linear regression using plot(mod.grow). But we prefer to do it ourselves, so that we can use some features from other packages.

- Residuals should have constant variance and no trend in mean; plotting residuals versus fitted values (panel A) provides a visual check on these properties. The plotted curve is a fitted nonparametric spline curve, using the gam function in the mgcv package (Wood 2011). It hints at the possibility of a small nonlinear trend, but this may be driven by a few points at the left (and since the data come from the IBM, we know that the underlying growth model really is linear).

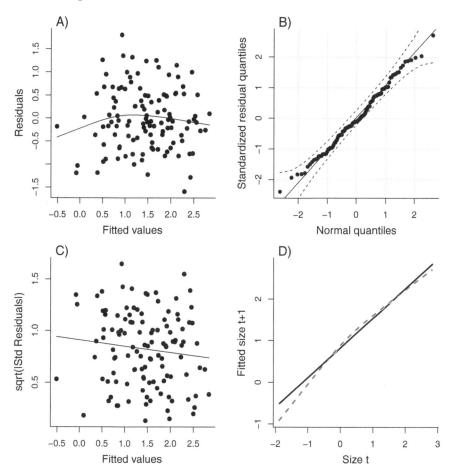

Fig. 2.9 Diagnostic plots for the monocarp growth function; see text for details. Source file: Diagnose Monocarp Growth Kernel.R

- Residuals should be Gaussian. A quantile-quantile plot (panel B), using the qqPlot function in the car library, supports this. Perfectly Gaussian residuals would fall on the 1:1 line (solid). The dashed lines are a 95% pointwise confidence envelope, so we should worry if more than 5% of points fall outside the envelope or if any points lie far outside it. This one looks OK. As a further check, we can test for statistically significant departures from Gaussian distribution:

```
> sresid <- rstandard(mod.grow); shapiro.test(sresid);
Shapiro-Wilk normality test
data: sresid
W = 0.9909, p-value = 0.6288
```

This confirms what we know: the growth distribution is Gaussian.

- A better check for constant error variance is a scale-location plot (panel C). The plotted points are the square-root magnitude of the standardized residuals, and the curve is again a fitted spline. The spline hints at a possible weak trend so we test for significance (and find none):

```
> cor.test(zhat,sqrt(abs(sresid)),method="k");
Kendall's rank correlation tau
data:  zhat and sqrt(abs(sresid))
z = -1.1606, p-value = 0.2458
```

"Standardized" means that residuals have been adjusted for differences in leverage between points, which cause residual variance to be nonconstant even when the true error variance is constant. In large data sets standardization usually has little effect because no point has much leverage, but in smaller data sets standardization can remove spurious patterns in the magnitude of residuals.[4]

In a typical regression analysis the main goal is to estimate the trend represented by the regression line. It's OK if the residuals are "close enough" to Gaussian and the variance is "close enough" to constant. But in an IPM, the scatter around the mean growth rate is also an important part of the model. A smaller growth variance in large individuals might be important for predicting longevity because it keeps big individuals from shrinking, even if it is inconsequential for estimating the effect of size on mean growth rate.

- To follow up on the hint of nonlinearity in panel (A), we can compare the linear model with a spline fit to the same data (panel D, solid and dashed lines).

The spline in panel (D) suggests that growth is a weakly nonlinear function of size. We know this isn't true in this case – it's an accident of random sampling – but with real data we would have to decide between the linear and nonlinear models. The statistical evidence is equivocal: a significance test using anova(mod.grow,gam.grow) is marginally nonsignificant ($P = 0.063$), while AIC slightly favors the nonlinear model (AIC=243.3) over the linear model (AIC=244.9).

The AIC difference is small enough that most users would probably select the simpler model. However, its lower AIC means that the nonlinear model is expected to make more accurate predictions (recall that in the frequentist framework AIC is a large-sample approximation to out-of-sample prediction error, which is why frequentists, and agnostics like us, guiltlessly use both AIC and p-values). Moreover, Dahlgren et al. (2011) have shown that even weak nonlinearities can sometimes have substantial effects on model predictions, so the nonlinear model should be taken seriously. When the statistical evidence for one model over another is equivocal, unless one of the models is strongly favored based on some underlying biological hypothesis, we believe that the

---

[4] What we call standardized residuals are sometimes called Studentized residuals. We follow the terminology used in R.

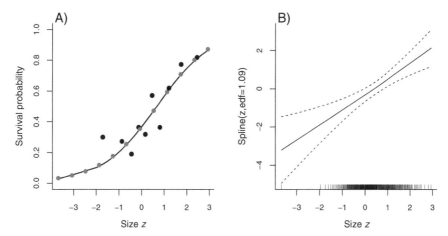

Fig. 2.10 Diagnostic plots for the fitted *Oenothera* survival function. (A) Survival as a function of size. The solid curve is the prediction of the fitted GLM (logistic regression); the red circles are the predictions of the fitted nonlinear GAM (nonparametric logistic regression); the black circles are survival fraction as a function of mean size for a series of size classes defined by percentiles of the size distribution. (B) The result of plotting the fitted GAM using plot(gam.surv), which shows the fitted spline regression (solid curve) on the scale of the regression model's "linear predictor." If this curve is a straight line (with 1 degree of freedom specifying the slope), the GAM is equivalent to the GLM. In this case the fitted GAM has 1.09 "effective degrees of freedom," and the linear model is well within the confidence bands on the GAM estimate. Source file: Diagnose Monocarp Survival.R

best approach is to try both models and attach most confidence to conclusions that the models agree on. Model averaging is another possibility. But current model averaging approaches are not always effective (Richards et al. 2011), and we think that it is more informative to show the degree of uncertainty by presenting results from the range of plausible models.

Residual plots are less informative for the survival or flowering models, because all observed values are either 0 or 1 and there is no expectation of Gaussian residuals. One visual check is to compare model predictions with survival rates within size classes (Figure 2.10A). And we can again compare the linear model with a nonlinear model using gam (Figure 2.10B). In this case the linear model is supported: it has lower AIC, and the difference between the linear and nonlinear models is minuscule. Further checks are to test for overdispersion, to test whether the fit is significantly improved by adding predictors other than size (e.g., individual age if it is known), and so on.

2.7.3 Implementation: choosing the size range

It seems natural that a model's size range should correspond to the range of observed sizes, perhaps extended a bit at both ends. Many published studies have used this approach. For *Oenothera*, we presented above a "data" analysis based on a sample of 1000 individuals. In a sample of 1000 individuals from the stable size distribution, the range of observed sizes is typically -2.55 to 3.9 (these were the median min and max sizes from 10000 replicate samples). Because size is on a natural-log scale, setting $[L, U] = [-2.65, 4.0]$ allows the IPM to include individuals $\approx 10\%$ smaller or larger than any in the sample.

However, it's important to check whether a size range chosen in this sort of way is really big enough that individuals aren't getting "evicted" from the model. *Eviction* refers to situations where the distribution of subsequent size z' extends beyond $[L, U]$, so that individuals in the tails of the growth or recruit size distributions aren't accounted for when the model is iterated (Williams et al. 2012).

Preventing eviction of new recruits is easy: make sure that $[L, U]$ is wide enough to include effectively all of the recruit size distribution. The fitted recruit size distribution is Gaussian with mean of ≈ -0.1 and standard deviation of ≈ 0.8. Some recruits are evicted because a Gaussian has infinite tails. But we can check that this number is small enough to ignore by computing the non-evicted fraction:

```
> pnorm(4,mean=-0.1,sd=0.8) - pnorm(-2.65,mean=-0.1,sd=0.8);
[1] 0.9992823
```

This is the fraction of recruits whose size is below 4, but not below -2.65. So fewer than one recruit in a thousand gets evicted – not a problem.

For older individuals, we need to look at the growth kernel $G(z', z)$. Figure 2.11A shows that with $[L, U] = [-2.65, 4]$ in the *Oenothera* IPM, the largest individuals lose a substantial part of their subsequent size distribution to eviction. So eviction is happening, and the next question is: does it matter? A large *Oenothera* is very likely to flower and die rather than continuing to grow, so what would happen if it lived is irrelevant, both in the model and in reality. To see if this argument really holds up, we can look at the fraction of individuals that the IPM sends to the "right place." For a real *Oenothera* there are three possible fates: flower and die, die without flowering, or survive with some new size z'. In the model there's a fourth possibility: eviction. The probability of going to the wrong place (eviction) for a size-z individual is the integral of $s(z)(1 - p_b(z))G(z', z)$ from $z' = U$ to ∞. Plotting this probability when $U = 4$ (solid curve in Figure 2.11B) turns up a surprise: eviction is most likely in the middle of the size range, where flowering is not a near-certainty. Still, the maximum risk of eviction is under 1%, and relative to other flaws in a data-driven model this one is probably not worth fretting over.

But what if you're not so lucky, and your IPM is sending too many individuals to the wrong place? Or if you're worried that eviction might still affect evolutionary analyses, because it causes reproductive value to be under-estimated?

A useful first step is to ask if eviction is a symptom of flaws in the growth model. Perhaps the data are better described by a different growth model with lower or no eviction: a model with nonconstant growth variance, or a bounded growth distribution (e.g., a shifted and stretched beta distribution). Or perhaps there's some reason why real individuals die before reaching their maximum potential size, and you need to make that happen in the model.

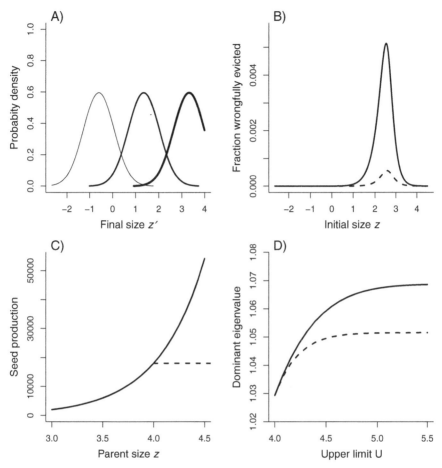

Fig. 2.11 Detecting and eliminating eviction of large individuals in the *Oenothera* IPM.
A) Plots of the growth kernel $G(z', z)$ for z at the lower endpoint, midpoint, and upper
endpoint of the size range $[L, U] = [-2.65, 4]$. B) Size-dependent probability of eviction
with $L = -2.65$ and upper size limit $U = 4$ (solid) and $U = 4.5$ (dashed). C) Per-capita
seed production as a function of parent size with no demographic ceiling (solid) and a
ceiling at $U_1 = 4$ (dashed). D) Dominant eigenvalue λ as a function of the upper limit
U for $L = -2.65$ with no demographic ceiling (solid) and a ceiling at $U_1 = 4$ (dashed).
Source file: MonocarpGrowthEviction.R

Again, what if you're not so lucky: the growth model fits the data well, but
eviction still happens. Williams et al. (2012) present several possible solutions,
but caution that the best approach is case-dependent: there's no universally
right cure for eviction. In *Oenothera*, as in many other cases, individuals near
the top of the size range tend to shrink, so setting $U = 4.5$ reduces the maximum
eviction risk to under one in a thousand (Figure 2.11B, dashed curve). However,
the model then extends far beyond the largest individual likely to be sampled,
and that kind of extrapolation is dangerous. Increasing U from 4 to 4.5 is

almost a doubling of the maximum size, and the maximum fecundity is tripled (Figure 2.11C, solid curve). To avoid creating such Titans, you can impose a demographic ceiling: a size U_1 such that individuals larger than U_1 have the same vital rates as individuals of size U_1. This approach, used by Easterling et al. (2000), is equivalent to adding a discrete class of "large" individuals (all those above the ceiling) who are demographically identical. The dashed line in Figure 2.11C shows per-capita seed production with a demographic ceiling at $U_1 = 4$.

In practice adding a demographic ceiling can be done very simply: a line like

```
K=h*outer(meshpts,meshpts,kernel.function,m.par)
```

becomes

```
K=h*outer(meshpts,pmin(meshpts,U1),kernel.function,m.par)
```

However, if you add a ceiling we recommend making its value one of the entries in the parameter vector m.par, and using it in each of the kernel's component functions, so that you can easily explore how sensitive your model's predictions are to the location of the ceiling. Imposing a ceiling is a modeling decision that should be guided by your substantive knowledge: what is the most reasonable way to extrapolate beyond the range of the data? Because the biggest are often the most fecund, how you extrapolate can matter even if the biggest are rare. Figure 2.11D shows how the dominant eigenvalue is affected by the upper size limit with and without a demographic ceiling at $U = 4$. With the ceiling, there's no meaningful gain from increasing the upper limit past $U = 4.5$, but that depends on what has been assumed about fecundity at unsampled sizes. As always, it's important to consider alternative plausible assumptions and how those affect your conclusions.

2.7.4 Implementation: the number of mesh points

The number of mesh points m required for accurate results depends on the kernel. An integral computed by midpoint rule is exactly right if the function being integrated is linear on each of the size classes of width $h = (U - L)/m$ centered at the mesh point. It will be close to right if the function is close to linear on intervals of length h.

For IPMs this means that h needs to be small compared to the standard deviation of offspring size, and to the standard deviation of growth increments $z' - z$. If neither of these is too small, then a simple but effective approach is to start with a reasonable value of m and increase it until numerical results stabilize to the accuracy you need. But if either of these is small, direct application of midpoint rule might lead to a very large matrix and to calculations that take a long time or crash when memory runs out.

There are a several possible ways around this problem. If the offspring size distribution is the problem and the variance of growth increments is much larger, the offspring size variance can be increased without substantially changing model outputs, because the variance in initial offspring size is swamped by the variance in their first growth increment. Suppose new offspring have standard deviation σ_0 and first year growth has standard deviation $\sigma_G \gg \sigma_0$. Their size at age 1 then has variance $\sigma_0^2 + \sigma_G^2$. Increasing σ_0 to a larger value $\tilde{\sigma}_0$ such that $\tilde{\sigma}_0^2$ is still $\ll \sigma_G^2$ will have a trivial effect on the variance in their size at age 1. Again, there is a simple numerical check on this. Find a value of $\tilde{\sigma}_0$ large enough that model predictions stabilize with increasing m while m is still small enough for your computer to handle. Then increase $\tilde{\sigma}_0$ to $2\tilde{\sigma}_0$, and if the change in model predictions is very small then you're safe. This approach not likely to fail unless small size differences at birth are amplified by a largely deterministic pattern of growth, rather than washed out by variance in later growth increments.

Small variance in growth increments is a more difficult problem. In practice it arises for long-lived, slow-growing species. If growth increments are small relative to the range of individual sizes in the population, the standard deviation of growth increments is necessarily small too. The kernel can't be "fixed" in that case without changing model forecasts. One option then is to use computational methods that can deal with large matrices. For example, the dominant eigenvalue and corresponding eigenvectors can be found by iteration (Ellner and Rees 2006) instead of using `eigen` to compute all of the eigenvalues and eigenvectors. A second option is to use more efficient ways of evaluating integrals, which we discuss in Section 6.8. The third, and likely best, option is to use methods for sparse matrices. These methods (such as those in R's `spam` and `Matrix` libraries) only store and work with the nonzero entries in a matrix. If growth is nearly deterministic, the IPM kernel is nearly zero in large regions representing impossible size transitions. In that case, if you zero out the corresponding entries in the iteration matrix (e.g., set to exactly zero all matrix entries below some threshold, such that the sum of each column is reduced by less than 0.1% of its total), the resulting matrix will be sparse and computations will be much faster with sparse matrix methods. Sparse matrix calculations can also be useful when individuals are cross-classified by multiple traits (Ellner and Rees 2006). We return to these issues and illustrate the methods in detail in Section 6.6.

2.8 Looking ahead

This chapter has covered the tools for building a basic size-structured IPM from data, make projections with it, and ask some probing questions about whether the model is really doing what you want it to do. But there's much more to do than projection. An IPM is not just a population model, it's a model for the lives of individuals, and we can extract a lot of information about individual life histories and their variation. We take this up in the next chapter.

2.9 Appendix: Probability densities versus probabilities, and moving from one scale of measurement to another

One important difference between matrix and integral projection models is that demographic transitions in the matrix model are described by *probabilities*, where an IPM uses a *probability density* for the same transition. For example, in a size-structured matrix model, the projection matrix entry a_{32} representing transitions from size-class 2 to size-class 3 is the product of the survival probability s_2 for size-class 2, with the probability g_{32} that a survivor grows into size-class 3: $a_{32} = s_2 g_{32}$. In the basic size-structured IPM, we typically have instead

$$P(z', z) = s(z)G(z', z).$$

Here $s(z)$ is the size-dependent survival probability, just like in a matrix projection model. But $G(z', z)$ is the *probability density* of size z' at the next census for survivors having initial size z.

Probability densities are enough like probabilities that you can often ignore the difference. But sometimes you can't. In past writings we have sometimes glossed over the difference, for example calling $G(z', z)$ the probability of growing to size z'. But conversations and emails with IPM builders have convinced us that this shortcut created *much* more confusion than it avoided, so in this book we try to be more accurate.

The density of water is the mass per unit volume of water, such as g/cm^3. To get the mass, you multiply density by volume. Similarly, to get the probability for a given range of body sizes, we have to take the probability density and multiply it by the "volume" of the size range. Because size is one-dimensional, the volume of a size range is its length. For a size range of length h we therefore have

$$\text{Probability that new size } z' \text{ is in } [z_1, z_1 + h] \approx G(z_1, z)h. \qquad (2.9.1)$$

Why do we have $\approx$ rather than $=$ in equation (2.9.1)? Why is it just an approximation? The reason is that the probability density for z' is generally not constant, so the density at z_1 doesn't apply over the whole size range $[z_1, z_1+h]$. To be exact, we have to use the right density for each size, which is accomplished by integration:

$$\text{Probability that new size } z' \text{ is in } [z_1, z_1 + h] = \int_{z_1}^{z_1+h} G(z', z) \, dz. \qquad (2.9.2)$$

However, one way you *won't* go wrong is by thinking of equation (2.9.1) as being exactly right, so long as you remember that it only holds for small h (narrow size ranges).

From equation (2.9.1) we see one way in which probability densities are like probabilities: they give us the *relative* likelihood of different outcomes. In this case,

$$\frac{\text{Probability that new size } z' \text{ is near } z_1}{\text{Probability that new size } z' \text{ is near } z_2} \approx \frac{G(z_1, z)h}{G(z_2, z)h} = \frac{G(z_1, z)}{G(z_2, z)} \qquad (2.9.3)$$

so long as the same definition of "near" is used at z_1 and z_2. This is the best way to think about the intuitive meaning of probability density: it tells us the *relative* probability of two equal-length (small) ranges of the variable, but not the absolute probability of any particular value.

In the same way, regardless of how often we or anybody else has said otherwise, the fecundity kernel $F(z', z)$ is *not* the number of size-z' offspring produced by a size-z parent. Rather, as in equation (2.9.1), $F(z', z)h$ is (for small h) the number of offspring in the size range $[z', z'+h]$ produced by a size-z parent, and $F(z_2, z)/F(z_1, z)$ is the relative frequency of offspring with sizes near z_2 and z_1.

The difference between probability and probability density is crucial when you need to move a kernel from one scale of measurement to another.

Suppose that your data lead you to fit a growth model using a linear ("untransformed") size measure u, but for everything else log-transformed size works better, so you decide to use $z = \log(u)$ as your state variable. Then you need to take the growth model $G(u', u)$ and express it as a growth model on log-scale, $\tilde{G}(z', z)$.

If we think of $G(u', u)$ as the probability of growing to size u', the answer is easy. Going from log-size z to log-size z' is the same as going from size $u = e^z$ to size $u' = e^{z'}$. So this misguided thinking leads us to

$$\tilde{G}(z', z) = G(e^{z'}, e^z) \qquad \longleftarrow \textbf{REMEMBER: this is wrong!}$$

The misguided approach was right in one respect: we need to find $\tilde{G}$ by computing the same probability two different ways. But the right way to compute probabilities is by using equation (2.9.1), which really does involve probabilities. To simplify the calculations we can assume that h is small, and drop terms of order h^2 or smaller.

- $\tilde{G}(z', z)h$ is the probability that subsequent log size z' is in $[z', z' + h]$, starting from log size z.
- In terms of untransformed size $u = e^z$, this is the probability that subsequent size u' is in the interval $[e^{z'}, e^{(z'+h)}]$, starting from size $z = e^u$. For small h, the Taylor series for the exponential function ($e^x = 1 + x + \frac{x^2}{2} + \cdots$) implies that $e^h \approx 1 + h$. Thus

$$[e^{z'}, e^{(z'+h)}] = [e^{z'}, e^{z'} e^h] \approx [e^{z'}, e^{z'}(1 + h)] = [e^{z'}, e^{z'} + e^{z'} h],$$

a size range of length $e^{z'} h$, whose probability is therefore

$$G(e^{z'}, e^z) \times e^{z'} h.$$

We therefore have

$$G(e^{z'}, e^z)e^{z'} h = \tilde{G}(z', z)h$$

and therefore

$$\tilde{G}(z', z) = e^{z'} G(e^{z'}, e^z). \qquad (2.9.4)$$

The general recipe is as follows. Let f be the function that gives u (the scale on which G or some other kernel was fitted) as a function z (the scale where the IPM works). In the example above, $z = \log(u)$ so $u = e^z$ and the function f is $f(z) = e^z$. Then the kernel on the scale of the IPM is

$$\tilde{G}(z', z) = f'(z')G(f(z'), f(z)) \qquad \longleftarrow \textbf{This one is right!} \qquad (2.9.5)$$

where $f' = df/dz$. To implement this in an IPM, we recommend writing a function that computes $G(u', u)$ on the scale where the kernel was fitted, and a second function that computes $\tilde{G}$ by using equation (2.9.5) and calling the function that computes G.

Equation (2.9.5) applies also to functions of a single variable. The situation is especially simple for a function of just initial size z: $\tilde{s}(z) = s(f(z))$. For a function of just z' such as a parent-independent offspring size distribution, $\tilde{C}(z') = f'(z')C(f(z'))$.

The same approach works for moving the size distribution from one scale to another. At the end of Section 2.2 we gave the example of using an IPM to project $n(z, t)$ where z is log-transformed size, and then wanting to plot the distribution $\tilde{n}(u, t)$ of untransformed size $u = \exp(z)$. Again, we change scales by computing the same thing two ways. $\tilde{n}(u, t)h$ is the number of individuals in the interval $[u, u + h]$, for small h. That corresponds to the size interval from $z = \log(u)$ to $z = \log(u + h)$. To find the width of that interval we use the Taylor series

$$\log(u + h) = \log(u) + h/u + \cdots .$$

The z-interval width is therefore h/u (for small h), so the number of individuals in the interval is $n(z, t)h/u = n(\log(u), t)h/u$. We conclude that

$$\tilde{n}(u, t)h = n(\log(u), t)h/u$$

and therefore $\tilde{n}(u, t) = n(\log(u), t)/u$.

The general recipe is this. Let g be the function taking the scale u on which you want to plot, to the scale z where the IPM operates. Then $\tilde{n}(u, t) = g'(u)n(g(u), t)$. In the case $z = \log(u)$, $g = \log$ and $g'(u) = 1/u$.

2.10 Appendix: Constructing IPMs when more than one census per time year is available

Here we outline a general approach for arriving at the "right" IPM in situations where more than one census (defined as any period of data collection) is performed each year. Using our Soay IPM as a case study, we show how this approach should be applied when one or more censuses are imperfect, in the sense that not every class of individual is measured. This is exactly the situation we face in the Soay system. Only lamb masses (measured shortly after birth) are acquired in the spring, whereas information about individuals of all ages are gathered in the late summer catch. As with models based on a single census per year, our aim is to show how to derive a model that correctly reflects the life cycle and census regime, and that can be parameterized from the available data.

The key to this methodology is to initially assume we have all the data we need at each census, and specify the corresponding model. We then "collapse" this model down in stages based upon our knowledge of which data are really available and our modeling objectives. In the Soay system, this means we first

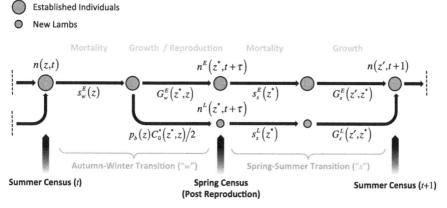

Fig. 2.12 Expanded life cycle diagram for the Soay sheep. Two census points are shown: the summer census of size z individuals and a post reproduction spring census of size z^* individuals. The diagram shows the fate of established individuals, $n^E(\ldots)$, and new lambs, $n^L(\ldots)$ over the autumn-winter transition (w subscript) and the spring-summer transition (s subscript). The life cycle is conceptualized as occurring in 4 phases: (1) an autumn-winter mortality phase; (2) an autumn-winter growth and reproduction phase; (3) a spring-summer mortality phase; (4) a spring-summer growth phase. These processes are described by survival functions $s_\circ^\circ(\ldots)$, growth kernels $G_\circ^\circ(\ldots)$, a recruit size-distribution kernel $C_0^*(z^*, z)$ and a probability of reproduction function $p_b(z)$.

need to construct a model that projects the dynamics from late summer to spring (post reproduction), and then from spring to late summer again. To do this, we will need to keep track of both established individuals (denoted with a superscript E) and new lambs (denoted with a superscript L). The established individuals' class includes every individual that survives to their first August catch. We distinguish winter and spring components of the demography with subscripts w and s, respectively. Spring size is denoted z^*. The remaining notation follows the conventions introduced in the main text unless otherwise stated (Figure 2.12).

As before, we aim to construct a model that projects the total August (summer) population size distribution, $n(z,t)$, between years. All individuals present at an August census are, by definition, established individuals. We begin with the equations for the spring size distributions of established individuals, $n^E(z^*, t+\tau)$, and new lambs, $n^L(z^*, t+\tau)$, produced by the August population at time t. These involve a winter survival-growth kernel, $P_w^E(z^*, z)$, and a winter fecundity kernel, $F_w^E(z^*, z)$, such that

$$n^E(z^*, t+\tau) = \int_L^U P_w^E(z^*, z)n(z,t)dz,$$

$$\text{where } P_w^E(z^*, z) = s_w^E(z)G_w^E(z^*, z)$$

$$n^L(z^*, t+\tau) = \int_L^U F_w^E(z^*, z)n(z,t)dz$$

$$\text{where } F_w^E(z^*, z) = s_w^E(z)p_b(z)C_0^*(z^*, z)/2$$

(2.10.1)

These expressions are very similar to the two components of the kernel from the Soay case study with one census per year, though everything in them now refers only to the winter transition. The fecundity component of the model does not contain a recruitment probability, $p_r(z)$, because this pertains to the next transition. We are also working with a different offspring size kernel, denoted C_0^* (not C_0), which describes the spring (not summer) mass of lambs.

The spring transition only involves survival and growth. The life cycle diagram tells us that the spring components of the model are

$$P_s^E(z', z^*) = s_s^E(z^*)G_s^E(z', z^*)$$

$$n^E(z', t+1) = \int_L^U P_s^E(z', z^*)n^E(z^*, t+\tau)dz^*$$

$$P_s^L(z', z^*) = s_s^L(z^*)G_s^L(z', z^*) \tag{2.10.2}$$

$$n^L(z', t+1) = \int_L^U P_s^L(z', z^*)n^L(z^*, t+\tau)dz^*$$

The total August density function next year is then

$$n(z', t+1) = n^E(z', t+1) + n^L(z', t+1).$$

We now have a *consistent* model that properly accounts for the life cycle and known census times. This model projects from one summer to the next using two transitions: summer to spring and then spring to summer. It is therefore the IPM equivalent of a seasonal (or more generally, periodic) matrix projection model.

If we really had measured the size of all individuals in the spring, we could stop the model derivation here and begin to parameterize the various component functions from the data. However, in reality this is not a *feasible* model because only lambs were measured in the spring. There is no simple way to parameterize the established individuals' spring survival function, $s_s^E(z^*)$, and growth kernel, $G_s^E(z', z^*)$, in terms of spring size, z^*.

The solution to this problem is to collapse the survival-growth components of established individuals into a single transition. Instead of two survival-growth transitions governed by $P_w^E(z^*, z)$ and $P_s^E(z', z^*)$, we have to model the summer-summer transition in a single step, such that

$$n^E(z', t+1) = \int_L^U P^E(z', z)n^E(z, t)dz \tag{2.10.3}$$

$$P^E(z', z) = s_w^E(z)s_s^E(z)G^E(z', z)$$

In this new formulation we have combined the two growth phases into one summer to summer growth kernel, $G^E(z', z)$, that can be parameterized from the summer catch data. We are still separating the winter and summer survival processes, though now they must be expressed as functions of the preceding summer mass, z. Information about mortality is gathered over most of the year from regular visual censuses and by searching for dead individuals. Capture-mark-recapture methods could therefore potentially be used to estimate $s_w^E(z)$ and $s_s^E(z)$. However, this type of analysis is time-consuming and considerable expertise is needed to deal with missing individual size data (Langrock and King 2013).

What else can be done to simplify the model? Some more knowledge of the system helps here. We know that virtually all the mortality of established individuals is experienced during the winter transition. This means that it is reasonable (and convenient) to set $s_s^E(z) = 1$. After combining equation (2.10.3) and the lamb components of equations (2.10.1) and (2.10.2), following a little rearranging we get

$$n(z', t+1) = \int_L^U [P(z', z) + F(z', z)]n(z, t)dz$$

$$P(z', z) = s_w^E(z)G^E(z', z) \tag{2.10.4}$$

$$F(z', z) = s_w^E(z)p_b(z) \int_L^U [s_s^L(z^*)G_s^L(z', z^*)C_0^*(z^*, z)/2]dz^*$$

Now let $s_s^L(z^*) = p_r$ (i.e., a constant), $s_w^E(z) = s(z)$, and $G^E(z', z) = G(z', z)$; and define $C_0(z', z) = \int G_s^L(z', z^*)C_0^*(z^*, z)dz^*$. This is essentially the same model we described in our example. The only difference is that here we have incorporated lamb spring-summer growth into the mathematical details of the model, whereas in the original example we estimated $C_0(z', z)$ directly and allowed the data to effectively "do the integration" for us.

An important idea revealed by the derivation of the example model is that it will often be possible to construct a range of models of different complexity, particularly if more than one census is available. The choice of final model obviously depends upon the aims motivating its construction. If a model is being developed primarily to project the dynamics, e.g., to compare the population growth rates at different sites or explore transient dynamics, then it is reasonable to adopt the simplest possible model. This minimizes the number of parameters we have to estimate and keeps the implementation simpler. If, on the other hand, the model is to be used to understand selection on the life history, e.g., via the calculation of parameter sensitivities, then it makes more sense to keep the components of lamb demography separated as in equations 2.10.4. This provides more insight into the effect of reproduction on population growth rate. The formulation we outlined assumes that there are no maternal effects on lambs that play out over spring and summer. This assumption could be relaxed, though it results in a much more complicated model because we have to keep track of the bivariate mother-offspring size distribution. However, this effort might be warranted if the model is going to be used to understand how phenotypic maternal effects impact the dynamics.

If we really had gathered size data for individuals of all ages at both censuses it would be natural to construct a seasonal IPM. Alternatively, we could "collapse" the model to project the dynamics among years. We would then have to make a decision about which census point to use: spring to spring or summer to summer. The kernels would be different, but the resultant dynamics would be essentially the same. In both cases the model would be of the post-reproductive census variety, because both reproduction kernels contain survival functions of the established individuals. What if we had collected spring size data prior to, instead of after, reproduction? In this case we could construct either a post-reproductive census model (by projecting from summer to summer) or a pre-reproductive census model (by projecting from spring to spring). This illustrates another feature of IPMs constructed from multiple censuses. The pre- versus post-reproductive census distinction is no longer necessarily a feature of the data collection methodology. Instead it reflects the life cycle, the data, and our modeling decisions.

Chapter 3
Basic Analyses 1: Demographic Measures and Events in the Life Cycle

Abstract A fitted IPM describes how a population as a whole changes in total abundance and in its size or state distribution. The case studies in Chapter 2 illustrated how the familiar measures of population change (stable population structure, long-term growth rate, reproductive value, etc.) can be calculated from an IPM. But the kernels that comprise an IPM also describe how individuals change in size or other state variables over their lifetime. In this chapter, we take a closer look at population growth, and then show how a fitted IPM can be used to calculate statistics summarizing the life cycle, such as age-specific fecundity and survival probability, expected lifespan, lifetime reproductive output, and size at death. For many of these it is also possible to calculate the expected variation among individuals that results just from chance events, such as the stochasticity of who lives and who dies in a given year, without any inherent differences among individuals.

3.1 Demographic quantities

Age-structured demographic analysis is a fundamental tool in population ecology and evolutionary ecology. The life table describes how the chance of survival and expected fecundity change over the individual's lifetime. That information allows us to project the population forward in time and provides essential information about selection on survival and fecundity at different ages.

The basic IPM is structured by size rather than by age.[1] But each individual still has an age, so the demographic analyses based on the life table are still possible in an IPM. However, there's an extra layer to the analysis because individuals may differ in size at birth. As a result, many demographic variables become a function of size at birth as well as of age, or need to be generalized to allow for variation at birth.

---

[1] As before, we refer to the individual-level state variable as size, but it could be any continuously distributed trait such as egg-laying date or thermal optimum, or a multivariate trait.

© Springer International Publishing Switzerland 2016
S.P. Ellner et al., *Data-driven Modelling of Structured Populations*,
Lecture Notes on Mathematical Modelling in the Life Sciences,
DOI 10.1007/978-3-319-28893-2_3

For simplicity, we restrict ourselves to the basic size-structured IPM introduced in the last chapter, with a smooth power-positive kernel. However, all of
the results still hold for the multidimensional models (e.g., age-size, size-quality,
etc.) described Chapter 6, so we will use $\int_{\mathbf{Z}}$ rather than $\int_L^U$ to denote an integral
over all possible states.

Our convention about age is that an individual is age 0 at the first census
after its birth. *Lifespan* is defined as the number of census times at which an
individual is alive, so that an annual plant (with one census per year) is assigned
a lifespan of 1. This is one less than the definition of lifespan in Cochran and
Ellner (1992), so some of the formulas here differ from that paper.

3.1.1 Population growth

For both of the examples introduced in Chapter 2 population growth is density
independent, and so after some initial fluctuations the populations grow at a
constant rate, termed the long-term population growth rate and denoted by
λ. This is a key demographic quantity which summarizes how all the state-
dependent processes occurring in the life cycle combine to determine how rapidly
a population grows. This allows us to determine whether a population will
increase ($\lambda > 1$) or decline to extinction ($\lambda < 1$). λ is therefore is important
in conservation, where we want to ensure that $\lambda > 1$ so a population does not
go extinct, and also in pest and disease management where we would like the
opposite. λ is also used a great deal in evolutionary biology, to determine fitness
by calculating the rate of invasion of a new mutation into a population, and to
measure the strength of selection operating on a trait (see Chapter 9).

In an unstructured population, when population growth is density independent the population density $n(t)$ grows according to $n(t) = n(0)\lambda^t$, and so
$\lim_{t\to\infty} \frac{n(t)}{\lambda^t} = n(0)$. For an IPM we have a similar result that

$$\lim_{t\to\infty} \frac{n(z,t)}{\lambda^t} = Cw(z) \tag{3.1.1}$$

where $w(z)$ is the stable state distribution and $C > 0$ is a constant that depends
on the initial state of the population (Ellner and Rees 2006). Equation (3.1.1)
holds for all z, so if we look at any particular state, the population of individuals in this state will grow at the same rate, λ. The rate of convergence to the
stable state structure is asymptotically exponential, with error decreasing in
proportion to $(|\lambda_2|/\lambda)^t$, where $|\lambda_2|$ is the maximum magnitude of the nondom-
inant eigenvalues. An IPM typically has infinitely many eigenvalues, but the
mathematical theory insures that there is a single dominant eigenvalue λ, and
a nonzero gap between the magnitude of λ and that of any other eigenvalue.

A second widely used measure of population growth is R_0, which is the
generation-to-generation population growth rate or *Net Reproduction Rate*.
Consider first our unstructured model

$$N(t+1) = (s + p_b p_r b)N(t) \tag{3.1.2}$$

(equation 2.3.2). Here all individuals are the same, so R_0 can be defined as the average number of offspring that an individual produces over their lifetime.[2] That is, each individual in the present generation contributes R_0 individuals, on average, to the next generation. We can calculate R_0 by adding up all the offspring produced, on average, in each year of life. In the first year of life (age 0) $f = p_b p_r b$ offspring are produced. The probability of survival to age 1 is s, so on average sf offspring will be produced in year 2. Multiplying this by s again gives the average number produced in year 3, $s^2 f$, and so on. Adding all this up gives us R_0, so

$$R_0 = f + sf + s^2 f + s^3 f + \cdots = f(1 + s + s^2 + \cdots) = f(1-s)^{-1}. \quad (3.1.3)$$

In an IPM and other structured models, defining R_0 as the average number of offering produced over a female's lifetime is ambiguous, because age-0 individuals can differ in state and therefore can differ in their expected future breeding success. R_0 is therefore defined as the long-term per-generation rate of increase. Imagine marking g_1 living individuals and calling them "generation 1." All of their offspring make up generation 2. All the offspring of generation 2 are generation 3, and so on. If g_k is the number of offspring in generation k, then R_0 is defined as

$$R_0 = \lim_{k \to \infty} (g_{k+1}/g_k). \quad (3.1.4)$$

We find R_0 by adding up the expected offspring production in each year of life, just as we did for the unstructured model. If $n_0(z)$ is the state distribution of a cohort at birth, then the offspring that they produce in their first year are a population with state distribution

$$Fn_0(z') = \int_Z F(z', z) n_0(z) \, dz \quad (3.1.5)$$

and the survivors to age 1 from the cohort are

$$Pn_0(z') = \int_Z P(z', z) n_0(z) \, dz \quad (3.1.6)$$

The left-hand sides of equations (3.1.5, 3.1.6) are examples of the *operator notation* introduced in Section 1.4. We use operator notation a lot in this chapter, because Fn_0 and Pn_0 are easier to read than the integrals they represent and they look like the corresponding formulas for matrix models. The operator notation that we use is summarized in Table 3.1. Operator notation also translates directly into computer code when an IPM is implemented with a matrix representing the kernel, and vectors representing size distributions and other functions of z. So when you see something like Fn_0 it's safe to think of it as F%*%n0, a matrix F multiplying a vector n_0, and $v \circ w$ can be thought of as element-by-element product v*w (Table 3.1).

---

[2] In a female-based life table, this would be the number of female offspring produced over a female's life.

Table 3.1 Operator notation for IPMs. Here A, B denotes any kernels (i.e., A is a real-valued function $A(z', z)$), f and g are any functions of z or z', and $\mathbf{Z}$ is the set of possible values for z. The operations $f \circ g, f/g, f^k, A \circ B$ defined below act like the element-by-element operations `f*g`, `f/g`, `f^k`, `A*B` in R, while fA, Af, AB act like matrix-vector and matrix-matrix multiplication `f%*%A`, `A%*%f`, `A%*%B`.

Notation	Meaning and/or formula
Af	"Right-multiplication" $Af(z') = \int_Z A(z', z)f(z)\,dz$, e.g., IPM iteration $n_{t+1} = Kn_t$.
fA	"Left-multiplication," $fA(z) = \int_Z f(z')A(z', z)\,dz'$; if A is a transition probability (from z to z') then $fA(z)$ is the expected value of $f(z_{t+1})$ given $z_t = z$.
$A \circ B$	Pointwise product of kernels, $(A \circ B)(z', z) = A(z', z) \times B(z', z)$.
AB	Composition of kernels, $AB(z', z) = \int_Z A(z', y)B(y, z)\,dy$, analogous to matrix multiplication.
A^2	Second iterate of kernel A, $A^2 = AA$ and similarly A^3, A^4, etc.
$f \circ g$	Pointwise product of functions, the function $(f \circ g)(z) = f(z) \times g(z)$.
$\langle f, g \rangle$	Inner product of functions, the number $\int_Z (f(z) \times g(z))dz$
f^2	Pointwise square of a function, $f^2(z) = f(z)^2$; similarly f^3, f^4, etc.
f/g	Pointwise ratio, the function $(f/g)(z) = f(z)/g(z)$.

Getting back to R_0: we've found the offspring that the focal cohort produces at age zero, namely Fn_0, and the number from the cohort that survive to age one, Pn_0. Those survivors then produce offspring, which have state distribution FPn_0. To calculate the survivors to age 2 and their state distribution, we need to consider all transitions $z \rightarrow y \rightarrow z'$ where y is the range of possible states at age 1. This is given by

$$\int_Z P(z', y)P(y, z)\,dy = P^2(z', z). \qquad (3.1.7)$$

The survivors to age 2 are therefore P^2n_0, producing offspring FP^2n_0. Carrying on, the transition $z \rightarrow z'$ in k time steps is given by $P^k(z', z)$, and the expected offspring production by the survivors in that year is $FP^k(z', z)n_0$. Adding up all years, the expected offspring from the cohort are

$$(F + FP + FP^2 + FP^3 + \cdots)n_0 = F(I + P + P^2 + \cdots)n_0$$
$$= F(I - P)^{-1}n_0 = FNn_0, \qquad (3.1.8)$$

where $N = (I - P)^{-1}$ is called the *fundamental operator*. The *Neumann series* $(I + P + P^2 + \cdots) = (I - P)^{-1}$ used in (3.1.8) corresponds to the geometric series $1 + r + r^2 + \cdots = (1 - r)^{-1}$ that is valid for real numbers r with $|r| < 1$, and it is valid so long as the survival probability is < 1 for all individuals (or more generally, if there is a k such that the probability of surviving k years is < 1 for all individuals).

Equation (3.1.8) is the IPM analog of equation (3.1.3). It shows that $R = FN$ is the kernel that projects the population from one generation to the next, so R is called the *next generation kernel*. $R(z', z)h$ is an individual's expected lifetime production of offspring with size at birth in $(z', z' + h)$, conditional on the individual's initial size z. If n_0 is the state distribution of a cohort at birth, then Rn_0 is the state distribution of all their offspring at birth, regardless of when the birth occurred.

The long-run generation-to-generation growth rate R_0 is therefore given by the dominant eigenvalue of R (so long as there is a unique dominant eigenvalue), and successive generations converge to the corresponding eigenvector. This will generally be the case unless size at birth has very long-lasting effects, for example if an individual who is small at birth never has large offspring, and vice versa. The "mixing at birth" condition that we discuss in connection with general IPMs (Chapter 6) is sufficient to rule out such cases, and R_0 is then well defined and unique. We then have $\lim_{k \to \infty} g_k/(R_0)^k = G$, where G is a constant that depends on the initial population distribution (Ellner and Rees 2006). So on the generation timescale, there is exponential population growth with finite growth rate R_0.

R_0 and λ are related, as in matrix models and other structured population models: $\lambda - 1$ and $R_0 - 1$ have the same sign (Li and Schneider 2002; Ellner and Rees 2006; Thieme 2009), so $\lambda > 1$ if and only if $R_0 > 1$. This means that R_0 can also be used to determine whether a population increases or decreases. The dominant eigenvector of R gives the stable state distribution of size at birth for a generation. In general this is not the size distribution of newborns in any particular year, unless offspring size is independent of parent size. Rather, it is the distribution of size at birth for all individuals in the same generation, regardless of when they are born.

Reassuringly, in a population structured by age the general definition of R_0 in terms of the next-generation kernel is equivalent to the familiar $R_0 = \sum_a l_a f_a$ (see Box 3.1). That is, R_0 is the expected lifetime reproduction by a randomly chosen newborn. The same is true in a size-structured model where the offspring size distribution $c_0(z')$ is independent of parent size. In that case c_0 is also the size distribution for all newborns in any generation, so R_0 equals the expected number in the next generation starting from a cohort with size distribution c_0, $R_0 = \int_Z Rc_0$. If we define $m(z)$ to be the expected current fecundity of a size-z individual (e.g., $m = p_r p_b b$ in our *Oenothera* IPM), then writing out the last integral we get

$$R_0 = \langle m, Nc_0 \rangle \tag{3.1.9}$$

when offspring size is independent of parental size.

The dominant eigenvector of R is the stable distribution of the state at birth for all individuals in a generation. In a population structured only by age, all individuals are age 0 at birth, regardless of when they are born. The dominant eigenvector of R is therefore $e_1 = (1, 0, 0, \cdots, 0)$.

Next, we need to find Re_1. F is the fecundity part of a Leslie matrix; it has $(f_0, f_1, f_2, \cdots)$ as its first row and zero everywhere else. P is the survival part; it has the age-specific survivorship's $(p_0, p_1, p_2, \cdots)$ on the subdiagonal and zero everywhere else, so $Pe_1 = (0, p_0, 0, \cdots) = (0, l_1, 0, \cdots)$, $P^2 e_1 = P(Pe_1) = (0, 0, p_1 l_1, 0, \cdots) = (0, 0, l_2, 0, \cdots)$, and so on. Then recalling $l_0 = 1$ we have

$$Re_1 = FNe_1 = F(I + P + P^2 + \cdots)e_1 = F \begin{pmatrix} 1 \\ l_1 \\ l_2 \\ \vdots \end{pmatrix} = \begin{pmatrix} \sum_a l_a f_a \\ 0 \\ 0 \\ \vdots \end{pmatrix}.$$

The dominant eigenvalue of R is therefore $\sum_a l_a f_a$.

Box 3.1: R_0 for an age-structured population.

3.1.2 Age-specific vital rates

In a size-structured model, survivorship to age a, l_a, and age-specific expected fecundity, f_a, are functions of the individual's size or state at birth. Because individuals are age zero at birth, $l_0(z) = 1$ for all z. If an individual is born with state z_0, the distribution of its state the next year is $p(z'|z_0) = P(z', z_0)$. Adding up the probability of all possible states at age 1 gives the probability that the individual is still alive, so the probability of surviving to age 1 is

$$l_1(z_0) = \int_Z P(z', z_0) \, dz'.$$

The state distribution at age 2 is given by $P^2(z', z_0)$ (equation 3.1.7), and similarly the distribution at any age a is given by $P^a(z', z_0)$ where P^a is the ath iterate of the kernel. To get the survivorship we have to add up the probability over all possible sizes, giving

$$l_a(z_0) = \int_Z P^a(z', z_0) \, dz'. \tag{3.1.10}$$

You can think of $l_a(z_0)$ as the probability of survival to age a for a single individual with state z_0 at birth, or as the expected fraction of individuals surviving to age a in a cohort of individuals having state z_0 at birth.

In the course of finding R_0 we have already computed that the state distribution of offspring produced at age a is $FP^a(z', z_0)$, for an individual born with state z_0. The average per-capita fecundity at age a is the total reproductive output at age a of survivors, divided by the number of survivors. This is computed as

$$f_a(z_0) = \frac{1}{l_a(z_0)} \int_Z FP^a(z', z_0) \, dz'. \tag{3.1.11}$$

This value is analogous to the age-specific fecundities in the top row of a Leslie matrix: the average number of offspring present at time $t+1$, per age-a individual present at time t.

More generally, we can compute the average survival and per-capita fecundity for a cohort of individuals with any initial state distribution $c(z)$:

$$\tilde{l}_a = \langle l_a, c \rangle = \int_{\mathbf{Z}} l_a(z)c(z) \, dz$$

$$\tilde{f}_a = \frac{1}{\tilde{l}_a} \int_{\mathbf{Z}} \int_{\mathbf{Z}} FP^a(z', z)c(z) \, dz \, dz' \tag{3.1.16}$$

Because c is a probability distribution (by assumption), $\langle l_a, c \rangle$ denotes an average value with respect to the distribution. In Box 3.2 we show how these formulas look using operator notation, and how they then translate into R code.

3.1.3 Generation time

There are several ways to define generation time. Our definition of R_0, as the per generation rate of increase, means that over one generation T the population increases by λ^T, and so $\lambda^T = R_0$. This gives us

$$T = \frac{\log(R_0)}{\log(\lambda)}. \tag{3.1.17}$$

Other definitions of T that demographers also use can be computed from the age-specific survival and fecundity functions. Another measure of generation time is the mean age of mothers at offspring production. In our conventions, an individual is age $a + 1$ when the offspring of "last year's" a-year-olds are first censused. This measure of generation time is therefore computed (under our conventions[3]) as

$$T = \frac{\sum(a+1)l_a(z_0)f_a(z_0)}{\sum l_a(z_0)f_a(z_0)} \tag{3.1.18}$$

where the denominator is there to ensure the probability distribution of age at offspring production, which is proportional to $l_a(z_0)f_a(z_0)$, sums to 1. This measure is not terribly useful as it applies to a particular state at birth, z_0. For a cohort we simply substitute average rates (equation 3.1.16) into the preceding equation. There are two obvious choices of cohort to average with respect to: 1) the offspring state distribution in species where this does not vary with parental state, and 2) the stable offspring-state distribution in species where parental state affects offspring initial state. If w is the stable state distribution, the stable offspring-state distribution is Fw scaled to have integral 1.

---

[3] As is often the case in age-structured demography, under different conventions the formulas look slightly different.

When we use operator notation (Table 3.1), the formulas for age-specific survival and fecundity become

$$l_a = \mathbf{e}P^a \tag{3.1.12}$$
$$f_a = (\mathbf{e}FP^a)/l_a \tag{3.1.13}$$
$$\tilde{l}_a = \langle l_a, c \rangle = \mathbf{e}P^a c \tag{3.1.14}$$
$$\tilde{f}_a = (\mathbf{e}FP^a c)/\tilde{l}_a \tag{3.1.15}$$

It's worth pausing to think about how to parse an expression like equation (3.1.13). P^a gives the state distribution at age a for each initial state z_0, and FP^a then gives the resulting state distributions for the offspring that those survivors produce at age a. Left-multiplication by the constant function $\mathbf{e} \equiv 1$ is equivalent to the integration over z' in equation (3.1.10): it sums up the total number of offspring (like summing one column of a matrix). The numerator in equation (3.1.13) is therefore a function of initial state. The same is true of the denominator. The right-hand side of (3.1.13) is thus the ratio of two functions, which defines f_a as a function of initial state.

These matrix-style formulas correspond very closely to code for the calculations if the IPM is implemented using midpoint rule or another bin-to-bin method (see Section 6.8). Suppose that P, F are iteration matrices representing the P and F kernels, for example using the monocarp IPM code from Chapter 2:

```
Mtrue <- mk_K(200,m.par.true,L=-2.65,U=4.5);
P <- Mtrue$P; F <- Mtrue$F;
```

Let e be a vector of all 1's whose length equals the dimension of the iteration matrices,

```
e <- matrix(1,nrow=1,ncol=nBigMatrix)
```

Then for example

```
l2 <- e%*%P%*%P;
f2 <- e%*%F%*%P%*%P/l2;
plot(Mtrue$meshpts,l2); plot(Mtrue$meshpts,f2);
```

computes vectors l2 and f2 vector containing $l_2(z_0)$ and $f_2(z_0)$ evaluated at the mesh points z_i. When you do these calculations yourself (you *are* running all of the code as you read the book, aren't you?) you will see that l_2 is maximized at intermediate values of the initial size z_0. The tiniest newborns are likely to die very young without flowering, and very large ones (which don't really occur) are likely to flower and die before reaching age 2.

Box 3.2: Calculating Age-specific Vital Rates.

3.2 Life cycle properties and events

One underlying premise of IPMs is that life is a stochastic process (our lives, at least, tend to support this assumption). The growth kernel $G(z', z)$ says that two individuals of the same size now are likely to differ in size next year. The survival function $s(z)$ says that some will die while others of the same size survive. Up to now we have focused on the population and cohort levels (population growth rate, average age-specific mortality, etc.). In this section we focus on individual life trajectories, the differing paths that individuals follow through life, growing, breeding, and dying according to the probabilities specified by the IPM kernels.

Individual state dynamics are specified by the survival kernel P. Analogous to Figure 2.1, if we hold z fixed and let z' vary, $P(z', z)$ is the probability density for subsequent size conditional on current size being z. Actually it's not quite a probability density because its integral is the survival probability $s(z)$, not 1. To fix that, we add a discrete point Δ representing death, with $1 - s(z)$ being the probability of going from z to Δ. Then because each individual's future prospects are completely determined by their current state z, their changes in state are a Markov chain. The properties of this chain tell us about the different paths that individuals take through life.

Markov chain theory was used to study individual life trajectories in human demography by Feichtinger (1971, 1973), then in ecological matrix models by Cochran and Ellner (1992), extended and simplified by Caswell (2001) making better use of Markov chain theory. Later applications and extensions include age-stage relationships in variable environments (Tuljapurkar and Horvitz 2006) and analyses of among-individual variation in lifetime descriptors such as lifespan and total fecundity (Caswell 2011a,b, 2012; Tuljapurkar et al. 2009; Steiner and Tuljapurkar 2012; Steiner et al. 2012). Metcalf et al. (2009b) applied some of these results to IPMs for tropical trees in a variable light environment, by using midpoint rule to approximate the IPM as a large matrix model. Feel free to follow their example: no sane individual will doubt that it works. But this may run into trouble because of matrix size when z is multidimensional, and there is no mathematical proof that such calculations converge to the right answer - or to anything at all - as the size of the matrix is increased. So here we take an IPM-centric approach, using continuous-state Markov chains. Many of the results here are new to ecology so we derive them below, but we have put the more technical derivations in the Appendix, Section 3.4.

Properties of individual life trajectories can always be estimated by repeatedly simulating the chain and recording the events in each individual's life, as we do later. One payoff from analytic formulas is the freedom to explore different scenarios and connect cause and effect by perturbation analyses. What happens to lifetime reproductive output if recruits are 10% larger, or if the relationship between size and fecundity gets steeper? For many questions like this, Markov chain theory gives exact answers in a few lines of code.

This section is the most technical in the book and uses some methods from Markov Chain theory that may well be unfamiliar to you. If you've studied finite-state Markov chains, you should be able to follow the whole section. If

Table 3.2 Notation for age-specific vital rates and life cycle events.

Notation	Meaning and/or formula
$\mathbf{e}$	Constant function $\mathbf{e}(z) \equiv 1$.
$\mathbf{i}$	Identity function $\mathbf{i}(z) = z$.
I	Identity operator: $fI = If = f$, $AI = IA = A$.
N	Fundamental operator $N = (I - P)^{-1}$.
R	Next generation kernel $R = FN$.
c, c_0	State distribution with integral= 1 (a probability distribution).
n, n_0	State distribution, not necessarily a probability distribution.
$\beta(z)$	Number of new recruits produced in the current year (a random variable with distribution depending on z).
$\bar{\beta}(z)$	Mean per-capita number of new recruits produced, $p_b(z)p_r b(z)$.
π_b, π_0	Survival-growth kernels for breeders and nonbreeders, respectively.
$\sigma_+^2(z)$	Variance in per-capita number of new recruits produced in the current year by a breeder.
$\sigma_b^2(z)$	Variance in per-capita number of new recruits produced in the current year, including breeders and nonbreeders, eqn. (3.2.14).

not, you can still use the main results, which are summarized in Table 3.3 using the notation defined in Tables 3.1 and 3.2. You can skim the rest of this section, or go directly to Section 3.3 where we illustrate the results and calculations using the basic size-structured *Oenothera* model.

The calculations and formulas in this section assume that the kernel has been written in the generic form $K(z', z) = s(z)G(z', z) + F(z', z)$. So for applying the results to the *Oenothera* kernel (2.5.1) where breeding implies death, $s(z)(1 - p_b(z))$ has to be used, instead of $s(z)$, for the survival probability, and $1 - s(z)(1 - p_b(z))$ is the chance of death.

3.2.1 Mortality: age and size at death

Age-specific mortality can be computed from the survivorship curve: the probability that death occurs between ages $a - 1$ and a is $(l_{a-1} - l_a)$, the probability of surviving to age $a - 1$ but no longer. If this happens, the individual's lifespan is a by our convention. The mean lifespan can then be computed in principle as $\sum_{a=1}^{\infty} a(l_{a-1} - l_a)$, but there's a simpler way. For a random variable X with possible values $\{0, 1, 2, \cdots, \}$, a basic result from probability theory is that the expected value of X is given by

$$E[X] = \sum_{k=0}^{\infty} \Pr[X > k].$$
(3.2.1)

Table 3.3 Main results from the analysis of individual life trajectories. The properties listed as functions of z_0 are conditional on the individual's state at birth being z_0. "Reproduction" and "breeding" mean an attempted breeding, regardless of whether any recruits survive to the next census. The formulas are for a kernel written in the form $K(z) = s(z)G(z', z) + P(z', z)$.

Notation	Formula	Meaning
$l_a(z_0)$	$\mathbf{e}P^a$	Probability of survival to age a.
$f_a(z_0)$	$(\mathbf{e}FP^a)/l_a$	Average per capita fecundity at age a conditional on surviving to age a.
$\overline{\eta}(z_0)$	$\mathbf{e}N$	Mean lifespan conditional on initial state z_0 = mean time until death for an individual with current state z_0.
$\sigma_\eta^2(z_0)$	$\mathbf{e}(2N^2 - N) - (\mathbf{e}N)^2$	Variance in lifespan conditional on initial state z_0 = variance in time until death for an individual with current state z_0.
$P_0(z', z)$	$(1 - p_b(z))\pi_0(z', z)$	Survival kernel for modified model in which reproduction is an absorbing state (the other absorbing state is death).
N_0	$(I - P_0)^{-1}$	Fundamental operator for P_0
$B(z_0)$	$p_b N_0$	Probability of reproducing at least once.
$\overline{\omega}(z_0)$	$(\mathbf{i} \circ (1 - s))N$	Mean size at death.
$\Omega(z', z_0)$	$(1 - s(z'))N(z', z_0)$	Size at death kernel.
$P_{(b)}(z', z)$	$P_0(z', z)B(z')/B(z)$.	Survival kernel P_0 conditioned on breeding at least once.
$\Omega_{(b)}(z', z_0)$	$(1 - \mathbf{e}P_{(b)})N_{(b)}(z', z_0)$	Size at first breeding kernel.
$\overline{a}_R(z_0)$	$\mathbf{e}(I - P_{(b)})^{-1} - 1$	Mean age at first breeding.
$\overline{r}(z_0)$	$\mathbf{e}FN$	Mean lifetime output of recruits.
$\sigma_r^2(z_0)$	$r_2 - \overline{r}^2$	Variance in lifetime output of recruits (see eqn. (3.2.15) for r_2).
S, S_2	eqn. (3.4.16)	First and second moments of age at last breeding.

where $\Pr[A]$ denotes the probability of outcome A. Let $\eta(z)$ denote the (random) lifespan of an individual given their initial state z. All individuals have $\eta > 0$; $\eta > 1$ if the individual survives to age 1, which happens with probability $l_1 = \mathbf{e}P$; $\eta > 2$ if the individual survives to age 2, which happens with probability $l_2 = \mathbf{e}P^2$; and so on. Applying equation (3.2.1) we get

$$\overline{\eta} = E[\eta] = \mathbf{e} + \mathbf{e}P + \mathbf{e}P^2 + \cdots$$
$$= \mathbf{e}(I + P + P^2 + P^3 + \cdots) \qquad (3.2.2)$$
$$= \mathbf{e}(I - P)^{-1} = \mathbf{e}N$$

For an individual with state z at birth, we want to know: how often do we expect the individual's state to be in the interval $J = (z', z' + h)$ before they die?

It's actually easier to think this through if we consider an arbitrary distribution of state at birth, $c_0(z)$. The state distribution at age k is then $c_k = P^k c_0$, so the probability of being in J at age k is $h c_k(z')$ for h small.

Now think of the individual's state at each age, starting with 0, as a series of zero-one coin tosses that have value 1 if the individual's state is in J, and 0 if the individual has state outside J or is already dead. The sum of those random coin tosses is the total number of times that the individual was in J at ages $0, 1, 2, 3, \cdots$. The expectation of a zero-one coin toss is the probability of getting 1. The expectation of the sum (which is the expected number of times in J) thus equals the sum of the success probabilities, which is h times

$$c_0 + c_1 + c_2 + c_3 + \cdots = (I + P + P^2 + P^3 + \cdots)c_0$$

evaluated at z'. Recalling that $I + P + P^2 + P^3 + \cdots = N$, the equation above says that $h N c_0(z')$ is the expected total number of times the individual is in $(z', z' + h)$. $N c_0$ is therefore the distribution function for the expected total time spent in each state, when the initial state has distribution c_0.

Writing out $N c_0(z') = \int_{\mathbf{Z}} N(z', z) c_0(z)$, we see that $N c_0(z')$ is the mean of $N(z', z)$ when z is chosen at random from the initial state distribution c_0. When the initial state is z with probability 1, the mean is just $N(z', z)$. Therefore $N(z', z)$ is the distribution function for the expected total time spent in each state, for an individual with state z at birth.

Box 3.3: The Meaning of the Fundamental Operator.

where I is the identity operator (i.e., $I f = f$ for any function $f(z)$) and $N = (I - P)^{-1}$ is the fundamental operator defined in the previous section. N is analogous to the fundamental matrix in the theory of finite-state Markov chains (Kemeny and Snell 1960; Caswell 2000, Chapter 5). The final expression in (3.2.2) is a function of z, the expected lifespan as a function of initial state. So for a cohort with initial state distribution $c(z)$, the average lifespan is $\tilde{\eta} = \langle \mathbf{e}N, c \rangle$.

The fundamental operator has an intuitive interpretation. Holding z fixed, $N(z', z)$ is the distribution function for the expected total time that an individual with initial state z spends at state z' during its lifetime (technically, $N(z', z)h$ is the expected total time in $(z', z' + h)$ for small h). If you aren't familiar with N, read Box 3.3 to see why this is true. $\mathbf{e}N$ then adds up the expected time spent in all states to give the expected lifespan as a function of initial state, and $\langle \mathbf{e}N, c \rangle$ averages the expected lifespan across a cohort's distribution of initial states.

To compute the variance of lifespan we just need the mean of η^2, because $Var(\eta) = E[\eta^2] - (E[\eta])^2$. By definition

$$E[\eta^2] = \sum_{a=1}^{\infty} a^2 (l_{a-1} - l_a).$$

Using (3.1.12) and simplifying (for details see the Appendix to this chapter, Section 3.4) gives

$$E[\eta^2] = \mathbf{e}(2N^2 - N) \qquad (3.2.3)$$

and therefore

$$\sigma_\eta^2 = Var(\eta) = \mathbf{e}(2N^2 - N) - (\mathbf{e}N)^2. \qquad (3.2.4)$$

Equations (3.2.2–3.2.4) are the IPM versions of equations 5.10–5.12 in Caswell (2001), which they greatly resemble.

σ_η^2, like $\bar{\eta}$, is a function of z, giving the variance in lifespan among a group of individuals with initial state z. The observed variance in lifespan among members of a cohort with initial state distribution $c(z)$ is given by

$$Var_c(\eta) = E_c[\eta^2] - (E_c[\eta])^2 = \langle \mathbf{e}(2N^2 - N), c \rangle - \langle \mathbf{e}N, c \rangle^2. \qquad (3.2.5)$$

Lifespan is age at death. In a size-structured IPM, an individual also has a *size* at death. The expected size at death $\bar{\omega}$ for a living individual of size z (e.g., a newborn of size z) can be calculated by considering the outcome of the next time-step. If the individual dies, its size at death is z. If it survives, its mean size at death is the mean conditional on its new size, thus

$$\bar{\omega}(z) = z(1 - s(z)) + \int_Z \bar{\omega}(z') P(z', z)\, dz' = (\mathbf{i}(z) \circ (1 - s(z))) + \bar{\omega}P. \quad (3.2.6)$$

This gives $\bar{\omega}(I - P) = \mathbf{i} \circ (1 - s)$ and therefore

$$\bar{\omega} = (\mathbf{i} \circ (1 - s))N = \int_Z z'\,(1 - s(z'))\,N(z', z)\, dz'. \qquad (3.2.7)$$

The mean size at death for a cohort of newborns with size distribution c is then $\tilde{\omega} = \langle \bar{\omega}, c \rangle$.

We can also calculate the entire distribution of size at death for a cohort, and from that the variance, skew, and so on. The answer is implicit in equation (3.2.7), but let's get there honestly. Start with a size-z newborn. The chance that it dies while its size is in $(z', z'+h)$ is the sum over all ages a of the chance that it dies at age a with size in $(z', z'+h)$. The chance at age a is $(1 - s(z'))P^a(z', z)h$, the mortality at size z' times the chance of being in $(z', z'+h)$ at age a. Adding all those up gives the size-at-death kernel,

$$\Omega(z', z) = (1 - s(z'))N(z', z). \qquad (3.2.8)$$

$\Omega(z', z)$ should be a probability distribution (as a function of z', for any fixed z) because everyone dies eventually, and we verify in Appendix 3.4 that this is true. The distribution of size at death for a cohort with initial state distribution c is then Ωc.

3.2.2 Reproduction: who, when, and how much?

In addition to measures of average reproductive output such as R_0, we can study individual variation in reproduction. We consider here a few measures of variation to illustrate the possibilities, but many others can be calculated by similar methods.

Some individuals succeed in reproducing before they die. How many? Size-structured IPMs have generally been constructed so that size-specific mean fecundity is positive at all sizes, for example fecundity is described by a generalized linear model with a log or logit link function. This contrasts with matrix models, where some size-classes typically have exactly zero fecundity and so an individual's age at first reproduction can be defined as the age when the individual enters a size-class with positive fecundity (Cochran and Ellner 1992; Caswell 2001). To study reproductive timing in an IPM, we have to assume that the fecundity kernel explicitly includes a size-dependent probability of reproducing, $p_b(z)$. To allow for possible costs of reproduction, we write the survival kernel P as

$$P(z', z) = p_b(z)\pi_b(z', z) + (1 - p_b(z))\,\pi_0(z', z) \qquad (3.2.9)$$

where π_b and π_0 describe survival and growth for those who attempt breeding in the current year and those who don't, respectively. The IPM kernel then has the form

$$K(z', z) = p_b(z)\,[p_r b(z)C_0(z', z) + \pi_b(z', z)] + (1 - p_b(z))\pi_0(z', z), \qquad (3.2.10)$$

where $p_r b(z)$ is the mean number of offspring that survive to the next census for individuals that attempt breeding, and $C_0(z', z)$ determines the offspring size distribution. The extreme case of (3.2.10) is a monocarpic life cycle, which has $\pi_b = 0$.

The kernel $P_0(z', z) = (1 - p_b(z))\pi_0(z', z)$ describes survival and growth without breeding. It is the survival kernel for a modified model in which breeding is a second kind of death (a second absorbing state). The fundamental operator for the modified model, $N_0 = (I - P_0)^{-1}$, then gives the distribution function for expected total time in state z' prior to either dying or reproducing, conditional on initial state z.

Finally we can answer the question, how many breed before dying? The chance that an individual breeds at least once is the probability that it leaves the pre-breeding state by reproducing, which is $B = p_b N_0$. This is a function of z giving the probability of breeding at least once, for a state-z individual that has not yet had any offspring.

In the calculations above (and in the rest of this section) "breeding" and "reproduction" include breeding attempts that do not yield any new recruits, due to a failed breeding attempt or early mortality of neonates if $p_r < 1$. To calculate the probability of leaving at least one recruit, instead of (3.2.9) break your P up into P^- for producing no recruits in the current year (by not attempting to breed, or attempting and producing no recruits), and P^+ for producing one or more recruits. The chance of leaving at least one recruit over the lifetime is then

$p^+(I - P^-)^{-1}$ where $p^+(z)$ is the state-dependent probability of producing one or more recruits in the current year.

If reproduction occurs, at what age and size does it start? The modified chain P_0 has two absorbing states, *reproduce* and *die*. To compute the mean age at reproduction (for those who reproduce) we first compute the transition probabilities conditional on absorption into *reproduce*. Using the definition of conditional probability, the conditional survival kernel is[4]

$$P_{(b)}(z', z) = P_0(z', z)B(z')/B(z). \qquad (3.2.11)$$

The conditional Markov chain defined by $P_{(b)}$ has one absorbing state – *reproduce* – and the time until this happens is the individual's "lifetime." The mean time to reproduction is then $\mathbf{e}N_{(b)}$ (equation 3.2.2) where $N_{(b)} = (I - P_{(b)})^{-1}$ is the fundamental operator for the conditional chain. This formula gives the mean number of censuses at which the individual appears before breeding, so we need to subtract one time-unit to get the mean age (according to our conventions) at first breeding, $\overline{a}_R = \mathbf{e}N_{(b)} - 1$.

The conditional chain $P_{(b)}$ has an associated size-at-death kernel $\Omega_{(b)}$, which gives the distribution of size at first reproduction conditional on initial size z (note that in equation (3.2.8) the survival function for the conditional chain is $s_{(b)}(z) = \mathbf{e}P_{(b)}$ rather than $s(z) = \mathbf{e}P$).

Both $\overline{a}_R$ and $\Omega_{(b)}$ are conditional on the individual breeding at least once, as well as on the individual's initial size. As a result, the mean age and size at death for a cohort can't be computed by averaging these over the size distribution of recruits. Different-size recruits have different probabilities of breeding at least once, and this has to be taken into account by weighting recruits according to their chance of breeding. If $c(z)$ is the cohort initial size distribution, then the distribution weighted by the probability of breeding is

$$c_b(z) = c(z)B(z)/\langle c, B \rangle. \qquad (3.2.12)$$

The mean age at first reproduction, for those in the cohort who breed at least once, is then $\langle \overline{a}_R, c_b \rangle$. The size-at-first-breeding distribution for the cohort is $\Omega_{(b)}c_b$, a function of size from which you can compute the mean, variance, etc. of size at first breeding for individuals in the cohort.

How often does reproduction occur? The expected number of times an individual is size z', conditional on its initial size z, is given by $N(z', z)$. The chance of reproducing when size z' is $p_b(z')$. Adding up over all possible sizes, the expected number of times that an individual reproduces, conditional on initial size z, is $p_b N$. The expected number of times for individuals who breed at least once is then $p_b N/B$. As above, this is a count of breeding attempts and includes attempts that produce no new recruits.

---

[4] To keep things simple we assume that, as in most size-structured IPMs to date, $B(z) > 0$ for all z: regardless of current state, there is some chance to reproduce later in life. If this is not true, then there is one additional step: remove states for which $B(z) = 0$, and then calculate $P_{(b)}$, etc. on the remaining states.

How much do individuals vary in reproductive output? Let $r(z)$ denote the total reproductive output (number of recruits) in the current and all future years for an individual with current state z. Above we found that the expected value of r is $\bar{r}(z) = \mathbf{e}R = \mathbf{e}FN$. To find the variance of r, we have to add one ingredient to our population model: the between-individual variation in the number of new recruits produced in one year. "New recruits" takes into account p_r, the recruitment probability in the Figure 2.2A life cycle; setting $p_r = 1$ in the formulas below gives the number of offspring prior to this mortality, or the number of recruits when p_r really equals 1 as in Figure 2.2B.

Let $\beta(z)$ denote the (random) total *recruit* number as a function of parent size (i.e., the number of recruits produced by a size-z parent is a random value drawn from the distribution of $\beta(z)$). Because $\beta(z)$ includes the possibility of not breeding, and pre-recruitment mortality, its mean is

$$\bar{\beta}(z) = p_b(z)p_r b(z) \qquad (3.2.13)$$

the probability of breeding times the mean recruits per breeder. Let $\sigma_+^2(z)$ denote the variance in the number of recruits produced by a breeder (again, taking p_r into account – this generates variance in recruit numbers as a result of mortality between birth and recruitment). Then the variance of β works out to be

$$\sigma_b^2(z) = Var(\beta(z)) = p_b(z)(1 - p_b(z))\left(p_r b(z)\right)^2 + p_b(z)\sigma_+^2(z). \qquad (3.2.14)$$

Having found $\bar{r}(z)$, to compute $Var(r(z))$ we need $E[r(z)^2]$, which we denote $r_2(z)$. In Appendix 3.4 we show that

$$\begin{aligned} r_2 &= \left(\sigma_b^2 + (p_r p_b \circ b)^2 + 2p_r p_b \circ b \circ (\bar{r}\pi_b)\right) N \\ &= \left(\sigma_b^2 + \bar{\beta}^2 + 2\bar{\beta} \circ (\bar{r}\pi_b)\right) N. \qquad (3.2.15) \end{aligned}$$

The variance in lifetime reproductive output, conditional on initial state z, is then $r_2(z) - (\bar{r}(z))^2$. A similar calculation can be used to get the third moment of total reproductive output. For the case $\sigma_b^2 = 0$, Steiner and Tuljapurkar (2012) derived the probability generating function of lifetime reproduction in a stage-structured matrix model, which can be used to compute all the moments of lifetime reproduction.

These formulas also tell us about the mean and variance of the number of times that an individual breeds. To do this, we modify the model so that each breeding attempt produces exactly one recruit. That is, we set $p_r = b(z) = 1$ and $\sigma_+(z) = 0$ for all z, which gives $\sigma_b^2(z) = p_b(z)(1 - p_b(z))$. Thankfully, that's what σ_b^2 should be, because in this modified model the number of new recruits produced is a 0-1 coin toss, with probability $p_b(z)$ of getting 1. The lifetime number of breeding attempts for the original model is the lifetime reproductive output for the modified model in which

$$\bar{\beta} = p_b, \quad \bar{r} = p_b N, \quad r_2 = (p_b + 2p_b \circ (\bar{r}\pi_b)) N. \qquad (3.2.16)$$

When does reproduction end? That is, what is the distribution for age at last breeding (or breeding attempt), among those that do at least once? In the Appendix to this chapter, we derive this distribution (equation 3.4.10) and we also give formulas for the first and second moments of the age at last breeding S among individuals that breed at least once, which we denote as $E_b[S]$ and $E_b[S^2]$ (equation 3.4.16). Both of these are functions of the individual's initial size z. To compute the mean and variance of S for a cohort with initial size distribution $c(z)$, we need to average these moments against the cohort's distribution weighted by the probability of reproducing at least once, $c_b(z)$ (equation 3.2.12). The mean and variance for the cohort are then

$$
\begin{aligned}
E_c[S] &= \langle E_b[S], c_b \rangle \\
Var_c[S] &= \langle E_b[S^2], c_b \rangle - E_c[S]^2.
\end{aligned} \tag{3.2.17}
$$

3.2.3 And next...

We could go on without limit computing properties of individual pathways through life.[5] But recalling the title of this chapter, we'll stop here. There are two obvious directions for future work on life cycle events:

1. Models with environmental variation, either temporal or spatiotemporal (as in megamatrix models); for some results see Tuljapurkar and Horvitz (2006); Metcalf et al. (2009a). Most of the results in this chapter apply to temporally varying environments just by cross-classifying individuals by their state and the environment state (Caswell 2009). If the environment is uncorrelated or Markovian the resulting model is still an IPM. We say more about this in Section 7.7.
2. Developing analytic perturbation formulas, analogous to recent developments for matrix models (Caswell 2007, 2008, 2011b, 2012; Steiner et al. 2012); we present a few in Sections 4.5 and 4.6.

Developing those ideas is beyond our scope here, but if you need them, an interested theoretician may find a way to compute what you need. If not, then "when all else fails, lower your standards."[6] The IPM implies an IBM that can be simulated repeatedly to generate a sample from the distribution of life cycles, and any statistic you want can be estimated from the sample. Distributions of last-event times are complicated and covariances are even harder. If you are interested in the covariances among lifespan, reproductive lifespan and total offspring number, simulating the underlying IBM may be your only option – or by now, as you read this, maybe not!

---

[5] If you don't believe us, take a look at Cochran and Ellner (1992).

[6] Sidney Saltzman, Professor Emeritus of City and Regional Planning, Cornell.

3.3 Case study 1B: Monocarp life cycle properties and events

To illustrate the application of these approaches we use the *Oenothera* IBM but set the probability of recruitment, p_r, so that $\lambda = 1$ which means we don't need to discount by population size, as in Soay sheep example (Section 2.6.5). Other than this change the R-code is as before (Section 2.5.2); 1) Monocarp Demog Funs.R contains the demographic function that describes how size influences fate, and also the functions used to define and make the iteration kernel; 2) Monocarp Simulate IBM.R simulates the individual-based model, and 3) Monocarp Events Calculations.R implements the various event calculations. Monocarpic plants are a useful illustration as reproduction is fatal and so there are two ways to die.

As we are interested in testing the various calculations we will use the true parameter values to define the IPM, and collect a large sample of individual fates as we need to calculate various metrics stratified by age. For many of the calculations we will need the survival and reproduction iteration matrices so having constructed the IPM using mk_K(nBigMatrix, m.par.true,L,U) we then store the iteration matrices in P, F, so we can keep close to the formulae in Section 3.2. We then define e=matrix(1,nrow=1,ncol=nBigMatrix) which we will use for summing down the columns of various matrices, and offspring.prob <- h*c_0z1(meshpts,m.par) which is the probability distribution of offspring sizes - we'll use this as the initial cohort $c(z)$.

3.3.1 Population growth

Both R_0 and λ, Section 3.1.1, describe changes in population size over time but on different time scales; λ deals with changes from one time step to the next, whereas R_0 is the generation-to-generation growth rate. Not surprisingly then, both can be calculated as eigenvalues of the appropriate kernel. For λ we can use the code described in Section 2.5.5, first constructing the iteration matrix using mk_K and then using eigen to get the eigenvalues, as a reminder here's the code

```
IPM.true <- mk_K(nBigMatrix, m.par.true, -3.5, 5.5)
lambda0 <- Re(eigen(IPM.true$K,only.values = TRUE)$values[1])
```

the only.values = TRUE option specifies that only the eigenvalues are to be calculated; the calculation of the eigenvectors is the slow part for large matrices. For the calculation of R_0 we need the fundamental matrix, which in R is given by

```
N <- solve(diag(nBigMatrix)-P)
```

where diag(nBigMatrix) creates the identity matrix, I, P is the survival iteration matrix, and solve then inverts the $(I - P)$ matrix. The kernel $R = FN$ projects the population from one generation to the next, and R_0 is the dominant eigenvalue of this, so in R the code needed is

```
> R <- F %*% N
> R0 <- abs(eigen(R)$values[1])
> cat("R0=",R0,"\n");
R0= 1.00026
```

In the *Oenothera* IPM offspring size is independent of parental size. This means we can calculate R_0 directly by averaging the per-capita fecundity with respect to the size at birth, see Box 3.1; the code looks like this

```
> R0 <- sum((e %*% R) * offspring.prob)
> cat("R0=",R0,"\n");
R0= 1.000257
```

This is doing some nBigMatrix+1=251 integrations, can you see why? If not read on.

3.3.2 Mortality: age and size at death calculations

For the calculation of l_a we need to implement equation (3.1.16) so we need to calculate $\tilde{l}_a = \langle \mathbf{e}P^a, c \rangle = \langle l_a, c \rangle$. For the age 0 individuals this is evaluated in R using

```
                sum((e%*%P)*offspring.prob)
```

which is the probability of surviving to age 1. There is a lot going on here and we need to be careful to ensure we implement the various integrations required, so let's step through it slowly. The term $\mathbf{e}P$ is defined as $\mathbf{e}P(z) = \int_{\mathbf{Z}} \mathbf{e}P(z', z) \, dz'$, which is a single integration. The P iteration matrix has already been multiplied by h, so the matrix multiplication e%*%P does the integrations over z' at each of the mesh points (z_i) in a single calculation. The result of the R calculation is a vector, with each entry containing $\int_{\mathbf{Z}} \mathbf{e}P(z', z_i) \, dz'$, which is why the operation $\mathbf{e}P$ is a function of z. Let's see it in action:

```
> round( e%*%P, 4)[1:10];
 [1] 0.0095 0.01 0.0104 0.0108 0.0113 0.0118 0.0123 0.0129 0.0134 0.014
```

Summing down the columns of P corresponds to summing over all the sizes an individual can become next year. As all individuals that survive have a size next year, $\int_{Z} \mathbf{e}P(z', z) \, dz'$ must equal the probability of survival. Remembering that flowering is fatal we can calculate the probability of being alive next year for each of the mesh points, this is given by

```
> round(s_z(meshpts,m.par.true)*(1-p_bz(meshpts, m.par.true)),4)[1:10];
 [1] 0.0095 0.01 0.0104 0.0108 0.0113 0.0118 0.0123 0.0129 0.0134 0.014
```

which agrees with our previous calculation. For a cohort with size distribution $c(z)$ the average of $\mathbf{e}P$, remembering this is a function of z, is $\int_{\mathbf{Z}} c(z) \mathbf{e}P \, dz$ which if $c(z)$ is the probability distribution of offspring sizes can be calculated in R using sum((e%*%P)*offspring.prob). There is no need to multiply by the integration step, h, as offspring.prob <- h*c_0z1(meshpts,m.par). So to answer the question posed at the end of the last section, the matrix multiplication e%*%R sums down the columns of R which is nBigMatrix integrations, and the final sum

then integrates this with respect to the distribution of offspring sizes, giving
nBigMatrix+1=251 integrations in total.

For the other age classes we need P^a, which is defined as the ath power of
matrix P, so we do the calculations recursively

```
Pa <- P
for(a in 2:12){
        Pa=Pa %*% P
        la[a]= sum((e %*% Pa)*offspring.prob)
}
la <- c(1,la)
```

Note, we finally add $l_0 = 1$, as all individuals are age zero at birth. The calculation of the age-specific fecundities proceeds in a very similar manner except
now we now need to calculate terms like $\langle eFP^a, c \rangle$, which for age 1 individuals
can be calculated in R using

```
sum((e %*% F %*% P)*offspring.prob).
```

To compare the theoretical age-specific survival terms ($p_a = l_{a+1}/l_a$) and fecundities with the individual-based simulation, we first need to 1) convert the
number of seed produced into recruits by multiplying seed production by the
probability of successful recruitment, p_r, as reproduction is calculated in the
IPM in terms of recruits, and 2) set the fecundities of plants that didn't flower,
stored as NAs, to zeros. The resulting calculations show that the theoretical
values are in excellent agreement with the simulation, Figure 3.1.

With $\tilde{l}_a$ and $\tilde{f}_a$ we can now calculate generation time using the various definitions introduced in Section 3.1.3, the R code is

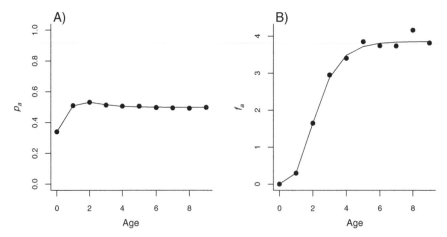

Fig. 3.1 Age-specific survival (A) and fecundity (B) for *Oenothera*. The points are from
a stochastic individual-based model, solid lines are the theoretical values calculated from
the IPM. Source file: Monocarp Event Calculations.R

```
> T1 <- log(R0)/log(lambda0)
> T2 <- sum((1:21)*fa*la)/sum(fa*la)
> cat("Generation time log(R0)/log(lambda) = ",T1,"\n")
Generation time log(R0)/log(lambda) = 4.160689
> cat("Generation time from age-specific rates = ",T2,"\n")
Generation time from age-specific rates = 4.215729
```

As is typical the differences between the different measures of generation time are rather small.

Moving onto the age and size at death, for average lifespan we need to evaluate $\tilde{\eta} = \langle \mathbf{e}N, c \rangle$, which if $c(z)$ is the offspring size distribution, is given by `sum((e %*% N)*offspring.prob)`. The `e%*%N` term sums down the columns of the fundamental matrix N, giving the expected lifespan at each of the mesh points, and then we average this, with respect to the probability distribution of offspring sizes. To calculate the variance in lifespan we evaluate $Var_c(\eta)$, equation (3.2.5), which in R is given by

```
sum((e%*%(2*N%*%N - N))*offspring.prob) - sum((e%*%N * offspring.prob))^2
```

Comparing theory with the simulation shows excellent agreement

```
> cat("Theory:mean=",mean.age.death," variance=",Var.nu,"\n")
Theory:mean= 0.7004 variance= 1.6878
> cat("Simulation:mean=",mean.age.death.sim," variance=",Var.nu.sim,"\n")
Simulation: mean= 0.6984 variance= 1.6795
```

We can also calculate expect longevity using $\sum \tilde{l}_a$ which was used to derive the preceding results, as a quick check we have

```
> sum(la)-1
[1] 0.7003955
```

By now you should be getting the hang of the operator notation summarized in Table 3.1. How would you evaluate the size at death kernel? The formula

$$\Omega(z', z) = (1 - s(z'))N(z', z) \qquad (3.3.1)$$

is an example of an element-by-element calculation. We have to be careful as there are two routes to death because flowering is fatal. The probability an individual dies is given by $p_b(z) + (1 - p_b(z))(1 - s(z))$ so in R we calculate $\Omega(z', z)$ using

```
Omega <- (p_bz(meshpts,m.par.true) +
         (1-p_bz(meshpts,m.par.true))*(1-s_z(meshpts,m.par.true))) * N
```

In R this multiplies each of the columns of the matrix N by the vector containing the probability of death at each of the mesh points. This is an example of the "recycling rule" in R where the vector of probabilities of death is repeatedly used in the element by element calculation. With Ω calculated we can then find the distribution of size at death for a cohort with initial state distribution c, which is given by Ωc. This is an example of "right-multiplication" (Table 3.1), and so we calculate the distribution of size at death using `dist.size.death <- (Omega %*% offspring.prob)`. Let's think this through: `offspring.prob` is the probability an individual is initially size z, and $\Omega(z', z)$ the chance an individual initially size z dies while being in the size interval $(z', z'+h)$ at some point in its life. So by averaging over the offspring distribution we can calculate the chance of dying in $(z', z' + h)$ for a randomly chosen individual. With the `dist.size.death` distribution we can then calculate any moment we require, for example the variance is given by

```
Var.size.death <- sum(dist.size.death*meshpts*meshpts) -
         sum(dist.size.death*meshpts)*sum(dist.size.death*meshpts)
```

As before the agreement between theory and simulation is excellent.

```
> cat("Theory: mean=",mean.size.death, "variance=",Var.size.death,"\n");
Theory: mean= 0.3505 variance= 1.3467
> cat("Simulation: mean=",mean.size.death.sim, " variance=",
  +    Var.size.death.sim, "\n");
Simulation: mean= 0.3508 variance= 1.3434
```

3.3.3 Reproduction: who, when, and how much?

The survival without breeding kernel, P_0, describes the transitions between different sizes for individuals that survive and don't breed, and so for the *Oenothera* IPM this is given by P, because reproduction is fatal and so the survival kernel has the form $(1 - p_b(z))s(z)G(z', z)$ as required. However, to keep as close to the text in Section 3.2 we define the matrices

```
# As reproduction is fatal the P kernel is the required P0 kernel
P0 <- P
N0 <- solve(diag(nBigMatrix)-P0)
B <- p_bz(meshpts,m.par.true) %*% N0
```

So `N0` is the fundamental matrix of the survival without breeding kernel, and B contains the probabilities of breeding at least once. The final quantity we need to calculate is the initial cohort distribution weighted by the probability of breeding, which is given by

```
# Breeding-weighted offspring size distribution
breed.offspring.prob <- B*offspring.prob/sum(B*offspring.prob)
```

as several of the calculations assume breeding occurs at least once. With these quantities we can calculate $P_{(b)}(z', z)$. There are many ways of doing this, and it's easy to muddle up the z's so we simply created two large matrices and then did the element-by-element multiplication and division

```
B.m.c <- matrix(B,nrow=nBigMatrix,ncol=nBigMatrix)
B.m.r <- matrix(B,nrow=nBigMatrix,ncol=nBigMatrix,byrow=TRUE)
P.b <- (P0 * B.m.c ) / B.m.r
N.b <- solve(diag(nBigMatrix)-P.b)
```

The B.m.c matrix is filled column-wise, the default, whereas B.m.r is filled by rows (the byrow=TRUE sees to that), the next line then does the element-by-element calculations for $P_{(b)}(z', z)$, equation (3.2.11). $P_{(b)}$ is the survival kernel conditioned on breeding at least once, so the fundamental operator (N.b) for this gives the number of censuses at which an individual is present before breeding, and so the mean age at first breeding, for those that breed at least once, is given by $\bar{a}_R = \mathbf{e}N_{(b)} - 1$ which a function of z. So to calculate the mean age at first breeding we need to average this with respect to the initial cohort distribution, weighted by the probability of breeding $\langle \bar{a}_R, c_b \rangle$, in R this is given by

```
sum((e %*% N.b) * breed.offspring.prob)-1
```

Comparing the simulation and analytical result again shows excellent agreement

```
> cat("Theory: mean=",mean.Repr.age,"\n");
Theory: mean= 2.985773
> cat("Simulation: mean=",with(sim.data,mean(age[Repr == 1])),"\n");
Simulation: mean= 2.983753
```

The size-at-death kernel, where death here is reproduction, can be calculated as $\Omega_{(b)}(z', z) = (1 - \mathbf{e}P_{(b)})N_{(b)}(z', z)$, and so if we want to know the mean size at flowering (which is the same as the mean size at first flowering as flowering is fatal) we need $\Omega_{(b)}c_{(b)}$, which is the probability of reproducing in the interval $(z', z' + h)$ for a randomly chosen individual from the offspring distribution that reproduces at least once. Now that we have the probability distribution for size at first reproduction, of those that reproduce, we can calculate the mean or variance, as in the variance in size at death calculation at the end of the previous section.

For monocarpic plants, where reproduction is fatal, all individuals that reproduce do so only once, so $p_b N/B$ must be 1 for all z, and it is

```
> breedingFreq <- p_bz(meshpts,m.par.true) %*% N0 / B
> cat("Mean breeding frequency (should =1):", range(breedingFreq),"\n");
Mean breeding frequency (should =1): 1 1
```

For the calculation of the variance in lifetime reproductive output we first need to calculate the variance in reproductive output each year. Seed production by a breeder follows a Poisson distribution with mean $b(z)$ and so the number of recruits (given that flowering occurs) also follows a Poisson distribution, with mean $p_r b(z)$. We therefore have $\sigma_+^2 = p_r b(z)$. Then from equation (3.2.14), with a bit of algebra we get

$$\sigma_b^2 = \bar{\beta}(z) + \frac{1 - p_b(z)}{p_b(z)} \bar{\beta}(z)^2 \tag{3.3.2}$$

(recall $\bar{\beta} = p_b p_r b$) which we evaluate using

```
mean.recs <- p_bz(meshpts,m.par.true) * m.par.true["p.r"] *
             b_z(meshpts,m.par.true)
```

```
sigma.b.2 <- mean.recs + (1 - p_bz(meshpts,m.par.true)) *
        mean.recs * mean.recs/p_bz(meshpts,m.par.true)
```

Mean lifetime reproductive output is straightforward to calculate ($\bar{r}(z) = \mathbf{e}R = \mathbf{e}FN$). That leaves us with $r_2(z)$, which is given by equation (3.2.15). For a monocarp $\pi_b = 0$ and hence

$$r_2 = \left(\sigma_b^2 + (p_r p_b \circ b)^2\right) N. \tag{3.3.3}$$

Here b is expected fecundity in terms of recruits, so in R

```
rbar <- e %*% R
r2 <- (sigma.b.2 + (m.par.true["p.r"] * p_bz(meshpts,m.par.true) *
      b_z(meshpts,m.par.true))^2) %*% N
var.b <- r2 - rbar*rbar
```

To estimate the mean and variance of reproductive output in the IBM, the simplest approach is to start many individuals at a common size, and then estimate the mean and variance of their lifetime reproductive output; this is what the function get_mean_var_Repr in Monocarp Demog Funs.R does. Using a for loop we can calculate these quantities at each of the mesh points.[7]

The results (Figure 3.2A and B) show excellent agreement between the simulation and analytical results, and as expected individuals that start life large produce many offspring. The variance in lifetime reproductive output is however rather more complicated. Initially when born very small, virtually all individuals die before reproducing and both the mean and variance are small. As size increases the variance initially increases as more individuals survive to reproduce. However, between sizes 2 and 3 the probability of flowering increases to ≈ 1 (Figure 2.4B) and so the variance that results from individuals flowering or not disappears (Figure 3.2D). As virtually all individuals now flower at the first opportunity the variance collapses back to the variance of the Poisson distribution of fecundity conditional on breeding (red line Figure 3.2B). Notice, despite following 100,000 individuals, our estimate of the variance in lifetime reproductive output is still highly variable, illustrating the considerable advantages in terms of computation time of having analytical results.

With two minor modifications we can use exactly the same code to calculate the mean and variance in the number of reproductive events. The changes we need are

```
b_z <- function(z, m.par){
    N <- 1/m.par.true["p.r"]   # seed production of a size z plant
    return(N)
}
sigma.b.2 <- p_bz(meshpts,m.par.true) * (1-p_bz(meshpts,m.par.true))
```

---

[7] Because this is slow we actually used mclapply, the multicore version of lapply within the parallel library. Going from 1 to 8 cores decreases the time required by 4.7-fold. In Windows mclapply is not available but parLapply has the same functionality.

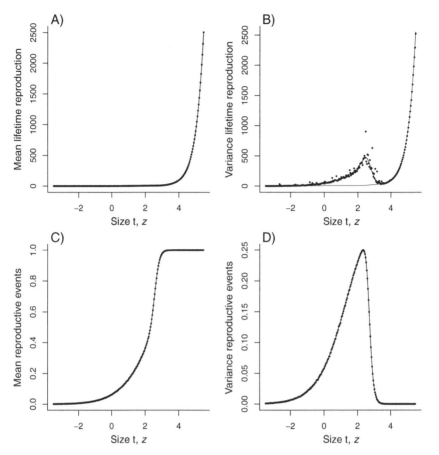

Fig. 3.2 A) Mean lifetime reproductive output from the IBM (dots), and expected value calculated from $\bar{r}(z) = \mathbf{e}R = \mathbf{e}FN$. B) Variance in lifetime reproductive output, from the IBM (dots) and black line the analytical result based on equation (3.2.15). Red line is the Poisson fecundity variance. C) Mean, and D) variance in the number of reproductive events, IBM (dots) and black line the analytical result. Source file: Monocarp Event Calculations.R

First we redefine the b_z function so $b(z) = 1$. As the b_z gives the expected seed production and we want the number of recruits per reproductive event to be 1, b_z returns 1/m.par.true["p.r"] which is multiplied by m.par.true["p.r"] when constructing the kernel, and so the number of recruits per reproductive event is 1. sigma.b.2 is then the Bernoulli variance that arises as a result of individuals flowering or not. We then have to remake the kernel using mk_K, as b_z has changed, and calculate the fundamental operator N. The results (Figure 3.2C and D) again show excellent agreement between the simulation and analytical results.

Finally we look at the mean and variance in the age at last reproduction. This for a monocarp is also the first time that reproduction occurs for which we've already calculated the mean, but let's go through the calculations anyway. First a bit of housekeeping, we need to undo our changes to b_z, and recompute the kernel and fundamental operator. The first thing to notice is that because $\pi_b = 0$ many of the formulae simplify. In particular $d_b(z) = 1$ and $A_b(z', z) = 0$ and so $f_b = 1$ so all we need is B and we're good to go:

```
> B <- p_bz(meshpts,m.par.true) %*% N
> E.bS <- ( p_bz(meshpts,m.par.true) %*% (N %*% N - N) )/B
> mean.age.last.breed <- sum(breed.offspring.prob*E.bS)
> mean.age.last.breed.sim <- with(sim.data, mean(age[Repr == 1]))
> E.bS2 <- (p_bz(meshpts,m.par.true) %*% (P + P%*%P) %*% (N%*%N%*%N))/B
> var.age.last.breed <- sum(breed.offspring.prob*E.bS2) -
+                       (sum(breed.offspring.prob*E.bS)^2)
> var.age.last.breed.sim <- with(sim.data,var(age[Repr == 1]))
> cat("Age last reproduction","\n")
Age last reproduction
> cat("Theory: mean=",mean.age.last.breed,
+                 " variance=",var.age.last.breed,"\n");
Theory: mean= 2.985773 variance= 2.821698
> cat("Simulation: mean=",mean.age.last.breed.sim,
+                 " variance=",var.age.last.breed.sim,"\n");
Simulation: mean= 2.983753 variance= 2.784852
```

And reassuringly the mean age at first reproduction equals that at last reproduction, as it should! Phew - put your feet up, pour yourself a cup of coffee/tea/a non-caffeinated beverage, it all works and you've earned it, now you can go apply it to your system - and perhaps more importantly when we say "Basic Analyses" you'll know what to expect next time.

3.4 Appendix: Derivations

1. Derivation of equation (3.2.3). The identity

$$\sum_{a=0}^{\infty} (a+1)P^a = N^2 \tag{3.4.1}$$

holds because $(I - P)$ times the kth partial sum of the left-hand side converges to N, and $(I - P)N^2 = N$. This is the IPM version of the identity

$$1 + 2r + 3r^2 + 4r^3 + \cdots = (1 - r)^{-2}$$

for real numbers $-1 < r < 1$, and holds so long as $||P|| < 1$ in some operator norm (the matrix version was proved by Cochran and Ellner (1992) and probably by others long before that). Then

$$E[\eta^2] = \sum_{a=1}^{\infty} a^2 (l_{a-1} - l_a) = \sum_{a=1}^{\infty} a^2 l_{a-1} - \sum_{a=1}^{\infty} a^2 l_a$$

$$= \sum_{a=0}^{\infty} (a+1)^2 l_a - \sum_{a=0}^{\infty} a^2 l_a = 2 \sum_{a=0}^{\infty} a l_a + \sum_{a=0}^{\infty} l_a \qquad (3.4.2)$$

$$= e(2 \sum_{a=0}^{\infty} a P^a + \sum_{a=0}^{\infty} P^a) = e(2 \sum_{a=0}^{\infty} (a+1) P^a - \sum_{a=0}^{\infty} P^a)$$

$$= e(2N^2 - N), \text{ as desired.}$$

2. Size at death is a probability distribution. We need to show that

$$\int (1 - s(z')) N(z', z) \, dz' \equiv 1. \qquad (3.4.3)$$

Using $N = I + P + P^2 + \cdots$ we can write the integral in (3.4.3) as the sum of terms,

$$I : \quad \int (1 - s(z')) I(z', z) \, dz' = 1 - s(z)$$

$$P : \quad \int (1 - s(z')) P(z', z) \, dz' = s(z) - sP(z)$$

$$P^2 : \quad (s - sP)P = sP - sP^2 \qquad \text{(because the } P \text{ term is } s - sP).$$

$$P^3 : \quad sP^2 - sP^3$$

$$\vdots$$

The kth partial sum is $1 - sP^{k-1}$ and $P^k \to 0$, so the sum is 1 as desired.

3. Derivation of equation (3.2.15). Total reproductive output from time t forward is the sum of the output in year t and the total from time $t + 1$ forward. Then conditioning on the first change in state from z_t to z_{t+1},

$$r_2(z) = E[r(z_t)^2 | z_t = z] = E[(\beta(z_t) + r(z_{t+1}))^2 | z_t = z]$$
$$= E[\beta(z_t)^2 | z_t = z] + 2E[\beta(z_t) r(z_{t+1}) | z_t = z] + E[r(z_{t+1})^2 | z_t = z]. \qquad (3.4.4)$$

The third term in the last line is $E[r_2(z_{t+1}) | z_t = z] = r_2 P$, so $r_2(I - P)$ equals the sum of the first two terms. For the first term we have

$$E(\beta(z)^2) = Var(\beta(z)) + E(\beta(z))^2 = \sigma_b^2(z) + p_r^2 p_b(z)^2 b(z)^2.$$

In the second term, $\beta(z_t) r(z_{t+1})$ can be nonzero only when reproduction occurs, which has probability $p_b(z)$ when $z_t = z$. Let $\mathcal{R}$ denote the event that reproduction occurs. The second term is then

$$2p_b(z)E[\beta(z_t)r(z_{t+1})|\mathcal{R}, z_t = z] = 2p_b(z)p_r b(z)E[r(z')|\mathcal{R}, z_t = z] \quad (3.4.5)$$
$$= 2p_r p_b \circ b \circ \bar{r}\pi_b. \quad (3.4.6)$$

We therefore have $r_2(I - P) = \sigma_b^2 + (p_r p_b \circ b)^2 + 2p_r p_b \circ b \circ (\bar{r}\pi_b)$, which gives equation (3.2.15).

4. Mean and variance of age at last reproduction. Consider an individual born at time 0, so (without loss of generality) age and time are equivalent. We first need the probability that breeding never occurs after time t. Let

$$f_0(z) = Pr[\text{No breeding at times } t+1, t+2, \cdots | z_t = z, \text{ No breeding at } t]$$
$$f_b(z) = Pr[\text{No breeding at times } t+1, t+2, \cdots | z_t = z, \text{ Breed at } t]. \quad (3.4.7)$$

If the individual dies between t and $t+1$, there is no more breeding after t. Let $d_b(z)$ and $d_0(z)$ be the probabilities of death conditional on size, after a time at which breeding does or does not occur, respectively, i.e.

$$d_b(z) = 1 - \int \pi_b(z', z)dz'$$

and similarly for d_0. If the individual survives, there is no more breeding after t if there is no breeding at time $t+1$ and then no breeding at any time thereafter. So conditioning on the change in state between t and $t+1$,

$$f_0(z) = d_0(z) + \int \pi_0(z', z)(1 - p_b(z'))f_0(z') \, dz'$$
$$f_b(z) = d_b(z) + \int \pi_b(z', z)(1 - p_b(z'))f_0(z') \, dz' . \quad (3.4.8)$$

Let A_0 be the kernel $A_0(z', z) = \pi_0(z', z)(1 - p_b(z'))$ and A_b the kernel $A_b(z', z) = \pi_b(z', z)(1 - p_b(z'))$. Equations (3.4.8) can then be written as

$$f_0 = d_0 + f_0 A_0 \quad \Rightarrow \quad f_0 = d_0(I - A_0)^{-1}$$
$$f_b = d_b + f_0 A_b = d_b + d_0(I - A_0)^{-1}A_b. \quad (3.4.9)$$

Now let S denote the age at last reproduction (with $S = \infty$ if the individual never breeds). Then $S = t$ if the individual breeds at time t (having some size z' at time t), and then never breeds again, i.e.

$$Pr[S = t|z_0 = z] = \int P^t(z', z)p_b(z')f_b(z') \, dz' = (p_b \circ f_b)P^t. \quad (3.4.10)$$

where P is given by equation (3.2.9). The probability (conditional on initial state z) that breeding occurs at least once is then the sum of (3.4.10) from $t = 0$ to ∞, which is

$$B = (p_b \circ f_b)N \quad (3.4.11)$$

where $N = (I - P)^{-1}$ is the fundamental operator.

To compute the mean time at last breeding, using (3.4.1) we have

$$
\sum_{t=0}^{\infty} t \Pr[S = t|z_0 = z] = \sum_{t=0}^{\infty}(t+1)\Pr[S = t|z_0 = z] - \sum_{t=0}^{\infty}\Pr[S = t|z_0 = z]
$$

$$
= \sum_{t=0}^{\infty}(t+1)(p_b \circ f_b)P^t - \sum_{t=0}^{\infty}(p_b \circ f_b)P^t \qquad (3.4.12)
$$

$$
= (p_b \circ f_b)(N^2 - N).
$$

For the variance we need the identity

$$
\sum_{t=1}^{\infty} t^2 P^t = (P + P^2)N^3, \qquad (3.4.13)
$$

derived as follows. We know Cochran and Ellner (1992, Appendix 2) that $\sum_{t=0}^{\infty}(t+1)^2 P^t = (I + P)N^3$. Therefore

$$
\sum_{t=1}^{\infty} t^2 P^t = \sum_{t=0}^{\infty}(t+1)^2 P^{t+1} = P\sum_{t=0}^{\infty}(t+1)^2 P^t = (P + P^2)N^3. \qquad (3.4.14)
$$

Then

$$
\sum_{t=0}^{\infty} t^2 \Pr[S = t|z_0 = z] = \sum_{t=1}^{\infty} t^2 (p_b \circ f_b)P^t = (p_b \circ f_b)(P + P^2)N^3. \qquad (3.4.15)
$$

The first and second moments of the time of last reproduction, conditional initial size and on reproduction occurring at least once, are then given by equations (3.4.12) and (3.4.15) divided by B (element-by-element),

$$
\begin{aligned}
E_b[S] &= (p_b \circ f_b)(N^2 - N)/B \\
E_b[S^2] &= (p_b \circ f_b)(P + P^2)N^3/B. \qquad (3.4.16)
\end{aligned}
$$

Chapter 4
Basic Analyses 2: Prospective Perturbation Analysis

Abstract A common motivation for constructing an IPM is to address the question: what will happen to the population if the underlying vital rates change? In Chapters 2 and 3 we saw how to calculate common statistics summarizing population growth and the life cycle. In this chapter we look at how to calculate functions and scalar quantities that describe how these statistics change when different components of the IPM kernel are subjected to a small perturbation. We begin by showing how to calculate the *sensitivity* and *elasticity* of λ to perturbations of the kernel, vital rate functions, and their parameters. We finish by briefly describing how to derive analogous expressions for life cycle statistics.

4.1 Introduction

Prospective analysis (Horvitz et al. 1997; Caswell 2000) refers to analyses that address the question: how much and in what direction does an emergent feature of the model change when we perturb the kernel in some way? The "emergent feature" can be anything of relevance that we know how to calculate, such as population growth rate, λ, generation time, T, or mean size at death. The aim is to understand how these properties of a model depend on its component parts. A population manager might want to know how improving components of reproduction and survival will influence λ, while someone interested in life history evolution might be more interested in understanding how generation time, T, depends on these processes.

Prospective analyses are forward looking. They tackle questions about what might happen if the life history or environment were to change. *Retrospective analysis* looks back in time (or across space) at observed patterns of variation in vital rates, to ask how those patterns have affected variation in a quantity like λ. Retrospective analyses are similar to statistical tools such as analysis of (co)variance, in that they partition variance in a "response" variable (λ, T, etc.) into contributions from different "explanatory" variables (e.g., time-varying model parameters). Prospective and retrospective analyses address different questions. A retrospective analysis cannot tell you much about the potential

S.P. Ellner et al., *Data-driven Modelling of Structured Populations*,
Lecture Notes on Mathematical Modelling in the Life Sciences,
DOI 10.1007/978-3-319-28893-2_4

effects of new management interventions (Caswell 2001). It explains the variation you observed, but says very little about sources of variation that have not yet been experienced. Retrospective analysis is usually done when parameters have been estimated separately for multiple years of a study; for an example see Section 7.6.

The focus of this chapter is on prospective analysis. Very many types have been developed for matrix models and are equally possible for IPMs. The purpose of this chapter is to provide an overview of the different kinds of calculations that can be done, and the different levels at which perturbations can be considered, using population growth rate and life cycle properties as the emergent features of interest.

4.2 Sensitivity and elasticities

Local prospective perturbation analysis (hereafter "perturbation analysis") amounts to calculating the partial derivative of a quantity with respect to a specific perturbation. Perturbations can be considered at several different "levels" of an IPM. For most IPMs we can consider at least three such levels: the projection kernel, the functions that comprise the kernel, or the parameters that define these functions. Other kinds of perturbation are certainly possible. For example, if we have expressed the parameters of vital rate functions in terms of an environmental covariate (e.g., soil moisture) then we might want to explore how λ changes as a function of this variable.

In the ecological literature, the term *sensitivity* is used for a partial derivative (e.g., the partial derivative of λ with respect to mean soil moisture). In many cases a more meaningful quantity is the *elasticity* or proportional sensitivity, the fractional change in the response relative to the fractional change in the quantity being perturbed: by what percentage does λ increase, if mean soil moisture goes up by 5%?

Sensitivities and elasticities can always be calculated numerically, by making small changes in the model and seeing how the response variable changes, see Box 7.2. Though straightforward, this may be computationally expensive; see Ellner and Rees (2006) for various tricks for speeding up the calculations. Fortunately, it is often possible to derive an analytical expression. Besides facilitating quick and accurate calculation, these may also provide insight into how particular model components influence the effect of a perturbation.

The aim of this chapter is to explain how to derive and use some common perturbation analyses. If you have only limited calculus experience, some of the calculations may look complicated at first glance. However, we mostly only rely on elaborations of the standard rules for derivatives (chain rule, product rule, quotient rule), which accommodate the fact that we are working with functions and kernels.

4.3 Sensitivity analysis of population growth rate

The most common perturbation analyses ask how the asymptotic population growth rate, λ, of a density-independent model changes when we perturb the kernel, $K(z', z)$. We can construct a fairly general perturbation analysis of λ

by considering the perturbation of $K(z', z)$ to $K(z', z) + \epsilon C(z', z)$ where ϵ is a small constant and $C(z', z)$ is a perturbation kernel - more on what exactly this is below. The sensitivity to this arbitrary perturbation is given by

$$\left.\frac{\partial \lambda(\epsilon)}{\partial \epsilon}\right|_{\epsilon=0} = \frac{\iint v(z')C(z', z)w(z)dz'dz}{\int v(z)w(z)dz} \tag{4.3.1}$$

where $\lambda(\epsilon)$ is the asymptotic population growth rate with the perturbed kernel, and $v(z)$ and $w(z)$ denote the dominant left and right eigenvectors (see Easterling et al. 2000, p.95, though this formula was already well known in the physics literature).

This one expression is all we need to calculate the sensitivity or elasticity of λ to perturbations at any "level" in an IPM: the projection kernel, the vital rate functions, or demographic model parameters. All of these are just a matter of identifying the perturbation kernel C, and working through some algebra. (Note C denotes a perturbation kernel whereas C_1 is an offspring state kernel).

Working with expressions like (4.3.1) quickly becomes cumbersome because we constantly have to write out the integrals, so at times we will use the operator notation summarized in Table 3.1. Formula (4.3.1) becomes

$$\left.\frac{\partial \lambda(\epsilon)}{\partial \epsilon}\right|_{\epsilon=0} = \frac{\langle v, Cw \rangle}{\langle v, w \rangle} \tag{4.3.2}$$

where $\langle f, g \rangle$ is the inner product $\int_Z f(z)g(z)\, dz$, and Af is $\int_Z A(z', z)f(z)\, dz$.

4.3.1 Kernel-level perturbations

Sensitivity analysis of λ at the level of the kernel is very similar to matrix models. The question is, how much does λ change when we apply an additive perturbation to the kernel $K(z', z)$ at some point (z'_0, z_0)?

We have to proceed carefully, because the kernel K is a continuous function. We can't perturb it at just one point and keep it continuous. The same is true for vital rate functions such as $s(z)$.

We want a way to compare the relative importance of perturbing the kernel at different places. One way (Easterling et al. 2000) is to perturb the kernel in a small region around (z'_0, z_0), and let the region size shrink to zero while the integral of the perturbation stays constant. Figure 4.1 illustrates this process for our size-structured ungulate model. Panel A shows the unperturbed kernel; the lower ridge is fecundity, with larger parents having larger offspring, and the higher ridge is survival and growth. In panel B a perturbation is added near $z = z' = 2$. The perturbation is proportional to a bivariate Gaussian kernel with each variable having a standard deviation of $\sigma = 0.1$. In panels C and D we localize the perturbation by decreasing σ to 0.05 and 0.02. We can compute the effect of each of these increasingly localized perturbations, using (4.3.1). The limiting value as $\sigma \to 0$ is what we mean when we say that the kernel was perturbed "at" $z = z' = 2$. This lets us compare perturbations to different transitions or vital rates.

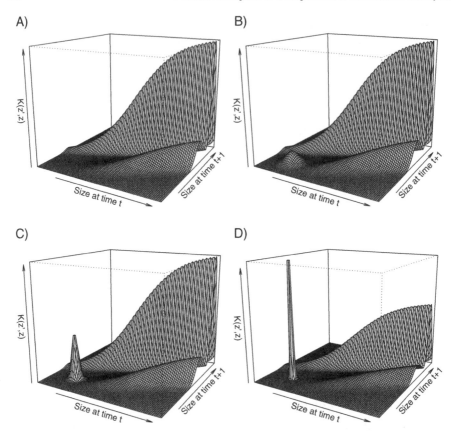

Fig. 4.1 Illustrating the process of perturbing a kernel at a single point. A) The unperturbed kernel for the size-structured ungulate model. B, C, D) Adding perturbations that are increasingly localized around the point $(z' = 2, z = 2)$, but all have the same integral. Note the change in vertical scale in D). Source file: `PerturbUngulate.R`

There's a useful shortcut to the process of shrinking σ that we introduce in Box 4.1, which is to use Dirac δ functions. These are not functions in the usual sense; they are an idealization of what you get as $\sigma \to 0$ in Figure 4.1. Introduced by physicist Paul Dirac, they proved so useful that mathematicians figured out how to justify them. For us, the key properties of δ functions are equations (4.3.3) and (4.3.5), which tell us how to work with one- and two-dimensional δ functions when they occur in a perturbation formula.

For example, if we increase the kernel at the point (z'_0, z_0) the perturbation kernel is $C = \delta_{z'_0, z_0}$. Then writing out (4.3.2) as a double integral (as in 4.3.1) and using (4.3.5), the right-hand side of (4.3.2) is the familiar eigenvalue sensitivity formula

$$\mathbf{s}(z'_0, z_0) = \frac{\partial \lambda}{\partial K(z'_0, z_0)} = \frac{v(z'_0)w(z_0)}{\langle v, w \rangle}. \tag{4.3.6}$$

(Note that the sensitivity function $\mathbf{s}$ is not the survival probability s).

For us, a δ function is shorthand for perturbing a demographic rate function in a region near a particular point, and letting the size of the region shrink to zero, like in Figure 4.1. You can think of δ functions as Gaussian probability distributions with a very small variance, which we then shrink to zero.

The one-dimensional Dirac δ function $\delta_{z_0}(z)$ is defined by the properties that its value is 0 except at $z = z_0$, and

$$\int_Z \delta_{z_0}(z)\, dz = 1$$

where Z is the set of possible values for z. Then for any smooth function $f(z)$,

$$\int_Z \delta_{z_0}(z)f(z)\, dz = f(z_0). \tag{4.3.3}$$

For example, increasing $s(z)$ to $s(z) + 0.01\delta_1(z)$ represents a perturbation in which survival probability s has been increased by 0.01 at $z = 1$. This perturbation increases the total number of survivors by

$$0.01 \int_Z \delta_1(z)n(z,t)\, dz \tag{4.3.4}$$

Using formula (4.3.3), we find that the value of (4.3.4) is $0.01n(1,t)$.

Is this the answer we should get? If we think of think of $\delta_1(z)$ as a Gaussian distribution with mean $z = 1$ and variance $\sigma^2 \approx 0$, the integral in (4.3.4) is then the expected value of $n(z,t)$ when z is Gaussian mean 1 and a very small variance. This is very close to $n(1,t)$, and it converges to $n(1,t)$ when $\sigma^2 \to 0$. So by using the properties of δ functions, we quickly get the same result as the entire process of making a smooth perturbation, and then localizing it to $z = 1$.

For perturbations to kernels we also need the bivariate δ functions $\delta_{z_0',z_0}(z',z)$. As in one dimension, $\delta_{z_0',z_0}(z',z)$ is zero except at $(z',z) = (z_0',z_0)$ and its integral over $Z \times Z$ is 1. As a result (and this is the only thing that matters) for any smooth function $A(z',z)$

$$\iint \delta_{z_0',z_0}A(z',z)\, dz'\, dz = A(z_0',z_0). \tag{4.3.5}$$

Again, it's safe to think of δ_{z_0',z_0} as a bivariate Gaussian distribution where z has mean z_0, z' has mean z_0', and z and z' are independent with very small variance. The integral in (4.3.5) is then the expectation of $A(z',z)$ with respect to this distribution, which converges to $A(z_0',z_0)$ as the variance goes to 0.

Box 4.1: Dirac δ functions.

The sensitivity formula also lets us write the general perturbation formula (4.3.1) in another way. Substituting (4.3.6) into (4.3.1) shows that for any perturbation $K(\epsilon) = K + \epsilon C$ we have

$$\left.\frac{\partial \lambda(\epsilon)}{\partial \epsilon}\right|_{\epsilon=0} = \iint \mathbf{s}(z', z)\frac{\partial K(z', z)}{\partial \epsilon}\, dz\, dz'. \tag{4.3.7}$$

This is an infinite-dimensional version of the chain rule for the sensitivity of λ, which is a useful shortcut through many calculations. For example, since the partial derivative of the kernel with respect to the survival and reproduction components at (z_0', z_0) is equal to 1, knowing (4.3.7) is sufficient to show

$$\frac{\partial \lambda}{\partial P(z_0', z_0)} = \frac{\partial \lambda}{\partial F(z_0', z_0)} = \mathbf{s}(z_0', z_0). \tag{4.3.8}$$

That is, the sensitivity functions associated with P and F are equal to the kernel sensitivity. This is a somewhat trivial example, since the kernel is just the sum of P and F, but in other calculations below rules like (4.3.7) often save a lot of time.

How do we compute the kernel elasticity function, $\mathbf{e}(z_0', z_0)$? Elasticity is the fractional change in λ resulting from a small perturbation, relative to the fractional change in the component of the model that is being perturbed. Because the derivative of $\log \lambda$ with respect to any parameter or kernel component θ is $(1/\lambda)(\partial \lambda/\partial \theta)$ (as promised, we didn't ask you to remember this from your calculus class), the derivative of $\log \lambda$ is fractional rate of change in λ. The same is true for whatever we perturb that results in λ changing. The kernel elasticity is therefore

$$\begin{aligned} \mathbf{e}(z_0', z_0) &= \frac{\partial \log \lambda}{\partial \log K(z_0', z_0)} = \frac{K(z_0', z_0)}{\lambda}\frac{\partial \lambda}{\partial K(z_0', z_0)} \\ &= \frac{K(z_0', z_0)}{\lambda}\mathbf{s}(z_0', z_0). \end{aligned} \tag{4.3.9}$$

The elasticity function integrates to 1 (Easterling et al. 2000), so $\mathbf{e}(z_0', z_0)$ can be interpreted as the proportional contribution of the kernel entry $K(z_0', z_0)$ to population growth. This is the same interpretation as the elasticity of λ to the matrix entries in a matrix model. The elasticity functions associated with P and F are derived in the same way, such that

$$\begin{aligned} \frac{\partial \log \lambda}{\partial \log P(z_0', z_0)} &= \frac{P(z_0', z_0)}{\lambda}\mathbf{s}(z_0', z_0), \text{ and} \\ \frac{\partial \log \lambda}{\partial \log F(z_0', z_0)} &= \frac{F(z_0', z_0)}{\lambda}\mathbf{s}(z_0', z_0). \end{aligned} \tag{4.3.10}$$

4.3.2 Vital rate functions

When considering perturbations to a vital rate function, two different types of perturbation are possible. The first describes how quickly λ changes when the vital rate function changes by a given amount or fraction *at all sizes*; this is a single number. The second considers the impact of altering the vital rate *near a particular size*, z_0, or *near a particular transition* from z_0 to z'_0; these are functions of z_0 or (z'_0, z_0). In either case, it is not possible to write down a completely general sensitivity formula, because the expressions depend on the form of the kernel. For example, the effect of perturbing survival depends on whether reproduction or mortality occurs first, because this determines where the survival function $s(z)$ appears in the kernel.

As an example of these two kinds of perturbations, consider the effect of perturbing the survival function in the kernel

$$K(z', z) = s(z)G(z', z) + p_b(z)p_r(z)C_1(z', z).\qquad(4.3.11)$$

To derive the sensitivity of λ to survival at a particular size, z_0, we want to know what happens to λ if we perturb $s(z)$ to $s(z) + \epsilon\delta_{z_0}(z)$ for some $z_0 \in [L, U]$. Because s is a function of one variable, the perturbation is a one-dimensional δ function (see Box 4.1). Substituting the perturbed survival into (4.3.11), the perturbed kernel is

$$K(z', z) + \epsilon\delta_{z_0}(z)G(z', z)\qquad(4.3.12)$$

so the perturbation kernel is $C(z', z) = \delta_{z_0}(z)G(z', z)$. Substituting this into (4.3.1) and using (4.3.3) with $f(z) = G(z', z)w(z)$, we find that the numerator of (4.3.1) is $\int v(z')G(z', z_0)w(z_0)dz'$. The sensitivity to perturbing the survival function at z_0 is therefore

$$\frac{\partial\lambda}{\partial s(z_0)} = \frac{\int v(z')G(z', z_0)w(z_0)dz'}{\int v(z)w(z)\,dz}.\qquad(4.3.13)$$

Using operator notation (Table 3.1) we can write this as $((vG) \circ w)/\langle v, w\rangle$, where vG is "left multiplication" of v and G, and $\circ$ denotes pointwise multiplication. Note that both sides of (4.3.13) are functions of z_0: the sensitivity of λ to survival at one size, as a function of the size at which survival is changed.

The formula makes sense. The change in λ induced by perturbing survival near z_0 is given by the product of the population density at z_0 with the expected reproductive value of surviving individuals, after accounting for their growth to different sizes, scaled by the total reproductive value of the population. That is, the effect of a local perturbation of the survival function depends on the fraction of individuals affected, where they end up after they grow, and their reproductive value at their new sizes. Given the sensitivity, the corresponding elasticity follows from its definition,

$$\frac{\partial\log\lambda}{\partial\log s(z_0)} = \frac{s \circ vG \circ w}{\lambda\langle v, w\rangle}.\qquad(4.3.14)$$

Like (4.3.13), the elasticity is a function of the individual size at which survival is perturbed. The right-hand side of (4.3.14) is a recipe for computing all at once the elasticities to survival at all sizes.

What about the sensitivity and elasticity of λ to survival *at all sizes*? That is, what happens to λ if $s(z)$ is perturbed to either $s(z) + \epsilon$ (sensitivity) or $s(z) + \epsilon s(z)$ (elasticity)? Using the chain rule (4.3.7) the sensitivity is

$$\frac{\partial \lambda}{\partial s} = \iint \mathbf{s}(z', z) \frac{\partial K(z', z)}{\partial s(z)} dz' dz. \tag{4.3.15}$$

Substituting $G(z', z)$ for $\partial K(z', z)/\partial s(z)$ as we did before, we find

$$\frac{\partial \lambda}{\partial s} = \frac{\iint v(z') w(z) G(z', z) \, dz' dz}{\int v(z) w(z) \, dz} = \frac{\langle v, Gw \rangle}{\langle v, w \rangle}. \tag{4.3.16}$$

You probably noticed that we did not really need to start from (4.3.15) to work out the formula for $\partial \lambda / \partial s$. We had already calculated $\partial \lambda / \partial s(z_0)$, so all we really needed to do was integrate over z_0 in (4.3.13).

We could go on to calculate sensitivities and elasticities of λ to the other vital rate functions in the example kernel (4.3.11). However, there are no new ideas, so we will leave perturbation analysis of vital rate functions until the ungulate case study below.

4.3.3 Parameters and lower-level functions

Much of the same logic is used to explore the impact on λ of perturbing an arbitrary low-level parameter, θ. We want to know what happens to λ if θ is perturbed to either $\theta + \epsilon$ (sensitivity) or $\theta + \epsilon\theta$ (elasticity). Not surprisingly, the sensitivity of λ to a low-level parameters depends on the details of the IPM. A good starting point for deriving these sensitivities is the chain rule

$$\frac{\partial \lambda}{\partial \theta} = \iint \mathbf{s}(z', z) \frac{\partial K(z', z)}{\partial \theta} dz' dz. \tag{4.3.17}$$

We will demonstrate an application using the example kernel (4.3.11). This time, instead of working with the survival function, we will consider the probability of reproduction function, $p_b(z)$. Let's assume $p_b(z)$ was estimated via a logistic regression on size using the *logit* link function. This means it has the form $p_b(z) = 1/(1 + e^{-\nu(z)})$, where ν denotes the linear predictor, such that $\nu(z) = \beta_0 + \beta_z z$. We refer to β_0 and β_z as the intercept and slope parameter, respectively.

One question we might ask is, what is the sensitivity of λ to the intercept of the reproduction function? To answer this we first expand (4.3.17), using the chain rule again, which gives us

$$\frac{\partial \lambda}{\partial \beta_0} = \iint \mathbf{s}(z', z) \frac{\partial K(z', z)}{\partial p_b(z)} \frac{\partial p_b(z)}{\partial \beta_0} \, dz' dz. \tag{4.3.18}$$

To apply equation (4.3.18) we need

$$\frac{\partial K(z', z)}{\partial p_b(z)} = p_r(z)C_1(z', z) \text{ and } \frac{\partial p_b(z)}{\partial \beta_0} = e^{\nu(z)}/(1 + e^{\nu(z)})^2.$$

Plugging these into equation (4.3.18), we find that the sensitivity is

$$\frac{\partial \lambda}{\partial \beta_0} = \frac{\iint v(z')w(z)a(z)p_r(z)C_1(z', z)\,dz'dz}{\int v(z)w(z)\,dz} = \frac{\langle vC_1, w \circ a \circ p_r \rangle}{\langle v, w \rangle}, \quad (4.3.19)$$

where $a(z) = \partial p_b(z)/\partial \beta_0$. The elasticity is now easy to write down, i.e., $\partial \log \lambda/\partial \log \beta_0 = (\beta_0/\lambda)\partial \lambda/\partial \beta_0$. Similar expressions apply if we are interested in the sensitivity associated with the slope parameter, but now $a(z) = \partial p_b(z)/\partial \beta_z = ze^\nu/(1 + e^\nu)^2$.

We can also consider the effect of perturbations that are in some sense intermediate to those acting on parameters and vital rate functions. For example, it may be useful to understand what happens to λ when the expected size of individuals next year, $m(z)$, is perturbed. As usual we can consider perturbations acting at all sizes, or just near a particular starting size z_0. For example, to calculate the sensitivity of λ to the expected size near z_0, we use

$$\frac{\partial \lambda}{\partial m(z_0)} = \int s(z', z_0)\frac{\partial K(z', z_0)}{\partial G(z', z_0)}\frac{\partial G(z', z_0)}{\partial m(z_0)}\,dz'. \quad (4.3.20)$$

If the growth kernel has been constructed using linear regression with Gaussian errors and size-dependent variance $\sigma^2(z)$, then

$$\frac{\partial G(z', z_0)}{\partial m(z_0)} = G(z', z_0)(z' - m(z_0))/\sigma(z_0)^2.$$

Plugging this expression along with $\partial K(z', z_0)/\partial G(z', z_0) = s(z_0)$ into (4.3.20), we find

$$\frac{\partial \lambda}{\partial \mu(z_0)} = \frac{\int v(z')w(z_0)s(z_0)G(z', z_0)(z' - m(z_0))\,dz'}{\sigma^2(z_0)\int v(z)w(z)\,dz}$$
$$= \frac{v[G \circ a] \circ s \circ w}{\langle v, w \rangle}, \quad (4.3.21)$$

where this time $a(z', z) = (z' - m(z))/\sigma^2(z)$, and again both sides are functions of initial size z_0. If you're not confident doing calculations with the chain rule, another approach is to use the general formula (4.3.1), with the perturbation kernel being

$$C = s(z)\frac{\partial G(z', z)}{\partial m}\delta_{z_0}(z) = s(z)G(z', z)a(z', z)\delta_{z_0}(z).$$

4.4 Case Study 2B: Ungulate population growth rate

To illustrate kernel sensitivity and elasticity analysis we use the Soay sheep case study from Chapter 2. So we assume you have already fitted the demographic analyses and implemented the IPM using the R scripts from that chapter. This means you have stored the estimated model parameters in m.par, stored the upper and lower integration bounds in max.size and min.size, and defined the functions used by mk_K to build the mesh points and iteration matrix.

The first step is to compute and store the iteration matrix, the mesh points and the mesh width, as we will use these over and over again.

```
IPM.sys <- mk_K(400, min.size, max.size, m.par)
K <- IPM.sys$K
meshpts <- IPM.sys$meshpts
h <- diff(meshpts[1:2])
```

If you have worked through Chapter 2 there will be nothing mysterious about this code. The only new idea here is that we have used the diff function to calculate the mesh width from the first two mesh points. Next we need to compute the dominant eigenvalue, λ, and the associated eigenvector, $w(z)$.

```
IPM.eig.sys <- eigen(K)
lambda <- Re(IPM.eig.sys$values[1])
w.z <- Re(IPM.eig.sys$vectors[,1])
```

We also need the dominant left eigenvector, $v(z)$. To do this we transpose the iteration matrix and use the eigen function again.

```
v.z1 <- Re(eigen(t(K))$vectors[,1])
```

4.4.1 Kernel-level perturbations

Calculating the kernel sensitivity function over a grid defined by the mesh points is straightforward. Only one line of code is needed to compute $\mathbf{s}(z_0', z_0)$ using (4.3.6):

```
K.sens <- outer(v.z1, w.z, "*")/sum(v.z1 * w.z * h)
```

Note, that there is no h in the numerator calculated by outer because the numerator in (4.3.6) involves values of v and w, but h is in the denominator because the inner product is an integral, and the sum is its approximation by midpoint rule. Figure 4.2A shows the calculated kernel sensitivity function. The shape of the sensitivity surface is governed by $w(z)$ and $v(z)$. A slice through this surface along a line of constant z' yields a curve that is proportional to the stable size distribution. A slice along a line of constant z produces a curve that is proportional to the relative reproductive value, which is an increasing function of body size. The resulting surface shows that absolute changes to the kernel which alter transitions into large size classes from the most abundant size class have the greatest relative impact on population growth rate.

The code to calculate the kernel elasticity function $\mathbf{e}(z_0', z_0)$ using (4.3.9) is also straightforward:

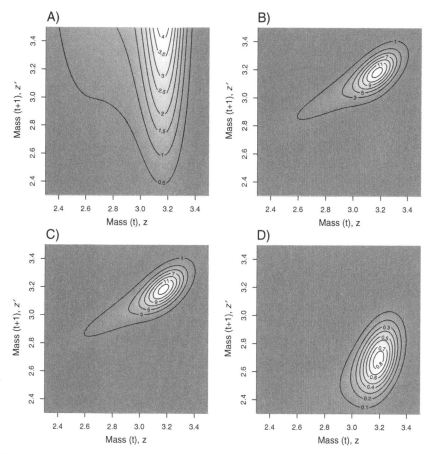

Fig. 4.2 Kernel-level perturbation analysis of the Soay IPM. (A) Sensitivity function of the kernel, K. (B) Elasticity function of the kernel, K. C) Elasticity function of the survival component, P. D) Elasticity function of the reproduction component, F. Source file: `Ungulate Pert Lambda Calc.R`

```
K.elas <- K.sens * (K / h) / lambda
```

We divided the iteration matrix `K` by the mesh width `h` because the formula involves the kernel *function* $K(z', z)$, while the iteration matrix entries are $hK(z_j, z_i)$. It is easy to lose track of where `h` belongs in these kinds of calculations. Fortunately, here we can use the fact that the integral of the kernel elasticity function is equal to one as a quick sanity check. Using the code snippet `sum(K.elas) * h^2` we find that the elasticity function does indeed "sum to one" (with h^2 in the sum here because it's a double integral with respect to z and z'). Figure 4.2B shows the resulting kernel elasticity function. This suggests that proportional changes to the survival-growth component of the kernel will have the greatest impact on λ, with transitions from the most common size class being most important.

The contour plot we used to summarize the elasticity surface does not reveal the contribution from reproduction. To better understand this we can use (4.3.10) to compute the separate elasticity surfaces associated with P and F. For example, the R code for P is:

```
Pvals <- IPM.sys$P / h
P.elas <- Pvals * K.sens / lambda
```

Figure 4.2C and D shows the P and F elasticity functions. In contrast to the sensitivity surfaces, the elasticity surfaces approach 0 at rare transitions. This makes sense because even a very large proportional perturbation of the kernel where it is near 0 still generates a very small absolute perturbation. Panel C clearly shows the survival-growth elasticity surface dominates the total elasticity $e(z'_0, z_0)$ - it is indistinguishable from panel B. The elasticity surface associated with F again shows that transitions from the most common size classes of ewes have the largest effect on λ, but that overall it makes only a small contribution to $e(z'_0, z_0)$.

To better quantify these differences we can integrate over each elasticity surface separately using `sum(P.elas)*h^2` and `sum(F.elas)*h^2`. These show that the relative contributions are 0.907 and 0.092, respectively, for survival-growth and reproduction. These sum to one (as they must) and demonstrate the relatively small effect that proportional changes to F would make to λ.

4.4.2 Vital rate functions

In order to see how to implement perturbations analyses involving vital rate functions it is worth first reminding ourselves what the kernel for the ungulate example looked like:

$$K(z', z) = s(z)G(z', z) + s(z)p_b(z)p_r C_0(z', z)/2. \qquad (4.4.1)$$

There are four functions we might consider in a perturbation analysis of this kernel: the survival function, $s(z)$, the growth kernel, $G(z', z)$, the probability of reproduction function, $p_b(z)$, and the offspring size kernel, $C_0(z', z)$. We will implement the sensitivity and elasticity calculations for each of these, first considering perturbations near z_0 or (z_0, z'_0), and then perturbations to the whole function.

Since we are primarily interested in the implementation details at this point, we will not step through the derivation of every expression. The sensitivity functions for perturbations near z_0 or (z_0, z'_0) are shown in Table 4.1. Notice that rather than expressing these sensitivity functions in a form that resembles (4.3.13) we have given these in terms of the kernel sensitivity function. This is convenient because it allows us to reuse our previous calculation of $s(z'_0, z_0)$. The term to right of $s(z'_0, z_0)$ in each case is $\partial K/\partial f$, where f is the focal function. Finally, it gets tedious to write "perturbations of f near z_0" over and over again, so we generally avoid this extra precision for the sake of brevity.

Table 4.1 Expressions for the sensitivity of λ to perturbations of functions used to construct the Soay sheep kernel, $K(z', z) = s(z)G(z', z) + s(z)p_b(z)p_rC_0(z', z)/2$. $\mathbf{s}(z', z)$ denotes the kernel sensitivity function, $s(z)$ is the survival function.

Sensitivity	Formula	Figure
$\dfrac{\partial \lambda}{\partial s(z_0)}$	$\displaystyle\int \mathbf{s}(z', z_0)(G(z', z_0) + p_b(z_0)p_rC_0(z', z_0)/2)\ dz'$	4.3A
$\dfrac{\partial \lambda}{\partial p_b(z_0)}$	$\displaystyle\int \mathbf{s}(z', z_0)(s(z_0)p_rC_0(z', z_0)/2)dz'$	4.3B
$\dfrac{\partial \lambda}{\partial G(z_0', z_0)}$	$\mathbf{s}(z_0', z_0)s(z_0)$	4.4A
$\dfrac{\partial \lambda}{\partial C_0(z_0', z_0)}$	$\mathbf{s}(z_0', z_0)s(z_0)p_b(z_0)p_r/2$	4.4C

To calculate the sensitivity of λ to perturbations of $s(z)$ we need to compute $G(z', z_0) + p_b(z_0)p_rC_0(z', z_0)/2$ over a 2d grid of (z_0', z_0) defined by the mesh points. This can be done by passing an anonymous function to `outer`:

```
dK_by_ds_z1z <- outer(meshpts, meshpts,
+        function (z1, z, m.par)
+          {g_z1z(z1, z, m.par) + pb_z(z, m.par) *
+          (1/2) * pr_z(m.par) * c_z1z(z1, z, m.par)}, m.par)
```

We then multiply this element-wise by the sensitivity function we calculated earlier and sum over the rows to do the integration, remembering to multiple by the mesh width:

```
s.sens.z <- apply(K.sens * dK_by_ds_z1z, 2, sum) * h
```

The sensitivity and elasticity of λ to perturbations of $p_b(z)$ are calculated in exactly the same way, this time passing the function $s(z_0)p_rC_0(z', z_0)/2$ to `outer`:

```
dK_by_dpb_z1z <- outer(meshpts, meshpts,
+        {function(z1, z, m.par) s_z(z, m.par) *
+        (1/2) * pr_z(m.par) * c_z1z(z1, z, m.par)}, m.par)
pb.sens.z <- apply(K.sens * dK_by_dpb_z1z, 2, sum) * h
```

The corresponding elasticity functions for $s(z)$ and $p_b(z)$ are found by scaling by the sensitivity function by $s(z_0)/\lambda$ and $p_b(z_0)/\lambda$, respectively:

```
s.elas.z <- s.sens.z * s_z(meshpts, m.par)/lambda
pb.elas.z <- pb.sens.z * pb_z(meshpts, m.par)/lambda
```

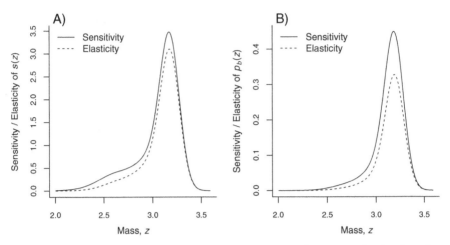

Fig. 4.3 Function-level perturbation analysis of the survival and reproduction functions of the Soay IPM. (A) Sensitivity (solid line) and elasticity (dashed line) of λ to local perturbations of the survival function, $s(z)$. (B) Sensitivity (solid line) and elasticity (dashed line) of λ to local perturbations of the reproduction function, $p_b(z)$. Source file: Ungulate Pert Lambda Calc.R

Figure 4.3 shows the sensitivity and elasticity functions associated with $s(z)$ and $p_b(z)$. In each case, the sensitivity and elasticity functions tell much the same story about the effect of perturbing these size-specific vital rate functions. Changing survival or the reproduction rate of the most common, reproductively active size classes has the greatest impact on λ. Accounting for their different scales, there is very little difference in the shape of the sensitivity or elasticity functions among each vital rate. The functions associated with $s(z)$ (Figure 4.3A) have a slightly larger "shoulder" near size classes corresponding to new (age 0) recruits. It seems survival of these young individuals influences λ more than their reproduction does, but what really matters are processes affecting the established ewes (age > 0).

We can use the same approach to compute sensitivity and elasticity functions for $G(z', z)$ and $C_0(z', z)$. The only difference is that now there is no need to use the apply function to integrate over z'. For example, the sensitivity of λ to changes in $G(z', z)$ near (z'_0, z_0) can be computed with:

```
dK_by_dg_z1z <- outer(meshpts, meshpts,
+                      function(z1, z, m.par) {s_z(z, m.par)}, m.par)
g.sens.z1z <- K.sens * dK_by_dg_z1z
```

The elasticity calculation is much the same as before, but now we also have to use the outer function to compute $G(z', z)$ on a grid:

```
g.elas.z1z <- g.sens.z1z * outer(meshpts, meshpts, g_z1z, m.par)/lambda
```

Exactly the same logic applies to the calculation of the $C_0(z', z)$ sensitivity and elasticity surfaces (R code for which is in Ungulate Pert Lambda Calc.R).

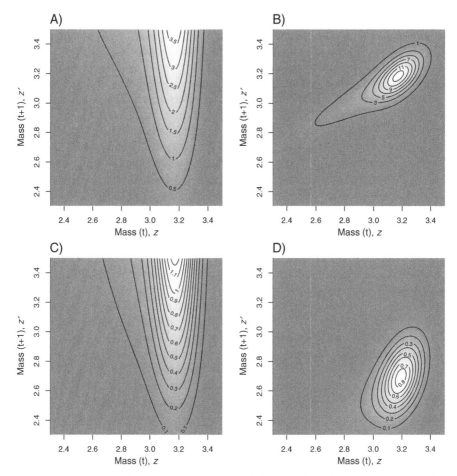

Fig. 4.4 Function-level perturbation analysis of the growth and offspring size kernels of the Soay IPM. (A) Sensitivity of λ to local perturbations of the growth kernel, $G(z', z)$. (B) Elasticity of λ to local perturbations of the growth kernel, $G(z', z)$. (C) Sensitivity of λ to local perturbations of the offspring size kernel, $C_0(z', z)$. (D) Elasticity of λ to local perturbations of the offspring size kernel, $C_0(z', z)$. Source file: `Ungulate Pert Lambda Calc.R`

Figure 4.4 shows the sensitivity and elasticity functions associated with $G(z', z)$ and $C_0(z', z)$. Panels A and C show the sensitivity surfaces of $G(z', z)$ and $C_0(z', z)$, respectively. In both cases it seems that population growth rate is most sensitive to rare transitions involving growth or reproduction into large size classes from the most abundant size class. We need to be careful about how we interpret the shape of these sensitivity surfaces. Because $\int G(z', z) \, dz' = \int C_0(z', z) \, dz' = 1$, an increase in a growth (or offspring size) transition must be compensated by a reduction somewhere else. One possible interpretation of panels A and C is that λ is increased by making either the expected value of z' larger or the skewness of the distribution of sizes around this value more positive. The only other conclusion we can draw from these two surfaces is that λ is relatively more sensitive (≈ 3 times) to changes in growth.

What about the elasticity functions associated with $G(z', z)$ and $C_0(z', z)$, shown in panels B and D, respectively? These plots are probably starting to feel a little repetitive by now; the $G(z', z)$ and $C_0(z', z)$ elasticity surfaces are identical to those derived from P and F, respectively. This is not surprising because $G(z', z)$ and $C_0(z', z)$ occur as a products with the other vital rate functions in P and F, and elasticities capture the impact on λ of proportional changes to these functions. In this example, there really was no need to compute the elasticity of λ to $G(z', z)$ and $C_0(z', z)$. We already knew the answer. Just as with the corresponding sensitivity surfaces, we need to be careful about how we interpret these elasticity functions because they describe conditional density functions. That said, we won't waste time looking at these in any more detail.

There is one more set of analyses we might want to carry out with the vital rate functions. We might like to know how λ responds to a perturbation of a vital rate function at all sizes. This kind of perturbation has generally not been done in matrix models, so it's worth pausing to make sure you understand their meaning. The sensitivity to $s(z)$ asks: what is the change in λ that results from changing $s(z)$ to $s(z) + \epsilon$ at all z (sensitivity $= \Delta\lambda/\epsilon$)? The elasticity asks: what is the *fractional* change in λ that results from changing $s(z)$ to $s(z) + \epsilon s(z)$ at all sizes (elasticity $= (1/\lambda)\Delta\lambda/\epsilon$)? In both cases the result is a single number, and the analytic formulas give us the limiting value as ϵ goes to 0.

We will start with elasticities. You should be able to guess their values from the work we have already done, but we will run through the calculations as a quick sanity check. These are easy because we already computed $s(z_0)$, $p_b(z_0)$, $G(z'_0, z_0)$, and $C_0(z'_0, z_0)$, so we just have to integrate the pointwise values over z_0 or (z'_0, z_0). The R code to calculate the elasticity to perturbing the entire survival function is just:

```
sum(s.elas.z * h)
[1] 1
```

This does the required integration by multiplying the computed $s(z_0)$ elasticity function by the mesh width, and summed up the resulting vector. Why does the result equal 1? The survival function is found in both P and F, so a 5% increase in s causes a 5% increase in P and F, hence a 5% increase in K and hence a 5% increase in λ. Thus, the elasticity equals one. What about elasticity of λ to perturbations of functions that influence only one component of the kernel, such as $G(z', z)$ or $C_0(z', z)$?

```
sum(g.elas.z1z) * h^2
[1] 0.90743
sum(c.elas.z1z) * h^2
[1] 0.09256998
```

The idea is the same as for s, though for these we multiply by h^2 to do the double integral. The resulting total elasticities to $G(z', z)$ and $C_0(z', z)$ perturbations are equal to those we calculated earlier for perturbations to the whole P and

F kernels, because they correspond to exactly the same increase in K. You can guess the value of the corresponding elasticity for $p_b(z)$. As we observed, these calculations are a sanity check because we know what the values should be, but it's reassuring to get the same number a different way.

The sensitivities of λ to perturbations of vital rate functions at all sizes are a little more informative, at least in the case of $s(z_0)$ and $p_b(z_0)$:

```
sum(s.sens.z * h)
[1] 1.221402
sum(pb.sens.z * h)
[1] 0.1332553
```

It seems that λ is about 10 times more sensitive to perturbations of the survival probability than the probability of reproduction. One useful interpretation of this result is that the strength of selection on the survival probability is about 10 times greater than that acting on the probability of reproduction (Chapter 9 provides a detailed explanation of this idea).

4.4.3 Parameters and lower-level functions

Each of the four size-dependent functions in the Soay kernel has two parameters that define the linear predictor: an intercept, β_0^V, and a slope term capturing the size dependence, β_z^V. We use the superscript to denote a particular vital rate (e.g., $V = s$ for the survival function parameters). The growth and offspring size kernels each involve one additional parameter that describes the variability in sizes next year, σ^V, resulting in a total of 10 possible parameters we can work with. We have implemented the λ sensitivity and elasticity calculations for these parameters in Ungulate Pert Lambda Calc.R.

The mechanics of these calculations are essentially the same for every parameter, so we will just look at one example in detail. How do we calculate the sensitivity and elasticity of λ to the intercept of the survival function, β_0^S? Start by taking a look at the chain rule expanded expression for the sensitivity with respect to β_0^S,

$$\frac{\partial \lambda}{\partial \beta_0^s} = \iint \mathbf{s}(z', z) \frac{\partial K(z', z)}{\partial s(z)} \frac{\partial s(z)}{\partial \beta_0^s} \, dz' dz. \tag{4.4.2}$$

We have already been calculated two of the three terms in the integrand, $\mathbf{s}(z', z)$ and $\partial K(z', z)/\partial s(z)$. This means we only need to calculate $\partial s(z)/\partial \beta_0^s$ and then we have everything we need. The survival function was derived from a logistic regression with a *logit* link function, which mean $\partial s(z)/\partial \beta_0 = e^{\nu(z)}/(1 + e^{\nu(z)})^2$, where $\nu(z) = \beta_0^s + \beta_z^s z$. The missing chunk of R code we need is therefore:

```
ds_by_dbeta0_z1z <- outer(meshpts, meshpts,
+          function(z1, z, m.par) {
+              nu <- m.par["surv.int"] + m.par["surv.z"] * z
+              exp(nu)/(1+exp(nu))^2
+          }, m.par)
```

As usual, we use outer even though the function only depends on z because we next want to use R's element-by-element multiplication to efficiently calculate

Table 4.2 Sensitivity and elasticities of λ to perturbations of parameters of the functions to construct the Soay sheep kernel, $K(z', z) = s(z)G(z', z) + s(z)p_b(z)p_r C_0(z', z)/2$. Source file: `Ungulate Pert Lambda Calc.R`

Vital rate	Parameter	Sensitivity	Elasticity
Survival, $s(z)$	Intercept	0.14	−1.34
	Size slope	0.42	1.54
Reproduction, $p_b(z)$	Intercept	0.026	−1.00
	Size slope	0.082	0.21
Growth, $G(z', z)$	Intercept	1.19	1.65
	Size slope	3.65	1.99
	Standard deviation	−0.0036	−0.00028
Offspring Size, $C_0(z', z)$	Intercept	0.25	0.089
	Size slope	0.79	0.55
	Standard deviation	0.056	0.0087

the double integral in (4.4.2) and then the elasticity. This just takes two very predictable lines of code:

```
s.int.sens <- sum(K.sens * dK_by_ds_z1z * ds_by_dbeta0_z1z) * h^2
s.int.elas <- s.int.sens * m.par["surv.int"]/lambda
```

The results of this sensitivity/elasticity calculation are given at the top of Table 4.2, along with the other results from the parameter-level perturbation analysis of λ implemented in `Ungulate Pert Lambda Calc.R`. The β_0^V sensitivities more or less recapitulate the insights derived from the sensitivity analysis of the vital rate functions. λ is more sensitive to the changes in the intercept of the survival function than the reproduction function, and more sensitive to changes in the growth function intercept than the offspring size function intercept. Comparing the sensitivities associated with the survival (or reproduction) intercepts to those associated with the growth (or offspring size) intercepts is not particularly informative because they work on different scales.

What about the sensitivity of λ to the slope terms? In all four functions these are all bigger than the corresponding intercept sensitivity. This makes sense if you keep in mind that the stable size distribution is to the right of zero, i.e., all individuals have positive (log) mass. When a slope is increased the associated vital rates of all individuals are increased. If the size distribution were to the left of zero we should expect to see negative sensitivities associated with slopes. The slope sensitivities are larger than those associated with the corresponding intercept sensitivities because perturbing a slope parameter affects larger individuals more than smaller individuals in this case, and larger size classes have higher rates of survival and reproduction.

One obvious feature of Table 4.2 is that some of the intercept elasticities are negative. This occurs when the parameter is negative and the associated sensitivity is positive: a proportional change in a negative parameter makes it more negative, reducing growth rate. This is an issue of scale again. Parameter elasticities are probably not very informative when they apply to scales that encompass the whole real line. This is often the case when vital rate functions have been fitted using a GLM with a nonlinear link function.

The expected size of an individual after growth (denoted $\mu_G(z)$) is linearly dependent on the intercept parameter of $G(z', z)$. This means a perturbation analysis of this intercept parameter describes how λ changes when the expected size of individuals changes at all sizes. What if we wanted to know the sensitivity or elasticity of λ to perturbations of the expected subsequent size near z_0? We need to apply expression (4.3.20). However, rather than compute $\partial G(z', z_0)/\partial m(z_0)$, we can reuse our prior calculation of $\partial G/\partial\beta_0^g$ over the grid of mesh points, because these are identical. The R code is then:

```
mu.g.sens.z <- apply(K.sens * dK_by_dg_z1z * dg_by_dbeta0_z1z, 2, sum) * h
mu.g <- m.par["grow.int"] + m.par["grow.z"] * meshpts
mu.g.elas.z <- mu.g.sens.z * mu.g / lambda
```

Analogous code can be used to calculate the sensitivity and elasticity of λ to perturbations of the expected size of offspring, $\mu_{C_0}(z)$, near z_0. The results are given in Figure 4.5.

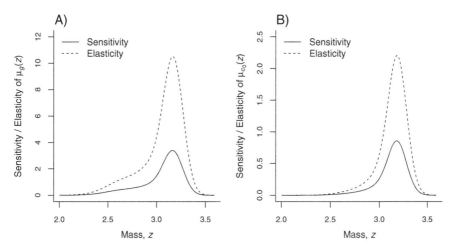

Fig. 4.5 Function-level perturbation analysis of the expected size functions of the Soay IPM. (A) Sensitivity (solid line) and elasticity (dashed line) of λ to local perturbations of the expected size function associated with growth, $\mu_g(z)$. (B) Sensitivity (solid line) and elasticity (dashed line) of λ to local perturbations of the expected size function associated with offspring size, $\mu_{C_0}(z)$. Source file: Ungulate Pert Lambda Calc.R

4.5 Sensitivity analysis of life cycle properties and events

Sensitivity analysis is also possible for the many life cycle measures that can be calculated from an IPM (Table 3.3). This section is technical, so you can skim it or go directly to Section 4.6 where we apply the results to the ungulate case study. Just two basic tools are needed: the sensitivity formulas for products and for inverses.

Products: If A and B are two kernels, then

$$\frac{\partial(AB)}{\partial\theta} = A\frac{\partial B}{\partial\theta} + \frac{\partial A}{\partial\theta}B. \qquad (4.5.1)$$

$\frac{\partial A}{\partial\theta}$ is the kernel whose value at (z', z) is $\frac{\partial A(z', z)}{\partial\theta}$ and AB denotes composition of the kernels rather than element-by-element multiplication. Equation (4.5.1) is like the product rule from calculus, except that the order of multiplication matters: both terms on the right-hand side have A before B. This product rule also holds for vector-kernel multiplications such as $p_b N_0$.

Inverses: If A has an inverse then

$$\frac{\partial(A^{-1})}{\partial\theta} = -A^{-1}\frac{\partial A}{\partial\theta}A^{-1}. \qquad (4.5.2)$$

This is like the quotient rule from calculus, $\frac{d}{dx}\left(\frac{1}{f(x)}\right) = -\frac{df/dx}{f^2}$. It follows by differentiating both sides of $AA^{-1} = I$, using the product rule for the left-hand side, and the right-hand side is constant so its derivative is zero. Solving the resulting equation gives (4.5.2).

The same formulas are also applicable to entry- or function-level perturbations. For example, a perturbation of $A(z_2, z_1)$ is represented by $\partial A/\partial\theta = \delta_{(z_2, z_1)}$. Using this in (4.5.2) the sensitivity of A^{-1} to this perturbation is $-A^{-1}(z', z_2)A^{-1}(z_1, z)$, which is a kernel because A^{-1} is a kernel.

An epic list of perturbation formulas could be produced by factorial combination of different perturbations with different life cycle properties. We don't have that much time on our hands, so we will give just two examples. First, consider the sensitivity of mean lifespan $\bar{\eta} = \mathbf{e}N$ to increases in survival at a particular size z_0, as in equation (4.3.12). Applying (4.5.2) to $N = (I - P)^{-1}$ we get the general sensitivity formula

$$\frac{\partial N}{\partial\theta} = -N\left(\frac{\partial(I - P)}{\partial\theta}\right)N = N\frac{\partial P}{\partial\theta}N.$$

The sensitivity of $\bar{\eta}$ is therefore

$$\frac{\partial\bar{\eta}}{\partial\theta} = \mathbf{e}\frac{\partial N}{\partial\theta} = \bar{\eta}\frac{\partial P}{\partial\theta}N. \qquad (4.5.3)$$

For a perturbation of survival at z_0, the perturbation kernel is $\partial P/\partial\theta = \delta_{z_0}(z)G(z',z)$, as in equation (4.3.12). Substituting this into the rightmost expression in (4.5.3), writing out the integrals and using the properties of the δ function, the sensitivity of $\overline{\eta}(z)$ to size-specific survival is

$$\frac{\partial\overline{\eta}}{\partial s(z_0)} = N(z_0,z) \times \int_{\mathbf{Z}}\overline{\eta}(z')G(z',z_0)\,dz'.$$

This expression makes a surprising amount of sense. For an individual born with state z, $N(z_0,z)$ is the expected time spent at z_0. The integral is the expected future lifetime of the additional survivors who are "created" each time z_0 is visited due to the increased survival, which depends on the size of those survivors the following year.

Second, consider the sensitivity of the variance in lifespan to an increase in survival at a particular size z_0. We start by considering a general perturbation. From equation (3.2.4) we have

$$\sigma_\eta^2 = \mathbf{e}(2N^2 - N) - (\mathbf{e}N)^2 = 2\mathbf{e}N^2 - \mathbf{e}N - (\mathbf{e}N)^2$$

so we need to compute and add up the sensitivities of the three terms on the right. The second and third are not difficult. The sensitivity of the second term is $-\frac{\partial\overline{\eta}}{\partial\theta}$ which we already calculated. The sensitivity of the third term (ignoring the sign) is

$$\frac{\partial(\mathbf{e}N)^2}{\partial\theta} = \frac{\partial\overline{\eta}^2}{\partial\theta} = 2\overline{\eta}\circ\frac{\partial\overline{\eta}}{\partial\theta}$$

This leaves the first term; ignoring the factor of 2,

$$\frac{\partial(\mathbf{e}N^2)}{\partial\theta} = \frac{\partial(\overline{\eta}N)}{\partial\theta} = \overline{\eta}\frac{\partial N}{\partial\theta} + \frac{\partial\overline{\eta}}{\partial\theta}N = \overline{\eta}N\frac{\partial P}{\partial\theta}N + \frac{\partial\overline{\eta}}{\partial\theta}N.$$

The sensitivity of σ_η^2 to a general perturbation is therefore

$$\frac{\partial\sigma_\eta^2}{\partial\theta} = 2\overline{\eta}N\frac{\partial P}{\partial\theta}N + 2\frac{\partial\overline{\eta}}{\partial\theta}N - \frac{\partial\overline{\eta}}{\partial\theta} - 2\overline{\eta}\circ\frac{\partial\overline{\eta}}{\partial\theta}. \tag{4.5.4}$$

We now want to derive the sensitivity to an increase in survival at a particular size z_0. The last three terms in (4.5.4) all involve simple calculations with terms we've already encountered. And the first term can also be written in a form that makes the calculations easy. It's convenient to do the right-most product first,

$$\frac{\partial P}{\partial\theta}N = \int_{\mathbf{Z}}G(z',u)\delta_{z_0}(u)N(u,z)\,du. \tag{4.5.5}$$

$\frac{\partial P}{\partial\theta}N$ is therefore the kernel H whose values are $H(z',z) = G(z',z_0)N(z_0,z)$. This means that we can express the first term in (4.5.4) as

$$2\overline{\eta}N\frac{\partial P}{\partial\theta}N = 2\overline{\eta}(NH). \tag{4.5.6}$$

This example illustrates that a numerical calculation might be much simpler than it seems. Looked at the right way (the right-hand side of 4.5.6), computing $2\bar{\eta}N\frac{\partial P}{\partial \theta}N$ for one value of z_0 is a simple for-loop to compute the entries of H from those of G and N, plus one vector-matrix-matrix multiplication. Wrap a for-loop on z_0 around that, and it's done.

4.6 Case Study 2B (continued): Ungulate life cycle

Here we demonstrate how to implement sensitivity analyses of $\bar{\eta}$ and σ_η^2 in R using the simple ungulate case study. Sensitivity analysis for life cycle measures is a little more challenging to implement than those associated with λ, because additional book keeping is required to keep track of the different kinds of "z." For example, we have to keep track of the initial size at birth, the size at which a perturbation is applied, and the intermediate terms that appear inside multiple integrals in (4.5.4).

To calculate the required sensitivities we first need to compute $\bar{\eta} = \mathbf{e}N$. The R code for doing this is no different from that given in Chapter 3:

```
e <- matrix(1, nrow = 1, ncol = nBigMatrix)
N <- solve(diag(1, nBigMatrix)-P)
eta.bar <- (e %*% N)[,,drop = TRUE]
```

The only new trick here is that we use the square brackets with drop = TRUE to remove the dimension attribute from eta.bar. This ensures the result is just an ordinary vector rather than a matrix object.

Two lines of R code are needed to calculate the sensitivity of $\bar{\eta}$ to an increase in survival at size z_0:

```
g.z1z0 <- outer(meshpts, meshpts, g_z1z, m.par) * h
eta.sens <- N * (eta.bar %*% g.z1z0)[,,drop = TRUE]
```

The first line computes the growth kernel over a grid of mesh points, remembering to scale by h so that the integrations work. The second line implements equation (4.5.3) with respect to the survival perturbation. We use the drop = TRUE option again to make the result of eta.bar %*% g.z1z0 a vector. The recycling rules used by R then ensure each column of N is multiplied by the intermediate result, which is a function of the perturbed size, z_0. This is exactly what we need, because each row of N corresponds to a different perturbed size class.

The calculations for the sensitivity of σ_η^2 to an increase in survival at size z_0 can be broken down into several steps. First, we calculate the sensitivity of $\bar{\eta}^2$, which is the last term in (4.5.3), ignoring the factor of -2. This requires one line of R code:

```
eta2.sens <- t(eta.bar * t(eta.sens))
```

Remember, the eta.sens object is a matrix in which columns correspond to different states an individual was born into and rows correspond to different sizes at which the perturbation occurs. This means we need to multiply each row of eta.sens by the eta.bar. Since R recycles vectors column-wise when they are multiplied by a matrix, we use the double transpose trick to ensure we get the recycling right and leave the designation of rows and columns unchanged.

The third term of (4.5.3) is the sensitivity of $\bar{\eta}$. We have already implemented this calculation so we can move to the second term. This is (ignoring the factor of 2) a simple integration that we calculate with one line of R code:

```
eta.bar.N <- eta.sens %*% N
```

The next calculation we need to implement corresponds to the first term of (4.5.3). The simplest way to compute this is to loop over z_0 values, calculating the H kernel with (4.5.5) and applying (4.5.6) at each iteration:

```
eta.bar.H.N <- matrix(NA, nBigMatrix, nBigMatrix)
for (i.z0 in seq_along(meshpts)) {
  H <- outer(g.z1z0[,i.z0], N[i.z0,], function(g.z1, N.z) g.z1 * N.z)
  eta.bar.H.N[i.z0,] <- eta.bar %*% N %*% H
}
```

Finally, to get the sensitivity of σ_η^2 we just add together four components of (4.5.3), remembering to scale each term as required:

```
eta.var.sens <- 2*eta.bar.H.N + 2*eta.bar.N - eta.sens - 2*eta2.sens
```

Chapter 5
Density Dependence

Abstract Density dependence has been one of the ecology's central topics and also one of its most contentious, from Nicholson versus Andrewartha and Birch, through the interspecific competition debates of the 1980s in community ecology, and up to the present day as the availability of long-term data has allowed meta-analysts to assess the prevalence and strength of density dependence across multiple taxa and habitats. In this chapter we take the perspective of an ecologist who believes that density affects some demographic processes in their study population, and who wants to examine the consequences of density dependence. We first present some examples of formulating, parameterizing, and implementing a density-dependent IPM. We outline the general theory of density-dependent IPMs, and then look at what the theory tells us about the Soay sheep IPM when we add intraspecific competition, and how the dynamics depend on demographic parameters.

5.1 Introduction

A general guide to building density-dependent IPMs is impossible, because there are too many possibilities. Any demographic rate can be affected by competition, and the dependence can take many possible forms: under- or over-compensating, Allee effects, and so on. So instead we present several examples of formulating, parameterizing, and implementing a density-dependent IPM, going from simple to complex. With these examples in mind we then review the general theory of density-dependent IPMs; there isn't much yet, but important first steps have been taken and more should be coming soon. To close the chapter, we look at what the theory tells us about the nonlinear population dynamics of Soay sheep with intraspecific competition, and study how the dynamics can change as a function of vital rate parameters.

© Springer International Publishing Switzerland 2016   111
S.P. Ellner et al., *Data-driven Modelling of Structured Populations*,
Lecture Notes on Mathematical Modelling in the Life Sciences,
DOI 10.1007/978-3-319-28893-2_5

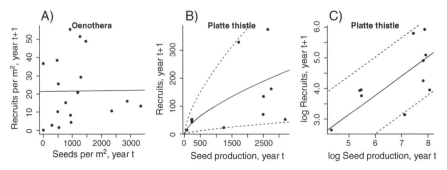

Fig. 5.1 Data on the relationship between total seed production and the number of new recruits the following year. (A) *Oenothera*, data from Kachi (1983). The solid line is the fitted linear regression; the slope is not significantly different from 0 ($P > 0.5$). (B) Platte thistle, data and fitted model from Rose et al. (2005). The solid line is Recruits=Seeds$^{0.67}$, fitted by negative binomial regression, and the dashed lines are the 10th and 90th percentiles of the fitted negative binomial distribution. (C) Fitted linear regression without intercept on log-log scale. Solid line is the fitted mean, and dashed lines are the 10th and 90th percentiles of the fitted Gaussian distribution. Data for this figure are in OenotheraRecruit.csv and PlatteThistleFig4.csv. Source file: PlotRecruitmentData.R

5.2 Modeling density dependence: recruitment limitation in *Oenothera* and Platte thistle

In a density-independent IPM for a plant that reproduces by seeds, the number of new recruits in a given year is linearly proportional to the number of seeds produced the previous year (all else being equal, and assuming that there is no recruitment from a long-lasting seed bank). But for *Oenothera* (Figure 5.1A), data on recruitment suggest that recruitment is independent of how many seeds were produced the previous year (Kachi 1983). This can't literally be true: with no seeds, there could be no new recruits. But apparently enough seeds are produced each year to saturate all the microsites that have come available for seedling establishment. As a result recruitment probability is not constant, but rather inversely proportional to the number of competing seeds.

With $\mathcal{R}$ total new recruits[1] the seedling contribution to the subsequent year's population is $\mathcal{R}c_0(z')$, the number of recruits multiplied by the seedling size distribution. The IPM iteration is then

$$n(z', t+1) = \mathcal{R}c_0(z') + \int_L^U P(z', z)n(z, t)\, dz \qquad (5.2.1)$$

with $P(z', z) = (1 - p_b(z))s(z)G(z', z)$ as in Chapter 2.

The situation is less simple for another monocarpic perennial, Platte thistle, *Cirsium canescens*. Rose et al. (2005) modeled *C. canescens* to study the effects of herbivory by the inflorescence-feeding weevil *Rhinocyllus conicus* that was introduced as a biocontrol agent for nonnative thistles. Figure 5.1B shows the recruitment data and the model fitted by Rose et al. (2005),

---

[1] Not to be confused with R, the next-generation kernel.

Recruits=Seeds$^{0.67}$. The variance appears to increase very rapidly with the mean, so Rose et al. (2005) used negative binomial regression to fit the model (Poisson regression also has variance increasing with the mean, but not fast enough: a Poisson regression fitted to these data has very high overdispersion). The Rose et al. recruitment model is linearized by log-transformation (i.e., log Recruits = 0.67 × log Seeds). So having set up a data frame Platte in PlotRecruitmentData.R with variables seeds and recruits, the model can be fitted by regressing recruits on log seeds with a log link function:

```
PlatteNB1 <- glm.nb(recruits ~ log(seeds),link="log",data=Platte)
PlatteNB2 <- glm.nb(recruits ~ log(seeds)-1,link="log",data=Platte)
```

The intercept coefficient is nonsignificant in PlatteNB1, which leads to PlatteNB2 and the Rose et al. model with exponent coef(PlatteNB2)[1]= 0.67. The dashed curves show the 10th and 90th percentiles of the fitted distribution, which should (and do) contain most of the observations.

But let's back up and ask: is that a reasonable model? The eye is pretty good at telling whether or not a relationship is linear. So let's look at a log-log plot, Figure 5.1C, and there's no sign of nonlinearity. When we fit a linear regression on log-log scale, the intercept is small and not significantly different from zero, so we re-fit without an intercept:

```
PlatteLN1 <- lm(log(recruits) ~ log(seeds)-1,data=Platte)
```

The fitted regression is log Recruits= 0.62×log Seeds, which is equivalent to Recruits=Seeds$^{0.62}$. And again, the percentiles of the fitted model encompass most of the data. Given the small number of observations, it seems prudent not to consider more complicated models.

The log-log and negative binomial regression curves have different exponents, 0.62 and 0.67, respectively. The larger exponent predicts 25% to nearly 50% more recruits for Seeds between 100 and 3000. That's a big enough difference to be worth thinking about. It happens because the log-log regression predicts the mean of log Recruits, while the negative binomial regression predicts the log of mean Recruits, and those are not the same thing (link=log specifies how the mean depends on the linear predictor in a GLM, but it doesn't transform the response). If log Recruits is Gaussian with mean μ and variance σ^2, then Recruits has mean $e^{\mu+\sigma^2/2}$. The log-log regression has residual variance $\sigma^2 \approx$ 0.6, and $e^{0.3} \approx 1.35$. So the log-log regression *really* implies that the mean of Recruits is 1.35 Seeds$^{0.62}$, which is numerically much closer to the Rose et al. model. All in all, then, the Rose et al. recruitment model seems reasonable.

The rest of the IPM was developed in the usual way, using 13 years of data on marked individuals in five 144 m^2 plots in the Sand Hills of Nebraska. Additional flowering plants outside the plots were sampled destructively each year to determine the relationship between size and seed set. The size measure was the log-transformed root-crown diameter (maximum value on two censuses in May and June). This is a pre-reproductive census, so the life cycle and IPM structure are the same as for *Oenothera*. The fitted demographic models are summarized in Table 5.1. The somewhat complicated seed set function $b(z)$ is

Table 5.1 Demographic rate functions for Platte thistle, *Cirsium canescens*, from Rose et al. (2005). The size measure z is the log-transformed maximum root crown diameter in the two annual censuses. ε is the mean number of *Rhinocyllus conicus* eggs oviposited on a plant of size z, and the intercept parameter e_0 varied over time as the weevil infestation developed (Rose et al. 2005, Table 2).

Demographic rate	Formula
Flowering probability $p_b(z)$	logit $p_b = -10.22 + 4.25z$
Seed set $b(z)$	$b = \exp(-0.55 + 2.02z) \times (1 + \varepsilon(z)/16)^{-0.32}$
Mean *R. conicus* eggs/plant	$\varepsilon(z) = \exp(e_0 + 1.71z)$
Seedling size $c_0(z')$	$z' \sim Norm(\mu = 0.75, \sigma^2 = 0.17)$
Survival $s(z)$	logit $s = -0.62 + 0.85z$
Growth $G(z', z)$	$z' \sim Norm(\mu = 0.83 + 0.69z, \sigma^2 = 0.19)$

the result of two submodels: the expected seed set as a function of plant size and the number of *R. conicus* eggs, and a negative binomial distribution for the number of *R. conicus* eggs with mean depending on plant size.

Now let's see how these fit together to give us an IPM with density-dependent recruitment. The survival kernel P is density-independent, with exactly the same form as for *Oenothera*,

$$P(z', z) = (1 - p_b(z))s(z)G(z', z). \tag{5.2.2}$$

We could work out the fecundity kernel, but there's a simpler way to calculate next year's new recruits. The total number of seeds is

$$S(t) = \int_L^U p_b(z)b(z)n(z, t) \, dz$$

and these produce $S(t)^{0.67}$ recruits whose size distribution is $c_0(z')$. Combining recruits with survivors, we get the complete iteration

$$n(z', t + 1) = \underbrace{c_0(z') \left(\int_L^U p_b(z)b(z)n(z, t) \, dz \right)^{0.67}}_{\text{New recruits}} \tag{5.2.3}$$
$$+ \underbrace{\int_L^U (1 - p_b(z))s(z)G(z', z)n(z, t) \, dz}_{\text{Survivors}}.$$

When the model (5.2.3) is iterated over time, the population settles down to a steady-state size distribution that depends on the degree of weevil infestation (Figure 5.2A). Weevil infestation reduces the equilibrium number of thistles, but thistles are projected to persist even at much higher weevil levels than were observed during the study period (Figure 5.2B, solid curve). Two factors contribute to this. First, establishment probability goes up very quickly when there are very few seeds. Second, the negative binomial distribution of weevil

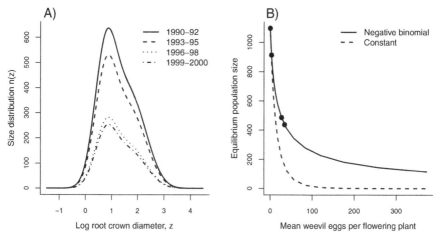

Fig. 5.2 Predicted effects of *R. conicus* weevil infestation on Platte thistle. (A) Equilibrium size distributions as weevil abundances increase over time. The four curves correspond to four values of the intercept e_0 in the relationship between plant size and mean number of weevil eggs. Those e_0 values are the average of the estimated values in Table 2 of Rose et al. (2005) for the time periods indicated. (B) Projected changes in equilibrium total thistle population as a function of mean weevil eggs per plant; curves were drawn by increasing e_0 beyond the empirical estimates and computing projected total population and mean weevil eggs per flowering plant at steady state. Solid curve uses the estimated negative binomial distribution of eggs per plant, as in (A); the four open circles correspond to the size distributions in (A). Dashed curve assumes that all plants have the expected number of weevil eggs given their size. Source files: `Platte Demog Funs.R` and `Platte Calculations.R`

eggs means that some fraction of the large plants have low weevil load and produce many seeds. From the weevil's perspective, their highly clumped distribution means that most weevils are sharing a plant with a lot of other weevils, which reduces the mean damage per weevil. If we remove this effect of clumping from the model by giving all thistles the average weevil load for a plant of their size, the impact of weevils becomes much larger (Figure 5.2B, dashed curve).

5.3 Modeling density dependence: Idaho sagebrush steppe

While plants are standing still and waiting to be counted they can also be measured, so the data to build an IPM are often straightforward to get (although it might involve a great deal of hard work). But a plant's neighbors also sit still, and a plant's fate is often strongly affected by the size, location, and species of nearby plants. Our final and most complex case study of density dependence is neighborhood competition among the four dominant plant species in Idaho sagebrush steppe, based on work by Peter Adler and collaborators (including SPE). This model includes density dependence in multiple vital rates and a "mean field" approximation for neighborhood competition that circumvents the need for a spatially explicit model. It also exposes some deficiencies in our

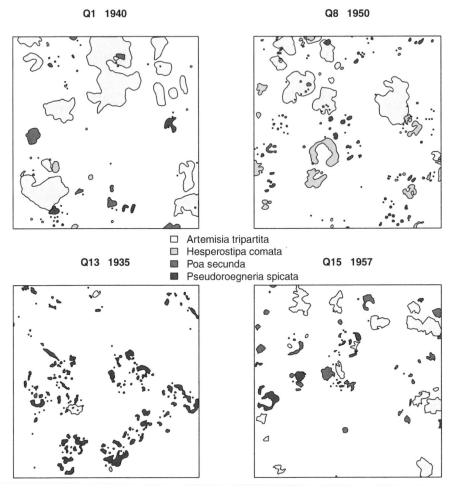

Fig. 5.3 Four examples of mapped quadrats from the Idaho sagebrush steppe data set (Zachmann et al. 2010). The heading on each panel gives the quadrat number and year. Shapefiles for these maps were provided by Peter B. Adler (*personal communication*). Source file: PlotShapefiles.R

current toolkit for demographic modeling. Instead of following our example, we encourage you to think about how you could do better.

A primary goal of the modeling was to compare the strengths of inter- and intra-specific competition among the dominant species. Whether these are similar or different in strength is the central difference between "neutral" and "niche" theories of community structure, but there are very few cases where this comparison has actually been made (Adler et al. 2010). The dominant species in the community are a shrub (*Artemisia tripartita*) and three C4 bunch grasses (*Hesperostipa comata, Poa secunda, Pseudoroegneria spicata*). Twenty-six permanent 1 m² quadrats were mapped with a pantograph during most growing seasons from 1923 to 1957, and again in 1973, by scientists at the US Sheep Experiment Station. These data (Zachmann et al. 2010) were digitized to GIS

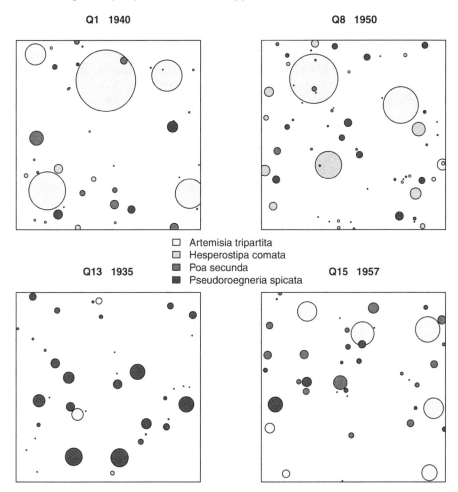

Fig. 5.4 Representation of genets in four quadrats from the Idaho sagebrush steppe data set (Adler et al. 2010). The heading on each panel gives the quadrat number and year. Data files were provided by Peter B. Adler (*personal communication*). Source file: MakeOverlapMaps.R

shape files.[2] Figure 5.3 shows four examples. The polygons represent mapped canopy cover for the shrub and basal cover for the grasses. A computer program grouped the digitized polygons into genets and classified genets as survivors or new recruits, based on spatial location and overlap with genets present in the previous year, allowing genets to fragment or coalesce (Adler et al. 2010). For modeling, each genet was represented by a circle having the same total area as the genet, centered at the centroid of the genet (Figure 5.4).

The richness of this data set made it possible to detect and model density dependence at all stages: survival, growth, and recruitment.

_____

[2] Available presently at knb.ecoinformatics.org/knb/metacat/lzachmann.6.36/ knb.

For each species, each vital rate was modeled as a function of log genet area z, competitive pressure W, and quadrat group (a categorical variable identifying a cluster of nearby quadrats). Competition was modeled on the assumption that a focal plant is affected by close plants more than distant plants, by large plants more than small plants, and by some species more than by others. The specific model was that the competitive pressure $w_{ij,km}$ on genet i in species j, from genet k in species m, was given by

$$w_{ij,km} = a_{jm}e^{-\alpha_{jm}d_{ij,km}^2}A_{km} \tag{5.3.1}$$

where d is the distance between the centers of the two genets, A_{km} is the untransformed area of genet km, and a_{jm} is the interspecific competition coefficient. The total competitive pressure on genet ij is then

$$W_{ij} = \sum_m \sum_k w_{ij,km}. \tag{5.3.2}$$

The form of (5.3.1) was the result of trial and error, comparing final results from several different equations for the effect of one genet on a neighbor. For example, an exponential competition kernel (i.e., d instead of d^2 in (5.3.1)) led to an individual-based model that behaved unrealistically and then crashed.[3]

Competitive pressure was then included as a covariate in all of the demographic models. The model for genet survival probability s was

$$\text{logit}(s_{ij}) = \gamma_{jt} + \phi_{jg} + \beta_{j,t}z_{ij} + W_{ij} \tag{5.3.3}$$

where t=year and g=quadrat group. Between-year variation was incorporated by fitting γ and β as random effects. All genets in the quadrats were included in calculating W_{ij} but plants near quadrat edges were not used in model selection or estimating parameters.

We saw no way to estimate all of the parameters in (5.3.1) and (5.3.3) at once, so the model was fitted in a series of steps. Still focusing on survival: we first estimated the parameters α_{jj} determining the spatial scale of intraspecific competition by fitting logistic regression models like (5.3.3) for different values of α_{jj} and finding the value that gave the best fit. These models only included intraspecific competition, which makes sense if interspecific competition is much weaker, and a range of models was fitted with more or fewer interaction terms. Reassuringly, the best α_{jj} for each species was insensitive to model structure (see Figure 5.5 for an example), and the shrub had a smaller α value than the grasses, indicating a larger radius of competitive interaction. The optimal value of α_{jj} was very similar for survival and growth in most species, so the value which maximizing the combined survival and growth likelihoods was used for both.

That leaves 12 more of the 16 α_{jm} to estimate. That seemed like too many to estimate reliably, on top of the 16 competition coefficients a_{jm} that also need to be estimated. We reduced the number of parameters by fitting the multispecies model in three different ways.

---

[3] There are now more flexible and objective ways to fit a competition kernel; see Teller et al. (2016).

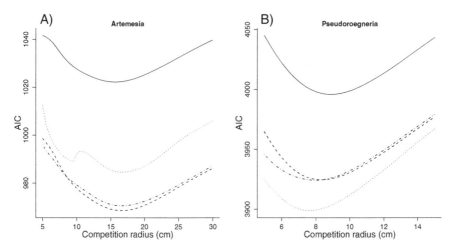

Fig. 5.5 Goodness of fit (AIC) for a set of logistic regression models of varying complexity for survival of (A) *Artemesia tripartita* and (B) *Pseudoroegneria spicata*, as a function of the intraspecific spatial scale parameter α_{jj}. The x-axis is $r_{jj} = 1/\sqrt{\alpha_{jj}}$, which we call the "competition radius" because the strength of competition between two genets depends on their distance d relative to r_{jj}. Source files: PSSP_survival_Gaussian_alpha.R and ARTR_survival_Gaussian_alpha.R

1. In the *Effects* model $\alpha_{jm} \equiv \alpha_{mm}$: each species has a characteristic zone of influence on other genets, regardless of their species.
2. In the *Response* model $\alpha_{jm} \equiv \alpha_{jj}$: each species has a characteristic zone within which it is affected by other genets, regardless of their species.
3. In the *Interaction* model $\alpha_{jm} = (\alpha_{jj} + \alpha_{mm})/2$.

These alternatives were compared using the same model structure with the best-fitting single species model for each species (e.g., the one with the lowest AIC in Figure 5.5), but including all four species in W. The difference between the alternatives was small (as we'll see, this is because interspecific competition is relatively weak) but the Effects model was best for most species, so we went with that.

Having specified all of the α_{jm} values, the rest is regression modeling, fitting and evaluating models with different subsets of the possible predictors and interactions, and different coefficients allowed to vary between years. Equation (5.3.3) is the final form for survival, and similar models were fitted for growth (a Gaussian distribution with size-dependent variance) and fecundity (negative binomial distribution of offspring per quadrat). We'll spare you the details and get on with the rest of the story.

With all of the demographic rates modeled for each species, it is then straightforward (in principle!) to implement a spatially explicit, individual-based model (IBM) for the community. But in that kind of model, each genet's demography depends on the location, size, and species of every other genet. With N genets, that means N^2 calculations at each time step just to get the competitive pressures $w_{ij,km}$, and $N(N-1)/2$ distances that need to be stored or recomputed - including every new seedling. As a result, the IBM could only be used to simulate small areas comparable to the study quadrats, not the full landscape.

Adler et al. (2010) therefore developed IPMs in which sums like (5.3.2) are replaced by their expectations under a "mean field" model for genet locations, so that the locations of individual genets don't have to be tracked. The first mean field model assumed that genet centers are distributed completely at random. This leads to a simple formula for the expected total competitive pressure on a species j genet,

$$\overline{W}_j = \pi \sum_m a_{jm} \Phi_m / \alpha_{jm}, \qquad (5.3.4)$$

where Φ_m is the fractional cover of species m (the total area of all genets divided by the total habitat area). This formula is derived in Section 5.7. The IPM using equation (5.3.4) generated species abundances much lower than those observed, indicating that the intensity of competition was overestimated. Spatial pattern analysis of the data then identified one small but important departure from spatial randomness: within each species, the distance between two large genets (the largest 1/3 in each species) was rarely less than 2–4 times the mean radius of large genets (Adler et al. 2010).

The second IPM therefore assumed that plants were distributed at random except that the circles representing conspecific genets cannot overlap. This calculation (Appendix 5.7) is still simple enough to implement in about a dozen lines of code (see Box 5.3), yet Adler et al. (2010) found that the IPM was effectively equivalent to a spatially explicit, individual-based simulation of the population.

Adler et al. (2010) used the IPM to evaluate the importance of niche differentiation in the Idaho sagebrush steppe community, with Steve Hubbell's "Unified Neutral Theory of Biodiversity" as their foil. The core tenet of neutral theories is that species similarities are more important than their differences: species are so evenly balanced that random drift to extinction is slow enough to be balanced by speciation. A central prediction, therefore, is that nothing much would change if species were (somehow) made less different from each other. The counter-prediction from niche-based theories is that species differences are essential to coexistence, so wiping out interspecific differences would change everything.

In the real world it would be impossible to test these predictions. But on the computer we can just change some parameters and seeing what happens. Part of fitting the model was estimating the interaction coefficients a_{jm} describing the impact of species m genets on species j genets. The signature of niche differentiation is that a_{mm} is larger than any of the other a_{jm}: a genet has the largest competitive impact on conspecifics (Chesson 2000). The interaction coefficients for all three density-dependent process (survival, growth, recruitment) showed exactly that pattern (Adler et al. 2010). Table 5.2 shows the estimated a_{jm} for recruitment rate. Each column gives the coefficients for the impact of one species on itself and the others, and the signature of niche differentiation is present: the diagonal entries are negative and much larger than all other values

Table 5.2 Estimated interaction coefficients a_{jm} for recruitment in the Idaho sagebrush steppe IPM, from Table 1 of Adler et al. (2010). Species codes ARTR:*Artemisia tripartita*, HECO:*Hesperostipa comata*, POSE: *Poa secunda*, PSSP:*Pseudoroegneria spicata*. The entry in row j, column m specifies the impact of species m cover on recruitment by species j. Asterisk indicates a coefficient that is different from zero at significance level $\alpha = 0.05$.

	ARTR	HECO	POSE	PSSP
ARTR	-0.0731*	-0.2425*	-0.2911	-0.0360
HECO	0.0224	-0.5471*	-0.2035	-0.0541
POSE	0.0041	-0.1155*	-1.1114*	-0.0032
PSSP	0.0389*	-0.1330*	-0.1576	-0.6007*

in the same column. These differences are much larger than what is required for all four dominant species to coexist stably in the model (Adler et al. 2010). Adler et al. called this "an embarrassment of niches," a pun[4] on "embarrassment of riches." Chu and Adler (2015) recently extended this analysis to five grassland and shrubland communities, again finding that niche differences (primarily at the recruitment stage) were far larger than needed to stabilize coexistence among common co-occurring grass species.

To which a Neutral Theory proponent might say: *so what?* We all know that species are different, but that doesn't mean that the differences actually matter. The Idaho steppe IPM said that the differences *do* matter. With niche differences removed by setting $a_{jm} = a_{mm}$ for all j and m, A. tripartita and P. spicata declined rapidly in the IPM, reaching near-zero cover in a century or less (Adler et al. (2010), fig. S10). Without the stabilization by interspecific niche differences, half of the dominant species would be quickly lost.

The differences also matter for forecasting the effects of climate variability and change (Adler et al. 2012). Some species have very strong negative frequency dependence in population growth, because intraspecific competition is much stronger than interspecific competition. Others have weaker frequency dependence. Perturbation analysis and numerical experiments showed that species with weaker frequency dependence are more sensitive to indirect effects of climate variation that directly affects other species in the community.

5.4 Theory

There is an enormous literature about nonlinear difference equations and nonlinear matrix projection models (see Caswell 2001, Chapter 16 for an introduction). But there is very little for continuous-state IPMs, and much of it was motivated by the Platte thistle model (5.2.3) in which competition only affects seedling establishment. To generalize somewhat, the model can be written in the form

$$n(z', t+1) = c_0(z')p_r(\mathbf{N}(t))\mathbf{N}(t) + \int_L^U P(z', z)n(z, t)\, dz \tag{5.4.1}$$

---

[4] Highly unsuccessful, in our experience.

where $\mathbf{N}(t)$ is the total seed production

$$\mathbf{N}(t) = \int_L^U p_b(z)b(z)n(z,t)\,dz$$

and p_r is the density-dependent recruitment probability. This is an IPM with density-independent survival kernel P and density-dependent reproduction kernel

$$F(z', z, \mathbf{N}) = p_r(\mathbf{N})c_0(z')p_b(z)b(z). \tag{5.4.2}$$

We make the biologically reasonable assumption that p_r is a decreasing function of $\mathbf{N}$. Nothing is drastically different when this isn't true, but some explanations would be a bit more complicated.

Two recent papers (Rebarber et al. 2012; Smith and Thieme 2013) analyzed model (5.4.1) to find conditions for population persistence versus extinction, and for convergence to a stable equilibrium. Persistence conditions have also been derived for a density-dependent plant model with an age-structured seed bank (Eager et al. 2014a), and for similar stochastic IPMs with density dependence (see Section 5.4.4). The stochastic IPMs all model spatially structured populations but they are important steps towards a more general theory.

The main conclusions from these analyses are exactly what we expect, once we spend a while thinking about what we ought to expect. The underlying math is beyond the scope of this book, and we keep it there. Below we explain the main conclusions without delving into the math, and then apply them to the Soay sheep IPM in Section 5.5.

5.4.1 Persistence or extinction?

The most basic question is whether the population will persist or go extinct. What do we expect?

A reasonable expectation is that the population will persist if a small population increases. Because population size only affects p_r, we can approximate the dynamics of a very small population by setting p_r to $p_r(0)$. This gives the density-independent kernel $K_0(z', z) = P(z', z) + p_r(0)c_0(z')p_b(z)b(z)$. Let's *add the assumption* that K_0 satisfies the assumptions for stable population theory (e.g., it is power-positive). Then a small population will eventually increase if K_0 has dominant eigenvalue $\lambda > 1$, or equivalently $R_0 > 1$. On the other hand, if K_0 has $R_0 < 1$, we expect that a small population will decrease to extinction; and it's hard to imagine how a larger population could increase, instead of declining to become smaller and eventually going extinct, because $p_r(\mathbf{N})$ can never be larger than $p_r(0)$.

It's useful to focus on R_0 rather than λ because R_0 is easy to compute when offspring state is independent of parent state, as it is in equation (5.4.1). Let $R_0(x)$ denote the value of R_0 for the density-independent IPM that results from (5.4.1) if p_r is held constant at $p_r(x)$, which has kernel $K_x(z', z) = P(z', z) + p_r(x)c_0(z')p_b(z)b(z)$. From equation (3.1.9) we have

$$R_0(x) = p_r(x)\langle p_b b, (I - P)^{-1}c_0\rangle. \tag{5.4.3}$$

It's worth pausing to realize that this expression is exactly what it ought to be. R_0 is the generation-to-generation population growth rate. The individuals in any generation have size distribution $c_0(z)$ at birth. So we know from Box 3.3 that $(I - P)^{-1}c_0$ is the distribution function for the expected total amount of time spent at each size during an individual's lifetime. The inner product with $p_b b$ then adds up the total seed production at each size, over the entire lifetime. Multiplication by $p_r(x)$ then gives the total number of recruits produced by an average individual – which is what we call R_0.

To summarize: If $R_0(0) > 1$ we expect the population to persist, and if $R_0(0) < 1$ we expect the population to go extinct. Then what? Any real population would eventually stop growing. In a model, density dependence can be so mild that increase without limits is still possible. Again, R_0 should tell the tale. Since $p_r(x)$ is decreasing but positive it must have a limit $p_r(\infty) = \lim\limits_{x \to \infty} p_r(x)$, and a corresponding $R_0(\infty)$. So if $R_0(\infty) > 1$ we should expect growth without limit, because a very large population will still have more and more individuals in each successive generation. But if $R_0(\infty) < 1$, population growth should be bounded.

These intuitive results are exactly what Rebarber et al. (2012) proved for (5.4.1), and Smith and Thieme (2013) and Eager et al. (2014a) extended this to more general models.[5] Just by asking what p_r does as the number of seeds goes to 0 or ∞, you can tell whether or not the population persists.

The simplest case is the constant recruitment model (5.2.1) that results from $p_r(\mathbf{N}) = \mathcal{R}/\mathbf{N}$. This has $p_r(0) = +\infty$ and so $R_0(0) = +\infty$, implying persistence. For this model we can go further. Writing the model as $n_{t+1} = \mathcal{R}c_0 + Pn_t$ we have the explicit solution $n_t = P^t n_0 + \mathcal{R}(I + P + P^2 + \cdots + P^{t-1})c_0$ implying that n_t converges to the steady state solution $\bar{n} = \mathcal{R}(I - P)^{-1}c_0$.

Determining the behavior of a persistent population is more complicated in general, because the possible outcomes include alternate stable equilibria, stable cycles, and chaotic dynamics. Nonequilibrium dynamics generally require some kind of overcompensation, so that under crowded conditions the population at time $t+1$ is a decreasing function of the population at time t: more parents imply fewer offspring when there are already too many parents for the environment to support. To rule out these possibilities, Rebarber et al. (2012) assume that $xp_r(x)$ (the number of Recruits) is increasing and convex down as a function of x (the number of Seeds); this is true in the Platte thistle model, which has $xp_r(x) = x^{0.67}$. When that is true, they proved that whenever $R_0(\infty) < 1 < R_0(0)$ there is a unique equilibrium population $\bar{n}(z)$. Moreover, the equilibrium is globally stable: $n(z, t)$ converges to $\bar{n}(z)$ as $t \to \infty$ so long as the initial population is nonzero. This matches our numerical findings for the Platte thistle model. It also shows that the behavior of model (5.4.1) doesn't depend on

---

[5] Actually these papers prove more, in greater generality, but you'll have to read them for the full story.

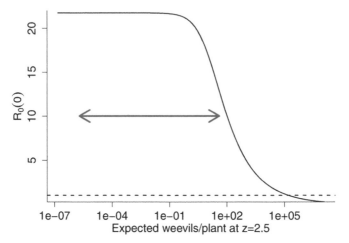

Fig. 5.6 $R_0(0)$ for the Platte thistle model as a function of the expected number of weevils on a typical-size flowering plant (log root crown diameter $z = 2.5$). The curve was computed by varying e_0 and computing $R_0(0)$ using equation 5.4.4. The arrow shows the range of e_0 values estimated for different years during the study period, with the largest values coming towards the end as the weevil infestation developed. Source file: `PlatteCalculations.R`

exactly how seedling competition is modeled. Whenever the number of new recruits each year is an increasing but concave-down function of the initial number of the seeds competing to join the population, the model will converge to a unique equilibrium from any initial population.

Using these results we can ask: what level of weevil damage would it take to wipe out Platte thistle? In that model $p_r(\mathbf{N}) = \mathbf{N}^{-0.33}$. This has the unrealistic property that $p_r \to \infty$ as $\mathbf{N} \to 0$. Let's modify it (for the moment) to $p_r(\mathbf{N}) = \min(\mathbf{N}^{-0.33}, 1)$ so that $p_r(0) = 1$. From (5.4.3) we then have

$$
R_0(0) = \langle p_b b, (I - P)^{-1} c_0 \rangle
$$
$$
= \int_Z \int_Z p_b(z') b(z') c_0(z)(I - P)^{-1}(z', z) \, dz' \, dz. \qquad (5.4.4)
$$

In Figure 5.6, we computed $R_0(0)$ for different values of the weevil burden intercept parameter e_0, and plot $R_0(0)$ as a function of the predicted mean weevil burden on a typical-size flowering plant ($z = 2.5$), given by $\exp(e_0 + 1.71 \times 2.5)$. Over the period of the Rose et al. study (indicated by the blue arrow) weevil damage began to have an appreciable impact on $R_0(0)$. However, the model projects that it would have to get much worse – more than a thousand-fold increase in mean weevil load at a typical flowering size – before $R_0(0)$ drops below 1 and the population becomes nonviable. This results from the high fecundity in the absence of weevil damage (~ 100 seeds per typical-size flowering plant) and the highly aggregated weevil distribution that spares many plants from serious damage unless mean weevil load is extremely high. It also results from the fact that (Recruits/Seed)$\to 1$ in the model as Seeds $\to 0$, which is an extrapolation beyond the range of the data. In the data, (Recruits/Seeds) is always below 0.25 (0.21 ± 0.02(sd) at the 4 lowest Seed densities). If we set the maximum value of p_r to be 0.25 rather than 1, this divides $R_0(0)$ by a factor of 4, so population extinction occurs when the $R_0(0)$ curve in Figure 5.6 drops to 4. It then takes a much lower mean weevil load (by a factor of ≈ 75) to obliterate the population. This kind of uncertainty about persistence is hard to avoid, unless data are available on the actual demography of uncrowded individuals.

Smith and Thieme (2013) extended the results of Rebarber et al. (2012) in several directions. First, they studied a generalization of (5.4.1) in which survival is not necessarily modeled with a smooth kernel. More importantly, they showed that the same conclusions about persistence and global stability still hold with overcompensating density dependence, so long as $x^2 p_r(x)$ is increasing. For example, $p_r(x) = A/(1 + Bx^2)$ with $A, B > 0$ is overcompensating because $x p_r(x)$ is not monotonically increasing, but $x^2 p_r(x)$ is monotonically increasing. In this case $R_0(\infty) < 1$ because $p_r(\infty) = 0$. So we can conclude that if $R_0(0) < 1$ the population goes extinct, while if $R_0(0) > 1$ the population converges to a unique stable equilibrium.

Persistence can occur in many ways besides convergence to equilibrium. But even without any assumptions about the shape of $p_r(x)$ it would be really surprising if a population with $R_0(\infty) < 1 < R_0(0)$ could do anything besides persisting within positive but finite limits on total population size. Smith and Thieme (2013) also provide some results in this direction. When $R_0(\infty) < 1$, there is a population size N^* such that the total population $\int_Z n(z,t)\, dz$ eventually drops below N^* and stays forever below N^*, for any initial population $n(z,0)$. And if it's also true that $R_0(0) > 1$, they show that the population is persistent in a somewhat complicated sense. Let $S(z)$ denote the expected total lifetime seed production of an individual born at size z, and $||n||_S = \int_Z n(z)S(z)\, dz$. $||n||_S$ is a measure of total population size in which some individuals count for more than others, based on their expected lifetime reproduction. Smith and Thieme (2013) showed that when $R_0(\infty) < 1 < R_0(0)$ there is an $\epsilon > 0$ such that for any initial population state, eventually $||n(z,t)||_S$ becomes larger than ϵ and stays there forever.

And that's all there is, folks – for now. It is reassuring that our intuitive expectations turn out to be right, but sobering to realize that mathematical proof is only available for the very special model (5.4.1). There are no published results (to our knowledge) for models where competition affects the growth or survival of established individuals or the size distribution of new recruits.[6] It seems safe to assume that if a small population grows and a large one shrinks, the population will persist within bounds. But we need to remember that this is just a conjecture, and check by simulations whether or not it's true for any model being studied.

---

[6] This is in part because such models are likely not to be monotone.

5.4.2 Local stability of equilibria

The dynamics near an equilibria are easier to study because near-equilibrium dynamics are approximately linear for any smooth dynamical system. For a finite-dimensional system the near-equilibrium linear dynamics are a density *independent* matrix model whose projection matrix is the Jacobian matrix for the nonlinear model evaluated at the equilibrium. The Jacobian matrix is easy to calculate because its entries are just partial derivatives of the functions that define the model, and its eigenvalues determine whether or not the equilibrium is locally stable.

The story for IPMs is similar, except that there is no easy recipe for the linear model that describes the near-equilibrium dynamics. It might not even be an IPM. So we limit ourselves to an important case where the linear model is an IPM and we can give a formula for its kernel. As above, we assume a density-dependent kernel $K(z', z, \mathbf{N})$ where $\mathbf{N}$ is some measure of total population size in which some individuals might count more than others,

$$\mathbf{N}(t) = \langle W, n \rangle = \int_{\mathbf{Z}} W(z) n(z, t)\, dz. \tag{5.4.5}$$

In the Rebarber et al. and Smith-Thieme models W equals the per-capita seed production $p_b b$ and only p_r is affected by $\mathbf{N}$. For this subsection we let W be any smooth nonnegative function, and the entire kernel can depend on $\mathbf{N}$.

For an $\mathbf{N}$-dependent kernel, the Jacobian kernel at an equilibrium $\bar{n}$ with corresponding "total population" $\bar{\mathbf{N}} = \langle W, \bar{n} \rangle$ is (Ellner and Rees 2006)

$$J(z', z, \bar{\mathbf{N}}) = K(z', z, \bar{\mathbf{N}}) + Q(z') W(z), \text{ where}$$

$$Q(z') = \int_{\mathbf{Z}} \frac{\partial K(z', z, \bar{\mathbf{N}})}{\partial \mathbf{N}} \bar{n}(z)\, dz. \tag{5.4.6}$$

The near-equilibrium dynamics are then approximated by the linear system

$$n(z', t+1) - \bar{n}(z') = \int_{\mathbf{Z}} J(z', z, \bar{\mathbf{N}})(n(z, t) - \bar{n})\, dz.$$

Under our assumptions about K, the Jacobian kernel J will have a dominant eigenvalue (or a set of codominant eigenvalues of equal magnitude). The equilibrium $\bar{\mathbf{N}}$ is locally stable if all eigenvalues of the Jacobian kernel at $\bar{\mathbf{N}}$ are < 1 in magnitude, and unstable if any are > 1 in magnitude.[7]

For equation 5.4.1, Ellner and Rees (2006) and Smith and Thieme (2013) give conditions on p_r which imply that an equilibrium (if one exists) is locally stable. But in general, local stability has to be checked numerically: iterating the model to find the equilibrium, and computing J and its eigenvalues.

---

[7] This follows from the proofs of Theorems 5.2 and 5.4 in Smith and Thieme (2013), under the assumption that the Jacobian kernel $J(z', z, \mathbf{N})$ is a continuous function. In that case the Jacobian defines a compact operator (on either $\mathcal{C}(\mathbf{Z})$ or $L_1(\mathbf{Z})$), and the proofs of Theorems 5.2 and 5.4 just use compactness of the Jacobian and not the special form of model (5.4.1).

5.4.3 Equilibrium perturbation analysis

When a density-dependent model has a stable equilibrium, perturbation analysis focuses on how that equilibrium is affected by changes in kernel entries, vital rate functions, or parameter values. If the kernel depends on a total density measure (5.4.5), then the Takada and Nakajima (1992, 1998) formula still applies.[8] Let $s_{\bar{N},\theta}$ denote the sensitivity of $\bar{N}$ to some aspect of the model, θ that is being perturbed (a kernel entry, a parameter, etc.). Then

$$s_{\bar{N},\theta} = -\frac{s_{\lambda,\theta}}{s_{\lambda,N}} \tag{5.4.7}$$

The numerator on the right-hand side is sensitivity of λ with respect to the perturbation, and the denominator is the sensitivity of λ with respect to $\bar{N}$. And λ, in this case, is the population growth rate (dominant eigenvalue) for the density-independent kernel $K(z', z; N)$ at $N = \bar{N}$. When crowding has only negative effects on vital rates $s_{\lambda,N}$ will be negative. In that case, (5.4.7) says that the sensitivity of $\bar{N}$ to the perturbation is proportional to the sensitivity of λ evaluated at the equilibrium.

Note that $s_{\lambda,N}$ is the same for all perturbations, and it can be computed from the general eigenvalue sensitivity formulas (4.3.1, 4.3.2),

$$s_{\lambda,N} = \left\langle \bar{v}, \frac{\partial K}{\partial N} \bar{w} \right\rangle \Big/ \langle \bar{v}, \bar{w} \rangle \tag{5.4.8}$$

where $\bar{v}$ and $\bar{w}$ are the left and right dominant eigenvectors, respectively, of the kernel $K(z', z, \bar{N})$ and $\partial K/\partial N$ is evaluated at $N = \bar{N}$.

For example, consider how perturbing the (z_2, z_1) kernel entry in $K(z', z; N)$ affects an equilibrium $\bar{N}$. The sensitivity of λ to this perturbation is the familiar eigenvalue sensitivity formula

$$s_{\lambda,K(z_2,z_1)} = \bar{v}(z_2)\bar{w}(z_1)/\langle \bar{v}, \bar{w} \rangle$$

Substituting this (5.4.8) into (5.4.7) we have

$$s_{\bar{N},K(z_2,z_1)} = -\bar{v}(z_2)\bar{w}(z_1) \Big/ \left\langle \bar{v}, \frac{\partial K}{\partial N} \bar{w} \right\rangle. \tag{5.4.9}$$

As in Caswell et al. (2004) the Takada-Nakajima formula can be generalized to IPMs that depend on several different total population measures. However, as we write there is nothing like Caswell (2008)'s comprehensive perturbation analysis for density-dependent matrix models, which gives sensitivity formulas for each entry in an equilibrium population vector, for measures of population dynamics along periodic orbits, and many other properties of stable equilibria or cycles. These results rely on matrix calculus, and to our knowledge there is no corresponding "function calculus" that can be used for IPMs. It is almost

---

[8] And the proof is the same: differentiate both sides of $\lambda(\bar{N}(\theta), \theta) = 1$ with respect to θ and solve for $d\bar{N}/d\theta$.

certainly safe to regard an IPM implemented with midpoint rule as a big matrix model and apply the formulas in Caswell (2008). But this is a conjecture, so it is essential to make sure that results are unaffected by the number of mesh points and the size range.

5.4.4 Density dependence and environmental stochasticity

Several papers have modified the constant recruitment model (5.2.1) to take into account of between-year variability in recruitment, replacing $\mathcal{R}$ by $\mathcal{R}(t+1)$, the number of new recruits in year $t+1$ (Childs et al. 2004; Metcalf et al. 2008; Rees and Ellner 2009). The special structure of this model makes it possible to solve it explicitly and show that the population converges to a unique stationary distribution, analogous to the stable equilibrium of the deterministic constant recruitment model (Ellner and Rees 2007, Appendix A). The stochastic model can be simulated by drawing $\mathcal{R}$ values at random from the set of observed recruit numbers, or by fitting a probability distribution to the observed numbers.

Otherwise, the only published IPMs with both density dependence and environmental stochasticity are models for spatial population structure. Hardin et al. (1988) added stochastic variation in population growth to the Kot and Schaffer (1986) model (8.1.3) for an unstructured population living on a bounded interval representing space (in one dimension). Eager et al. (2014b) modified the Platte thistle model to describe an annual plant with a seed bank. The population is censused in between growing seasons when it consists entirely of buried seeds, and $n(z,t)$ is the seed bank with z being the depth of burial in the soil. Disturbances occur at random times. These redistribute seeds vertically in the soil and allow some seeds to germinate, complete the aboveground life cycle, and replenish the seed bank.

With environmental stochasticity we still expect that a population will persist if it tends to increase when rare, and go extinct if a small population gets even smaller. However, the tendency to increase or decrease has to be measured by the stochastic growth rate λ_S (Chapter 7). If $\lambda_S > 1$ when all individuals have the maximum possible per-capita fecundity (which occurs in the absence of competition) we should expect persistence. Both Hardin et al. (1988) and Eager et al. (2014b) show that this is true in their models with undercompensating density dependence. Again, it is surprising that such an intuitive result has only been proved for a few special cases, and we expect to see progress soon.

5.5 Case study 2C: nonlinear dynamics in a Soay sheep model

The extensive long-term data on the Soay population makes it possible to identify effects of population density on several demographic processes. The full fitted models also include year-to-year variation (climate covariates such as NAO, environmental stochasticity, and temporal trends). To focus on density dependence, here we study a deterministic model representing a "typical" year in the middle of the study period, by setting all time-varying coefficients to their average value. As in Chapter 2 the model only tracks females, omits rare events such

```
## Recruit size distribution
c_z1z <- function(z1, z, Nt, m.par) {
    mean <- m.par["rcsz.int"] + m.par["rcsz.Nt"] * Nt +
            m.par["rcsz.z"] * z
    sd <- m.par["rcsz.sd"]
    return(dnorm(z1, mean = mean, sd = sd))
}


## Recruitment probability (survival from birth to first summer census)
## logistic regression
pr_z <- function(Nt, m.par) {
    linear.p <- m.par["recr.int"] + m.par["recr.Nt"] * Nt
                p <- 1/(1+exp(-linear.p))
    return(p)
}


## Survival probability: logistic regression
s_z <- function(z, Nt, m.par) {
    linear.p <- m.par["surv.int"] + m.par["surv.Nt"] * Nt +
                m.par["surv.z"] * z
    p <- 1/(1+exp(-linear.p))
    return(p)
}
```

Box 5.1: The density-dependent demographic functions in the Soay sheep model. Nt is the total number of females in year t. Other demographic functions are the same as in Chapter 3. The fitted values of the coefficients on Nt are all negative, so these are all decreasing functions of density.

as twinning, and simplifies the age structure by only distinguishing between new lambs and older females.

Density affects three processes in the kernel (equation 2.6.1): the probability that a newborn lamb successfully recruits into the population, p_r; the size distribution of new recruits, C_0; and the survival of adults s (Figure 5.7A). The demographic models are all simple extensions of the density-independent model in Chapter 2, in which the total number of females $\mathbf{N}(t) = \int_{\mathbf{Z}} n(z, t) \, dz$ is an additional predictor (Box 5.1). Nonetheless the model is beyond the reach of the mathematical theory reviewed above, because density affects survival and recruit size as well as reproduction. We have to study it computationally.

The first step is picking the size range to avoid eviction. Picking the lower limit L is trickier than usual because recruit size depends on their mother's size and on population density. Females of size $z \leq 1$ are unlikely to reproduce ($p_b(1) \approx 0.01$), and smaller adults have smaller lambs. We are safe if L is low enough to "catch" all lambs of a size-1 mother, at any population density that

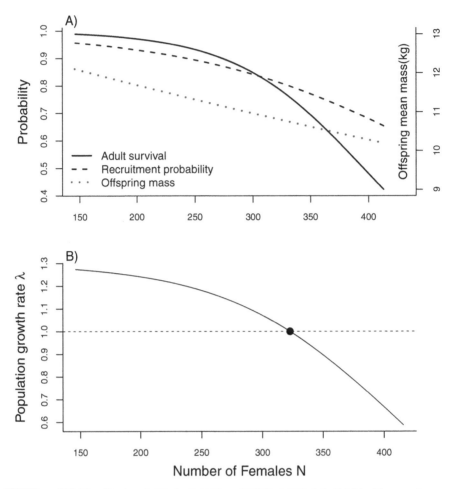

Fig. 5.7 (A) The density-dependent rates are adult survival (solid black), recruitment probability (dashed black), and mean offspring size (dashed blue), plotted here over the observed range of female densities. Note that offspring size is not log-transformed; adult survival and offspring size are plotted for a typical-size adult female, $z = 2.9$. (B) Long-term population growth rate $\lambda(\mathbf{N})$; the dashed line at $\lambda = 1$ locates the equilibrium population size $\bar{N} \approx 323$. Source files: Soay DD Demog funs.R and Soay DD IPM calcs.R.

we want to consider. With $\mathbf{N} = 520$ females, which is 125% of the highest population density in the data, the recruit-size distribution for a mother of size $z = 1$ is

```
rc <- function(z1) c_z1z(z1,1,520,m.par.true)
```

With a bit of trial-and-error, $L = 0.5$ is small enough:

```
> integrate(rc,-Inf,0.5)
0.004430661 with absolute error < 2.3e-07
```

So with $L = 0.5$ the size range will omit only a tiny fraction of lambs from females so small that they are unlikely to breed, and even fewer lambs from larger females, at any population density up to $\mathbf{N} = 520$. As an initial guess for U, we compute the fixed point of the mean growth function, and add 3 standard deviations of the growth variability plus an additional 20% (on un-logged scale):

```
m.par.true["grow.int"]/(1-m.par.true["grow.z"]) +
    3*m.par.true["grow.sd"] + 0.2
```

The result is 3.62, so let's try $U = 3.65$ and see what fraction of the largest individuals grow outside the size range, using the survivorship at $\mathbf{N} = 0$ to maximize the result.

```
> z1Dist <- function(z1) s_z(z1,0,m.par.true)*g_z1z(z1,3.65,m.par.true)
> integrate(z1Dist,3.65,Inf)
0.004796459 with absolute error < 3.8e-09
```

That's probably good enough, but to be safe we go out a bit more:

```
> z1Dist <- function(z1) s_z(z1,0,m.par.true)*g_z1z(z1,3.8,m.par.true)
> integrate(z1Dist,3.8,Inf)
0.0003110296 with absolute error < 1e-05
```

Now we can explore what the model predicts. The real population has remained relatively stable over decades of study, so if the model is at all realistic, it should exhibit bounded population growth and perhaps convergence to a stable equilibrium. As in Section 5.4.1 above, we can start by asking if a small population grows and a large population shrinks. Let $\lambda(x)$ denote the population growth rate (dominant eigenvalue) for the density-independent kernel K_x that results from holding $\mathbf{N}$ constant at value x (in this model the recruit size distribution depends on parent size, so there is no advantage to working with R_0 instead of λ). Survival and recruitment drop to 0 as $\mathbf{N} \to \infty$, so $\lambda(\infty) = 0$ and unbounded population growth is impossible. Computing $\lambda(N)$ across the range of observed female densities (Figure 5.7B) we find that $\lambda(0) > 1$ so a small population will grow. Because the effects of higher density are all negative, λ is a strictly decreasing function of $\mathbf{N}$ and so there is a unique population size $\bar{N} \approx 323$ at which $\lambda(\bar{N}) = 1$. $\bar{N}$ is an equilibrium population size, and reassuringly it is close to the average number of adult females over the study period (mean=280.2, median=286.5). The dominant eigenvector of $K_{\bar{\mathbf{N}}}$ is the population structure at the equilibrium.

The equilibrium appears to be stable: a small population increases monotonically and reaches the equilibrium within a few decades (Figure 5.8A). To confirm that the equilibrium is locally stable, we iterate the model until the population converges to the equilibrium, compute the Jacobian kernel (5.4.6) using the function in Box 5.2, and find its dominant eigenvalue λ_J:

```
> J <- mk_J(nbar=nbar, m=m, m.par=m.par.true, L=L, U=U, eps=0.01);
> lam_J <- eigen(J)$values[1]; cat(lam_J,"\n")
0.5509334+0i
```

```
## Midpoint rule iteration matrix for Jacobian at an equilibrium nbar
## eps is the increment for numerical differentiation of K
   mk_J <- function(nbar, m, m.par, L, U, eps=0.01) {
   h <- (U-L)/m; Nbar <- h*sum(nbar) # compute total population

   # first compute Q = dK/dN %*% nbar
   dK <- ( mk_K(Nbar+0.5*eps, m, m.par, L, U)$K
          - mk_K(Nbar-0.5*eps, m, m.par, L, U)$K )/eps
   Q <- dK%*%nbar

   # Next the kernel R(z',z) = Q(z')W(z), i.e. R[i,j]=Q[i]*W[j]
   # In this case W==1 so R[i,] = Q[i]
   R <- matrix(0,m,m)
   for(i in 1:m) R[i,] <- Q[i]

   # Finally, Q = R + K; note factor of h to give the iteration matrix
   # for J rather than kernel values, to parallel the output of mk_K
   J <- mk_K(Nbar, m, m.par, L, U)$K + h*R
   return(J)
}
```

Box 5.2: Jacobian kernel iteration matrix for the Soay sheep model.
The kernel construction function mk_K is in Soay DD Demog Funs.R

Since $|\lambda_J| < 1$, the equilibrium is locally stable. Once $n(z,t)$ gets close to $\bar{n}(z)$, the linear approximation predicts exponential convergence with $n(z,t) - \bar{n}$ decreasing by a factor of $|\lambda_J|$ between each year and the next. Figure 5.8B confirms this and provides a useful check on our calculation of the Jacobian kernel.

Is the equilibrium globally stable? This is a question about infinitely many initial conditions. The best we can do is try a few and see what happens (Figure 5.8C, D) – no surprises. The population always stabilizes with 323 females.

Well, that was certainly boring. Let's turn up the heat a bit and see what happens when the density dependence is stronger. A simple way of doing this is to add a parameter ρ that makes recruitment probability change more steeply near the equilibrium population size Nbar:

```
pr_z <- function(Nt, m.par) {
   linear.p <- m.par["recr.int"] + m.par["recr.Nt"] * Nt - rho*(Nt-Nbar)
   1/(1+exp(-linear.p))
}
```

This modification changes the intercept and the Nt coefficient in such a way that the equilibrium stays the same but the behavior near the equilibrium changes as a function of ρ.

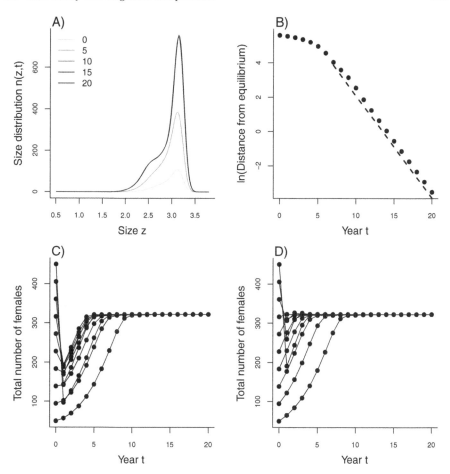

Fig. 5.8 (A) Growth of the population starting from 50 females at the steady-state size distribution for the kernel with $\mathbf{N} = 50$. Increasingly dark curves show $n(z, t)$ at $t = 0, 5, 10, 15, 20$. (B) Convergence to the equilibrium. Points show $\int_{\mathbf{Z}} |n(z, t) - \bar{n}(z)|\ dz$ as a function of t. The slope of the dashed curve shows the predicted rate of convergence based on the dominant eigenvalue of the Jacobian. (C) Dynamics of total population size starting from a range of initial population sizes with random initial size distribution. (D) As in panel C), but starting from the steady-state size distribution for the kernel with $\mathbf{N}$ equal to the initial number of females. Source file: Soay DD IPM calcs.R.

Now it gets more interesting. As ρ increases, a negative eigenvalue increases in magnitude until it becomes dominant at about $\rho = 0.02$ (Figure 5.9A). The steady state is still stable, but convergence is (as expected) oscillatory rather than monotonic (Figure 5.9B). At about $\rho = 0.05$ the equilibrium becomes unstable, when the dominant Jacobian eigenvalue passes through -1. In a one-dimensional map we would expect to see a period-doubling "flip" bifurcation, producing a period-2 limit cycle, and that's exactly what happens here (Fig-

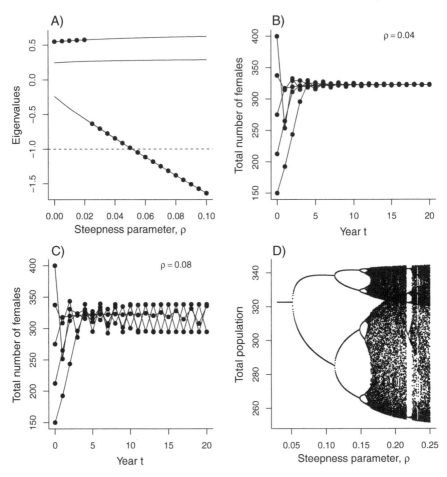

Fig. 5.9 (A) The three Jacobian kernel eigenvalues with largest absolute value as a function of the steepness parameter ρ that modifies the recruitment probability; all three are real numbers. Circles indicate the dominant eigenvalue and the dashed horizontal line at -1 indicates when the equilibrium becomes locally unstable. (B, C) Dynamics of total population size, as in Figure 5.8D, for two different values of ρ. (D) Bifurcation diagram, computed by iterating the model for 1000 time steps at each value of ρ and plotting the last 100 values of total population size as points above the corresponding ρ value. Source file: Soay Bifurcation Calcs.R

ure 5.9C). As ρ increases further, it's *déjà vu* all over again – the model exhibits a sequence of period-doubling bifurcations leading to chaos (Figure 5.9D) that's very familiar from one-dimensional difference equations like the discrete logistic model $x(t+1) = rx(t)(1 - x(t)/K)$.

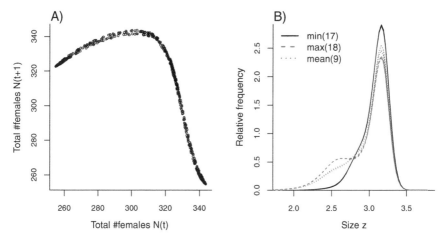

Fig. 5.10 (A) The dynamics of total female population $\mathbf{N}(t)$, for steepness parameter $\rho = 0.02$. The population was iterated for 1000 time steps, starting from 300 individuals. The graph shows $\mathbf{N}(t+1)$ as a function of $\mathbf{N}(t)$ for $t = 501$ to 1000. (B) Size distributions, normalized to total population $\mathbf{N} = 1$, at times when $\mathbf{N}(t)$ was within 1 of its minimum (black solid curve), maximum (red dashed curve), and mean (blue dotted curve) over $t = 501$ to 1000. Numbers in the figure legend are the number of overlaid curves in each case. Source file: `Soay Bifurcation Calcs.R`

The similarity to the discrete logistic is not coincidental. In fact, the model's dynamics are almost one-dimensional. After initial transients, the changes in total female population $\mathbf{N}(t)$ are almost perfectly described by a one-dimensional function giving $\mathbf{N}(t+1)$ as a function of $\mathbf{N}(t)$ (Figure 5.10A). This happens because the size structure is almost perfectly determined by the total population size (Figure 5.10B). The black curves are overlaid size frequency distributions at the 17 times (during the time interval $t = 501, 502, \cdots, 1000$) when $\mathbf{N}(t)$ was within 1 of its minimum value, and they are nearly identical. The same is true for the times when $\mathbf{N}(t)$ was within 1 of its maximum and within 1 of its mean.

These complex dynamics don't occur in the actual fitted model, so we won't pursue it any further. And to actually prove that there is a period-doubling bifurcation ending in chaos would be extremely difficult, though not necessarily impossible. In a remarkable paper, Day et al. (2004) give a rigorous computer-assisted proof of chaotic dynamics in a density-dependent spatial IPM. Day and Kalies (2013) develop a more general approach for rigorously computing dynamic properties of integrodifference equations, and apply it to demonstrate chaos in a density-dependent spatial IPM by computing a lower bound on the topological entropy of an attractor. The math is *way* above the level of this book (and its authors). But the seemingly impossible is starting to happen, and software implementing those methods is being developed. Before long there may be tools that let us all analyze complex dynamics in nonlinear IPMs.

5.6 Coda

The mathematical theory for density-dependent IPMs lags far behind the corresponding theory for matrix models and other structured population models. Sensitivity analysis, in particular, needs a lot of development and some innovative approaches to catch up with what's now available for matrix models (Caswell 2008).

But the lack of analytical machinery should not deter you, because there is nothing special about modeling how population density affects demographic rates. Asking how a plant's growth depends on the size and abundance of neighbors is the same as asking how growth depends on size and age, or on size and how much rain fell last year. And in many cases (such as our Platte thistle and Soay sheep models), numerical implementation doesn't require much more than an extra line or two to compute a measure of total population size. Density dependence is much less common in the IPM literature than it is in the real world, but it doesn't need to stay that way.

5.7 Appendix: Mean field approximations for neighborhood competition

The mean field approximations to neighborhood competition replace $\sum_k w_{ij,km}$ in equation (5.3.2) by its expectation under an assumed spatial distribution of genets. This produces a nonspatial IPM for the size distribution averaged across space.

Consider the region between the circles of radius x and $x + dx$ centered on a focal genet of species j. The area of this annulus is $2\pi x\,dx$ for $dx \approx 0$ (i.e., $\pi(x + dx)^2 - \pi x^2$ to order dx). According to equation (5.3.1), a species m genet whose center is in the annulus puts competitive pressure $e^{-\alpha_{jm}x^2}$ times its area on the focal genet. The expected total competitive pressure from all such genets is therefore $e^{-\alpha_{jm}x^2}2\pi x\,dx$ times the expected fractional cover of species m in the annulus (the total area of species m genets, as a fraction of the total area).

The first mean field model is that all genet centers are distributed uniformly and independently at random. The expected fractional cover of species m in the annulus is then its fractional cover in the habitat as a whole, which we denote Φ_m. Let $n_m(z, t)$ be the size distribution of species m, such that $n_m(z,t)h$ is the number of species-m genets with log-transformed genet area in $(z, z + h)$, per unit area of the habitat (e.g., n gives the population density in m^2 of genet cover per m^2 of habitat). The total area of those genets per unit area of habitat is $e^z n_m(z,t)h$, and so

$$\Phi_m = \int_L^U e^z n_m(z, t)\,dz. \qquad (5.7.1)$$

The expected competitive pressure on a species-j genet due to species m is then

$$\overline{w}_{jm}(t) = \int_0^\infty \Phi_m e^{-\alpha_{jm}x^2} 2\pi x\,dx = \pi\Phi_m/\alpha_{jm}. \qquad (5.7.2)$$

The expected total competitive pressure on a species j genet is then equation (5.3.4). The mean field IPM assumed that every species-j genet experienced competitive pressure $\overline{W}_j$.

The second mean field model adds the constraint that conspecific genets cannot overlap. For any one focal genet, this restriction on its neighbors affects only a negligibly small part of the habitat, so it is equivalent to distributing plants at random and then deleting any that violate the constraint. The closer the annulus, and the larger the focal plant, the higher the fraction of plants that are too big to be rooted in the annulus without overlapping the focal plant. Let $\Phi_m(u)$ be the cover of species m genets of radius u or smaller,

$$\Phi_m(u) = \int_L^{\log(\pi u^2)} e^z n_m(z, t) \, dz. \tag{5.7.3}$$

The expected competitive pressure by species m on a focal conspecific of radius r is then

$$\overline{w}_m(r) = 2\pi \int_r^\infty x e^{-\alpha_{mm} x^2} \Phi_m(x - r) \, dx \tag{5.7.4}$$

The computationally inconvenient infinite integral in (5.7.4) can be avoided as follows. Let r_U denote the radius corresponding to the upper limit U on log genet area, i.e., $r_U = \sqrt{e^U/\pi}$. Then whenever $x - r \geq r_U$, $\Phi_m(x - r)$ will equal the species' total cover $\Phi_m = \Phi_m(r_U)$, giving

$$\overline{w}_m(r) = 2\pi \int_r^{r+r_U} x e^{-\alpha_{mm} x^2} \Phi_m(x - r) \, dx$$
$$+ \ \pi \Phi_m e^{-\alpha_{mm}(r+r_U)^2} / \alpha_{mm}. \tag{5.7.5}$$

For each species m, an IPM implemented by midpoint rule will have a setup like

```
# boundary points b and mesh points v for log(genet area)
bigM <- 100; # number of mesh points
b <- L+c(0:bigM)*(U-L)/bigM
v <- 0.5*(b[1:bigM]+b[2:(bigM+1)])
```

A size-distribution vector nt, equal in length to v, represents values of the size (log genet area) distribution at the mesh points. The genet radii corresponding to the boundary and mesh points are

```
b.r <- sqrt(exp(b)/pi); v.r <- sqrt(exp(v)/pi); r.U <- sqrt(exp(U)/pi);
```

The sum of the first j entries in exp(v)*n represents the total cover of plants of transformed area b[j+1] or smaller, hence of radius b.r[j+1] or smaller. That lets us approximate $\Phi(r)$ as

```
Phi.r <- splinefun(b.r,h*c(0,cumsum(exp(v)*nt)),method="natural");
```

The size-dependent competitive pressure $W(r)$ (equation 5.7.5) can then be computed as follows:

```
Phi.tot <- h*sum(exp(v)*nt);
wr <- function(r,alpha,nt) {
 out <- integrate(function(x) x*exp(-alpha*(x^2))*Phi.r(x-r),r,r+r.U)$value
        + pi*Phi.tot*exp(-alpha*((r+r.U)^2))/alpha);
 return(2*pi*out);
}
Wr <- Vectorize(wr,vectorize.args="r");
```

The W values at all grid points are then Wr(v.r,alpha,nt). At each time-step, this function needs to be called once for each species to compute intraspecific competitive pressure as a function of genet size; the interspecific competitive pressures are pi*Phi.tot/alpha, equation (5.7.2).

Box 5.3: Computing neighborhood competition with midpoint rule.

Chapter 6
General Deterministic IPM

Abstract This chapter is where the IPM really comes into its own as a flexible and parsimonious framework for populations with complex demography, meaning that individual state is not described by a single number z. Instead z can be multidimensional (age and size, size and disease state, etc.); it can use different attributes at different stages of the life cycle (e.g., age-structured and size-structured breeding adults, depth-structured seeds, and stage-structured plants); or the future can depend on the past as well as the present. We explain here how everything about the basic IPM continues to work in this more general setting, and discuss numerical methods for dealing with (or avoiding) the larger iteration matrices that can result when z is multidimensional. The main case study is a model with size and age structure based on the Soay sheep system, and we also use a size-quality model, and a model for tropical tree growth, to illustrate different aspects of how a general IPM is implemented numerically.

6.1 Overview

Our focus so far has been on population structured by a single continuous variable, modeled by the basic IPM that was developed for populations with continuous variation in size. But in many species, demographic rates are affected by several attributes, such as size and age (e.g., Childs et al. 2003; Rees et al. 1999; Rose et al. 2002; Coulson et al. 2010), stage and size (Childs et al. 2011), age and sex (e.g., Coulson et al. 2001), or by measured attributes and additional unobserved variables that characterize individual "quality" (e.g., Rees et al. 1999; Cam et al. 2002, 2013; Pfister and Stevens 2002, 2003; Clark et al. 2004; Chambert et al. 2013, 2014). This chapter develops the computing methods and theory that allow IPMs to handle this situation, and others that can't be captured by the basic IPM, such as:

- When individuals are classified by different attributes during different parts of the life cycle, for example, if aboveground plants are classified by size while buried seeds are classified by depth of burial (Eager et al. 2013).

© Springer International Publishing Switzerland 2016

S.P. Ellner et al., *Data-driven Modelling of Structured Populations*,
Lecture Notes on Mathematical Modelling in the Life Sciences,
DOI 10.1007/978-3-319-28893-2_6

- When population structure is a mix of continuous and discrete variation (e.g., Rees et al. 2006; Hesse et al. 2008; Jacquemyn et al. 2010), such as a seed or seedling class in which you just count the total number of seeds and an "adult" population classified by size (Williams 2009; Salguero-Gomez et al. 2012).
- When there are time delays, so that the future depends on the past as well as the present (Kuss et al. 2008).

What unites these situations is that individual state is not specified by just one number. If a population is classified by weight w and quality q we need the bivariate distribution function at each time, $n(w, q, t)$. If weight and sex both matter we need a separate weight distribution for each sex, $n_m(w, t)$ and $n_f(w, t)$. So we need IPMs where z belongs to a more complicated individual-level state space $\mathbf{Z}$. For sex and weight, $\mathbf{Z}$ can be two copies of an interval $[L, U]$. For weight and quality, or any other pair of continuous variables, $\mathbf{Z}$ can be a rectangle. But $\mathbf{Z}$ could also be an set of functions specifying a state-dependent life history strategy such as reproductive effort as a function of age.

The approach we develop allows one general theory to cover many situations, so that the structure of the model can be adapted to the life cycle and to the results of demographic data analysis. If several measured variables affect the risk of death, you would fit a survival model with all of them as covariates. You can then build an IPM with individuals cross-classified by all the important covariates. We illustrate this in the next section, returning to the Soay sheep system, but now including the age-dependent demography seen in the real data. With that as motivation, we describe a general deterministic density-independent IPM, discuss some important examples of the general model, and outline the main results of stable population theory. We then demonstrate how models with discrete×continuous (e.g., age-size) or continuous×continuous (e.g., size-quality) classification of individuals can be implemented using the midpoint rule. Finally there are some technical sections, describing alternatives to the midpoint rule that are sometimes needed for multidimensional or spiky kernels, and detailing the precise mathematical assumptions and results.

6.2 Case Study 2D: Age-size structured ungulate, fitting the model, and projecting

Age is a good predictor of demographic performance in many animal populations. Unlike size, which typically exhibits a simple positive association with vital rates, the age-dependence of mortality and reproduction is often more complicated. Many processes interact to shape these relationships. For example, behavioral adaptations arising from past experience may improve performance with age, while senescence leads to a decline. If nothing else varies, selective mortality of lower "quality" individuals will result in a decline in age-specific mortality as poor quality individuals are removed from the population.

Age-structured IPMs therefore represent an important special case of the general IPM. In this section we show how to construct such models. The broader aim is to demonstrate how to build IPMs when individuals are cross-classified by a mix of continuous and discrete variables, as well as to provide a concrete example to illustrate some basic calculations. In order to do this, we will extend the ungulate case study introduced in Chapter 2.

The new case study is again motivated by the St. Kilda Soay sheep population, only now we want to accommodate both size- and age-dependent vital rates. As usual, "size" is used as a generic synonym for a continuous structuring variable, log body mass in this case. The basic assumptions are the same as in the simpler version: we still use the post-reproductive census life cycle in Figure 2.2, and we only consider the dynamics of females in a constant, density-independent environment.

6.2.1 Structure of an age-size IPM

The general form of a deterministic age-size IPM is

$$n_0(z', t+1) = \sum_{a=0}^{M} \int_L^U F_a(z', z) n_a(z, t) \, dz \tag{6.2.1a}$$

$$n_a(z', t+1) = \int_L^U P_{a-1}(z', z) n_{a-1}(z, t) dz \quad a = 1, 2, \ldots, M \tag{6.2.1b}$$

when M is an assumed maximum age beyond which no individuals survive. If there is no hard and fast age limit, then all individuals past some age can be grouped into a single "$M + 1$ or older" age-class. This adds one more equation to the model:

$$n_{M+1}(z', t+1) = \int_{\mathbf{Z}} [P_M(z', z) n_M(z, t) + P_{M+1}(z', z) n_{M+1}(z, t)] \, dz \tag{6.2.2}$$

and the sum in (6.2.1a) then needs to run up to age $M + 1$ instead of M. Adding the final age-class spares us from dealing with many more equations. It's also a good idea because sample size generally decreases with age and eventually the data are too sparse to parameterize age-specific kernels. Our age-size ungulate IPM includes this kind of "greybeard" age-class.

If we iterate the model defined by (6.2.1) it will converge to a stable age-size distribution, $w_a(z)$, with the population growing at a constant rate λ (assuming the conditions given in Section 6.5 are met). This means we can always calculate $w_a(z)$ and λ by iteration once we know how to implement the model. What if we need the reproductive value? Let $v_a(z)$ denote the size-dependent reproductive value for age a, and $P_a(z', z), F_a(z, z)$ the age-dependent survival and fecundity kernels. With a finite maximum age M, the transpose iteration is

$$v_a(t+1) = P_a^T v_{a+1}(t) + F_a^T v_0(t), \quad a = 0, 1, 2, \cdots, M \tag{6.2.3}$$

This is identical to iterating a transposed Leslie matrix, except for the transposition of P_a and F_a. Here P_a^T is the transpose kernel $P_a^T(z', z) = P_a(z, z')$, so

Table 6.1 Demographic processes, simulated variables, and demographic functions and their parameter values used in the age-size-structured IBM. The demographic functions and parameter values (stored in a vector, m.par.true) are defined in the R script Ungulate Age Demog Funs.R. The IBM simulation is implemented in Ungulate Age Simulate IBM.R.

Process	Simulated variable	Demographic function
Reproduction probability	$Repr \sim Bern(p_b(z,a))$	$\text{logit}(p_b) = -7.88 + 3.11z - 0.078a$
Recruitment probability	$Recr \sim Bern(p_r(a))$	$\text{logit}(p_r) = 1.11 + 0.18a$
Recruit size	$Rcsz \sim Norm(\mu_c(z), \sigma_c)$	$\mu_c = 0.36 + 0.71z$, $\sigma_c = 0.16$
Survival	$Surv \sim Bern(s(z,a))$	$\text{logit}(s) = -17.0 + 6.68z - 0.33a$
Growth	$Z_1 \sim Norm(\mu_g(z,a), \sigma_g)$	$\mu_g = 1.27 + 0.61z - 0.0072a$, $\sigma_g = 0.078$

in practice (6.2.3) is done with the transposes of the iteration matrices for P and F. For the greybeard age-class $M + 1$, if the model includes one,

$$v_{M+1}(t+1) = P_{M+1}^T v_{M+1}(t) + F_a^T v_0(t). \tag{6.2.4}$$

More generally, suppose $\mathbf{Z}$ consists of multiple components $\mathbf{Z}_i$ and the IPM is represented by component kernels, $n_i(t+1) = \sum_j K_{ij} n_j(t)$. The transpose iteration is then

$$v_j(t+1) = \sum_i K_{ij}^T v_i(t) \tag{6.2.5}$$

6.2.2 Individual-based model and demographic analysis

The case study is again based on a dataset constructed from an individual-based model (IBM), and to ensure that it is realistic, the model was parameterized using the real Soay field data. Essentially the same set of linear and generalized linear models was fitted to the dataset as before, but now the models for survival, growth, and reproduction include (log) body mass, z, and female age, a. [1] Rather than stepping through every aspect of the IBM we have summarized its components in Table 6.1. Review the ungulate case study in Chapter 2 if you need more detail about how the IBM is implemented in R. Survival, growth, and reproduction are size- and age-dependent, while lamb recruitment and birthsize only depend on maternal age and size, respectively. Beyond keeping track of

---

[1] The data analysis included year-to-year parameter variation; the parameter values reported here describe a typical year. We also simplified the nature of the age-dependence somewhat, ignoring the mild nonlinearities (on the linear predictor scale) when they occur. Likelihood ratio test was used to assess whether or not keep size and age in a particular demographic function.

both size and age of individuals, the only other significant change is a hard constraint on reproduction: we assume that individuals never conceive in their first year of life.

To generate the data we started with an initial population of 500 new recruits, simulated the population until the density reached 5000 individuals, and then selected a random sample of 3000 individuals to be used in the demographic analysis. The R code for the demographic functions and the IBM simulation can be found in `Ungulate Age Demog Funs.R` and `Ungulate Age Simulate IBM.R`, respectively. The code for running the IBM and carrying out the analysis we discuss next is in `Ungulate Age Calculations.R`.

The output of the simulation is again stored in an R data frame called `sim.data`. The only difference between this data frame and that generated by the IBM from the Chapter 2 ungulate case study is the addition of an extra column, a, that gives the age of the individual associated with each observation:

```
   z a Surv  z1 Repr Recr Rcsz
3.09 7   1 2.95    1    0   NA
3.17 3   1 3.19    1   NA   NA
3.08 0   1 3.11    0   NA   NA
3.23 7   1 3.19    1    1 2.61
3.25 5   1 3.01    0   NA   NA
3.23 7   0  NA   NA   NA   NA
3.28 7   1 3.29    0   NA   NA
2.19 0   0  NA   NA   NA   NA
```

A quick look at this data frame should convince you there are no surprises lurking inside it. For example, age 0 individuals never reproduce ($Repr = 0$ for all age 0 survivors).

The next step is to fit the regression models required to build the IPM. This step is no different from its counterpart for a simple IPM. We simulated the data assuming that survival, growth, and reproduction are both size- and age-dependent, and lamb recruitment and birth size only depend on maternal age and size, respectively. Let's fit these five models:

```
mod.surv.glm <- glm(Surv ~ z + a, family = binomial, data = sim.data)
mod.grow <- lm(z1 ~ z + a, data = grow.data)
mod.repr <- glm(Repr ~ z + a, family = binomial, data = repr.data)
mod.recr <- glm(Recr ~ a, family = binomial, data = recr.data)
mod.rcsz <- lm(Rcsz ~ z, data = rcsz.data)
```

For all but the first of these we first subsetted `sim.data` in the script to make sure we are only working with the appropriate cases. For example, we exclude new recruits when fitting the probability of reproduction model because we know that our simulated Soays never reproduce in their first year of life:

```
repr.data <- subset(sim.data, Surv==1 & a>0)
```

Leaving age 0 individuals in this regression would introduce a bias into the estimated functional relationship between reproduction and size for older age classes. The parameter values from these regressions are stored in m.par.est as before.

In this example we knew exactly how size and age influence each demographic process so we just went ahead and fitted the appropriate model to the data. It is worth stepping back and thinking about what to do with real data, when we never know the "truth" and we have to select among different possible models. When working with a discrete trait like life-stage, it might be reasonable to fit a separate model for each stage. For example, we expect the growth dynamics of immature and mature females to be different as a result of allocation to reproduction in the latter, so our model should reflect this. We might do this by including interaction terms in the models (e.g., age-size), or if we have enough data, by fitting a different regression for each stage.

Fitting a separate model for each *age* is probably not a good strategy in long-lived species, since this would involve estimating many parameters. One solution is to work with aggregate age classes. For example, we might classify female Soays as new recruits, yearlings, prime age individuals, and older individuals. However, inferring the right set of age classes is difficult, especially if we are serious about guarding against over-fitting (King et al. 2006). A more parsimonious option is to assume a nonlinear smooth relationship between a target vital rate and age, and use regression models to estimate the relationship. In our experience, age-fate relationships often *are* smooth and only weakly nonlinear. However, this approach only works if the species is sufficiently long-lived to provide data across a range of ages. Thus, different approaches might be preferable in short- and long-lived species.

6.2.3 Implementing the model

The $F_a(z', z)$ and $P_a(z', z)$ kernels in (6.2.1) associated with the age-size case study are structurally identical to the component kernels from the ungulate IPM from Chapter 2. They are

$$F_a(z', z) = s(z, a)p_b(z, a)p_r(a)C_0(z', z)/2 \qquad (6.2.6a)$$
$$P_a(z', z) = s(z, a)G(z', z, a). \qquad (6.2.6b)$$

How do we implement this model in R? One option is to embed iteration matrices associated with each $F_a(z', z)$ and $P_a(z', z)$ into one big matrix. However, this is not efficient for computing because most elements of this matrix will be zero, reflecting the fact that transitions from ages a to $a + 1$ are, tragically, the only ones possible - none of us are getting any younger.

So for age-size models we suggest that you work with a set of age-specific matrices stored in a list, rather than putting everything into one big matrix. This is more efficient computationally, and it's easier and safer to code the sums in (6.2.1) as a loop. The integrals are all one-dimensional, and can be handled in the usual way by the midpoint rule. λ, R_0, etc. can all be computed by

```
# Function to calculate the ''integration'' parameters.
mk_intpar <- function(m, L, U, M) {
  h <- (U - L) / m
  meshpts <- L + ((1:m) - 1/2) * h
  na <- M + 2
  return( list(meshpts = meshpts, M = M, na = na, h = h, m = m) )
}


# Function to calculate lists of iteration matrices.
mk_age_IPM <- function(i.par, m.par) {
  within(i.par, {
    F <- P <- list()
    for (ia in seq_len(na)) {
      F[[ia]] <- outer(meshpts, meshpts, F_z1z, a = ia-1, m.par = m.par)*h
      P[[ia]] <- outer(meshpts, meshpts, P_z1z, a = ia-1, m.par = m.par)*h
    }
    rm(ia)
  })
}
```

The within function is used to allow us to access elements of i.par directly and
store the F and P kernels in a list, along with the elements of the original i.par
list object.

Box 6.1: R functions used to construct an age-size structured IPM.

iteration, and when maximum speed is needed (e.g., to bootstrap an estimate
of λ) it's straightforward to use multiple cores (mclapply, parLapply or a dopar
loop) because the calculations for each age are independent of each other.

Box 6.1 shows how to build a set of age-size IPM kernels in R, assuming the
presence of a "greybeard" class. We defined two functions. The first, mk_intpar,
calculates the quantities needed to construct the iteration matrices and keep
track of age classes, returning these in a list object. We could do these calculations in the global workspace. The advantage of wrapping them up as a function
is that we can update everything we need at once, if for example, we decide to
change the number of mesh points or age classes. Keeping the results of these
calculations in a list also means that we only have to pass a single output (the
integration parameters i.par, in our script) to the functions we use to build and
analyze the model. When implementing more complicated IPMs it is safer and
easier to work like this.

The second function in Box 6.1, mk_age_IPM, iterates over the set of age classes,
building an iteration matrix for each F_a and P_a using the midpoint rule, storing each set in a list, and then returning these in another list along with the
implementation parameters. In order to use this we need to define R functions
that implement the component kernels and the underlying demographic func-

tions. Other than becoming functions of z and a, these are no different from those used to build the simple ungulate case study. For example, the R function implementing the survival-growth kernel and the survival function are:

```
P_z1z <- function (z1, z, a, m.par) {
  return( s_z(z, a, m.par) * g_z1z(z1, z, a, m.par) )
}


s_z <- function(z, a, m.par){
  linear.p <- m.par["surv.int"] + m.par["surv.z"]*z + m.par["surv.a"]*a
  p <- 1/(1 + exp(-linear.p))
  return(p)
}
```

We won't step through every function; all the kernel functions and the demographic functions are found in Ungulate Age Calculations.R and Ungulate Age Demog Funs.R, respectively.

A simple analysis of the IPM might begin by using iteration to compute λ, $w(z)$, and $v(z)$. Box 6.2 contains two functions to help us do this. The first, r_iter, implements (6.2.1) and (6.2.2). It takes a list of age-specific size distributions, reproduction kernels and survival kernels, and uses these to calculate the age-size distribution after one iteration of the model. The "r" in the name of this function refers to the fact that this is analogous "right" multiplication of a simple kernel. Not surprisingly, the second function, l_iter, implements (6.2.3) and (6.2.4); it performs the operation that is analogous to "left" multiplication of simple kernels. Both functions assume that the age-size distribution is stored as a list of vectors containing the age-specific size distributions.

To illustrate how we put all this together, let's calculate the population growth rate under the true and estimated parameter sets. We need to first set up the integration parameters by passing the number of mesh points, $m = 100$, the maximum age before entering "greybeard" age-class, $M = 20$, and the lower and upper bounds of the integration domain, $L = 1.6$ and $U = 3.7$, to the mk_intpar function. We then pass this output along with the true model parameters to the mk_age_IPM to build the IPM:

```
i.par <- mk_intpar(m = 100, M = 20, L = 1.6, U = 3.7)
IPM.sys <- mk_age_IPM(i.par, m.par.true)
```

To calculate λ we need to iterate the model until it converged on the stable age-size distribution. We found by trial and error that 100 iterations are sufficient when we start with a population comprised of only new recruits. The R code to set up the initial distribution of states is:

```
nt0 <- with(i.par, lapply(seq_len(na), function(ia) rep(0, m)))
nt0[[1]] <- with(i.par, rep(1 / m, m))
```

```
# Iterate the model one step using 'right' multiplication of the kernel.
r_iter <- function(x, na, F, P) {
  xnew <- list(0)
  for (ia in seq_len(na)) {
    xnew[[1]] <- (xnew[[1]] + F[[ia]]%*%x[[ia]])[,,drop=TRUE]
  }
  for (ia in seq_len(na-1)) {
    xnew[[ia+1]] <- (P[[ia]]%*%x[[ia]])[,,drop=TRUE]
  }
  xnew[[na]] <- xnew[[na]] + (P[[na]]%*%x[[na]])[,,drop=TRUE]
  return(xnew)
}

# Iterate the model one step using 'left' multiplication of the kernel.
l_iter <- function(x, na, F, P) {
  xnew <- list(0)
  for (ia in seq_len(na-1)) {
    xnew[[ia]] <- (x[[ia+1]]%*%P[[ia]] + x[[1]]%*%F[[ia]])[,,drop=TRUE]
  }
  xnew[[na]] <- (x[[na]]%*%P[[na]] + x[[1]]%*%F[[na]])[,,drop=TRUE]
  return(xnew)
}
```

Box 6.2: R functions used to implement equations (6.2.1–6.2.4).

To iterate the model 100 times, starting with this initial distribution, we use:

```
IPM.sim <- with(IPM.sys, {
  x <- nt0
  for (i in seq_len(100)) {
    x1 <- r_iter(x, na, F, P)
    lam <- sum(unlist(x1))
    x <- lapply(x1, function(x) x / lam)
  }
  list(lambda = lam, x = x)
})
IPM.sim$lambda
```

The with function used here is just for convenience; it allows us to avoid having to either first copy the elements stored in IPM.sys or access them explicitly via the $ operator. Running this code, we find that $\lambda = 1.015$. When we repeat these calculations with m.par.est, the model returns an estimate of $\lambda = 1.013$. With a different random number seed, the results (e.g., your results) will be slightly different. But repeating the exercise of generating the artificial data by running the IBM, estimating IPM parameters, and calculating λ, shows that

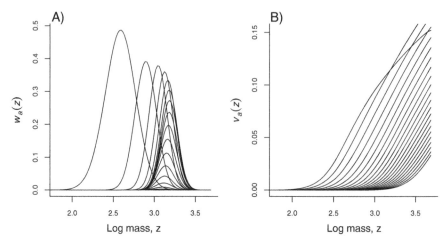

Fig. 6.1 A) Stable age-size structure. B) Reproductive value functions for each age-class. Source file: `Ungulate Age Calculations.R`

for this model 3000 observations from the IBM output are sufficient to arrive at a reasonably precise estimate of λ. Even with 1000 observations, the estimates are pretty good (1.0165 ± 0.016(standard deviation) in 250 replicates). This contrasts with the poorer results for the monocarp model in Figure 2.6 – data requirements are highly problem-dependent, and you won't know what they are until you do an exercise like Figure 2.6.

The code snippet used to calculate λ produces an estimate $w_a(z)$, stored in x. If we replace the `r_iter` function with `l_iter`, the same code can also be used to estimate $v_a(z)$. We'll spare you the details; the code is in `Ungulate Age Calculations.R`. Figures 6.1A and B show the results of these calculations. We can pick out a few features that arise as a result of age structure. The mean age-specific size reaches a maximum around age 5, and then declines slowly at later ages. This decline is driven by the negative age effect in the growth function. Ignoring the new recruits, we see that the size-specific reproductive value decreases steadily with age. This suggests that the decline in survival with age outweighs the improvement in reproductive performance that accompanies ageing.

Estimating λ, $w_a(z)$ and $v_a(z)$ as we have done here is not particularly efficient, because we have to use trial and error to decide how many iterations are sufficient for convergence. There is also much more we can do to understand the model. However, we will defer doing this until after we have reviewed the theory and numerical implementation of general deterministic IPMs.

6.3 Specifying a general IPM

We now describe a general density-independent IPM for species with complex demography. The description here is abstract, but in Section 6.4 we present a set of specific examples.

The space of individual states $\mathbf{Z}$ can include a set of discrete points $\mathbf{D} = \{z_1, \cdots z_D\}$ and a set of continuous domains $\mathbf{C} = \{\mathbf{Z}_{D+1}, \mathbf{Z}_{D+2}, \cdots \mathbf{Z}_{D+C}\}$. Each continuous domain is either a closed interval $[L_j, U_j]$ or a closed finite rectangle in d-dimensional space. Each set in $\mathbf{D}$ or $\mathbf{C}$ will be called a *component* and denoted $\mathbf{Z}_j, j = 1, 2, \cdots, M = D + C$. For example, $\mathbf{Z}_1$ and $\mathbf{Z}_2$ might represent different sexes or two different color morphs, within which individuals are classified by size.

The population state $n(z, t)$ also consists of a set of M components: discrete values $n_j(t), j = 1, 2, \cdots, D$ giving the total number of individuals in each discrete component, and continuous functions $n_j(z_j, t), j = D + 1, D + 2, \cdots, M$ giving the state distribution on the continuous components such that $\int_{\mathbf{Z}_j} n_j(z_j, t) \, dz_j$ is the total number of individuals in component j.

Transitions within and among components are described by a set of kernel components $K_{ij}, 1 \leq i, j \leq M$ with $K_{ij} \neq 0$ whenever individuals in $\mathbf{Z}_j$ at time t contribute to the population in $\mathbf{Z}_i$ at time $t + 1$. There are four different possible kinds of kernel components:

1. Discrete to Discrete: K_{ij} is a number, such as the fraction of diapausing *Daphnia* resting eggs that survive to the next year and remain in diapause.
2. Discrete to Continuous: $K_{ij} = k_{ij}(z')$, a state distribution that is the same for all individuals in $\mathbf{Z}_j$, such as the size distribution of seedlings produced by dormant seeds that germinate. Then $\int_{\mathbf{Z}_i} k_{ij}(z') \, dz'$ is the total number of seedlings at time $t + 1$ per dormant seed at time t.
3. Continuous to Discrete: $K_{ij} = k_{ij}(z)$, the state-dependent per-capita contribution of individuals in component j at time t to component i at time $t + 1$, such as the number of resting eggs produced by a *Daphnia* of age j and size z. Then $\int_{\mathbf{Z}_j} k_{ij}(z) n_j(z, t) dz$ is the total number of resting eggs produced by age-j parents.
4. Continuous to Continuous: $K_{ij}(z', z)$ is a genuinely bivariate function like the smooth kernels for the basic size-structured IPM in Chapter 2.

Many kernel components will be the sum of survival and reproduction kernels, $K_{ij} = P_{ij} + F_{ij}$. We assume that all of these are piecewise continuous. All types of kernel components can be made density dependent, or stochastic, but the theory in this chapter for density-independent, deterministic IPMs no longer will apply.

The only really restrictive assumptions in the general model are (i) that there are finitely many domains (rather than, for example, ages $0, 1, 2, \cdots$, without limit) and (ii) that all of the continuous domains are bounded. IPMs with these restrictions share many properties with matrix projection models, as we have seen in previous chapters. But an IPM with an unbounded domain can behave very differently, more like a spatial model on an infinite spatial domain (Chapter 8). Instead of converging to a stable state distribution, the population could spread forever as a traveling wave in "trait space."

But there is nothing biologically restrictive about bounded domains. Real traits in real populations have bounded distributions, and a realistic model should not produce individuals very different from any individuals that are

actually observed. So it should always be possible to find a finite range for each
trait that is large enough to prevent eviction, and not much larger than the
range of observed trait values.

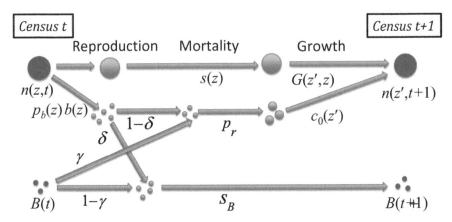

Fig. 6.2 Life cycle diagram for the monocarpic perennial model with a persistent seed
bank and pre-reproductive census. In the absence of a seed bank, this would be the
same as Figure 2.2A. Large and small circles represent established individuals and new
recruits, as in Figure 2.2A; tiny circles represent seeds. The two new parameters are γ,
the fraction of old seeds that germinate out of the seed bank, and δ, the fraction of new
seeds that go down into the seed bank.

6.4 Examples of the general deterministic IPM

6.4.1 Seeds and plants

Suppose that some seeds of our generic monocarpic perennial do not germinate
at their first opportunity, but instead remain dormant and enter a persistent
seed bank (e.g., Rees et al. 2006; Metcalf et al. 2008; Jongejans et al. 2011;
Eager et al. 2013). The simplest possible assumption is that all seeds within the
seed bank are demographically equivalent (despite differences in age, depth of
burial, and so on). To model that case, we just need to add a new variable $B(t)$
representing the number of seed bank seeds in year t, and add kernels for the
seed-to-rosette and rosette-to-seed transitions.

We have not encountered this kind of life cycle in previous chapters, so it's
best to start by drawing a life cycle diagram depicting our biological assumptions
and the timing of events (Figure 6.2). We distinguish between newly produced
seeds and older seeds that entered the seed bank in prior years. At the time t
census there are only older seeds and aboveground rosettes present. The census
is followed immediately by seed production, after which both new and older
seeds are present. Some of the seeds in each pool germinate (fraction γ of the
older seeds, and $1 - \delta$ of the newer seeds). These establish as rosettes with
probability p_r (we assume that p_r is the same for old and new seeds). As in the
original monocarp model, the size distribution of new recruits is independent
of parent size (another "all seeds are equal" assumption). Ungerminated seeds

remain in the seed bank and suffer mortality, with fraction s_B surviving until the next census.

Having drawn the diagram, it's now easy to write down the IPM. For the seed bank we have

$$B(t+1) = s_B \left[(1-\gamma)B(t) + \delta \int_L^U p_b(z)b(z)n(z,t)\,dz \right]. \qquad (6.4.1)$$

For rosettes there are now three inputs: survival of established rosettes, and new recruits from both new and old seeds:

$$n(z',t+1) = \int_L^U G(z',z)s(z)n(z,t)\,dz$$

$$+ c_0(z')p_r \left[(1-\delta)\int_L^U p_b(z)b(z)n(z,t)\,dz + \gamma B(t) \right]. \qquad (6.4.2)$$

Note, how the structure of the equations corresponds to the diagram, with the terms in square brackets representing two paths that merge into a common set of future demographic steps.

6.4.2 Susceptible and Infected

Bruno et al. (2011) constructed an IPM for the sea fan coral *Gorgonia ventalina* during a recent epizootic of the fungal pathogen *Aspergillus sydowii*. Demographic models were based on observations of ≈ 750 sea fans at three sites on the northern Belizean Barrier Reef in Mexico from 2002 to 2005. Fans were photographed annually, and the areas of uninfected tissue and of tissue with several levels of disease damage were quantified by image analysis software. The data analysis showed that future fates were affected by disease state (infected or uninfected) and by the area of uninfected tissue, but other covariates such as area of infected tissue or level of disease damage did not improve forecasts. The IPM therefore cross-classified individuals by size (continuous) and disease state (discrete: healthy or infected). Because sea fans have no adaptive immunity, past infection history was assumed not to matter for a currently disease-free individual.

The size measure was z =cube root of healthy tissue area, because that produced a growth model with linear mean and constant variance. The population state is described by $n_H(z,t)$ and $n_I(z,t)$, the size distributions for Healthy and Infected fans. Because larvae are widely dispersed, the model for a local population assumes a constant external input of recruits, which are uninfected because very small fans do not acquire the infection:

$$n_H(z', t+1) = \int_0^{20} P_{HH}(z', z) n_H(z, t) dz$$

$$+ \int_0^{20} P_{HI}(z', z) n_I(z, t) dz + \mathcal{R}c_0(z') \qquad (6.4.3a)$$

$$n_I(z', t+1) = \int_0^{20} P_{IH}(z', z) n_H(z, t) dz$$

$$+ \int_0^{20} P_{II}(z', z) n_I(z, t) dz \qquad (6.4.3b)$$

Here $\mathcal{R}$ is the recruitment rate and the P_{ij} are kernels describing survival and growth with initial disease state j and final disease state i. For example, the model for Healthy to Infected transitions was

$$P_{IH}(z', z) = \tau(z)(1 - m_H(z)) g_H(z'|z) \qquad (6.4.4)$$

where m_H is the probability of death for an uninfected individual, τ is the risk of infection, and g_H is the kernel describing changes in uninfected tissue area for initially uninfected individuals (fans that became infected did not appear to grow differently from ones that remained uninfected).

6.4.3 Time delays

Kuss et al. (2008) developed an IPM for the Alpine monocarpic perennial *Campanula thyrsoides* of the form

$$n(z', t+1) = \int_L^U P(z', z) n(z, t) \, dz + \int_L^U F(z', z) n(z, t-1) \, dz \qquad (6.4.5)$$

Fecundity is based on time $t-1$ because leaves mostly wither during the winter, and bolting occurs soon after snowmelt. Size in the previous year $(t-1)$ was therefore a better predictor of flowering than size in the year of flowering (t). The best predictor of flowering would be the size that the plant would reach in year t if it didn't flower, but that is an unobserved state (*sensu* Rueffler and Metz 2013) for flowering plants.

The individual state space $\mathbf{Z}$ is two copies of $[L, U]$, one for current size $n(z, t)$ and the other for lagged size $n_L(z, t) = n(z, t-1)$. $n(z, t+1)$ is given by the equation just above, and $n_L(z, t+1) = n(z, t)$. Schematically, the model can be written as

$$\begin{bmatrix} n(t+1) \\ n_L(t+1) \end{bmatrix} = \begin{bmatrix} P & F \\ I & 0 \end{bmatrix} \begin{bmatrix} n(t) \\ n_L(t) \end{bmatrix}, \qquad (6.4.6)$$

where I is the identity operator.

6.4.4 Individual quality and size

Individual quality refers to the net effect of unmeasured "latent" attributes, which result in persistent differences in performance between individuals whose measured attributes are very similar (Rees et al. 1999; Cam et al. 2002, 2013;

Chambert et al. 2013, 2014). Clark et al. (2004) refer to these as "random individual effects." It could be site quality (for a plant, or for an animal with a stable home range), lasting effects of events early in life, genotype or breeding value for a heritable trait affecting breeding success, immune function in a vertebrate, and so on. But the underlying cause is often unknown, and "quality" is just a name for observable differences with no observable cause.

In many cases quality variation is manifested as positive autocorrelation in growth or reproduction (e.g., Pfister and Stevens 2002, 2003; Haymes and Fox 2012). Individuals with rapid growth or high fecundity recently are likely to be "high quality" individuals who will continue to perform well in the future. If some individuals consistently grow fast for their size while others are consistently slow, a model that ignores these differences will under-predict the increasing variance in size that develops as some individuals steadily pull ahead while others fall behind. A positive correlation between parent and offspring performance (e.g., Chambert et al. 2014) suggests that observed variation is due to an unmeasured heritable trait – but it may also be that larger mothers have larger offspring. Negative autocorrelation can also occur, for example, if current growth or reproduction depletes unmeasured storage reserves (Ehrlén 2000).

To incorporate growth correlations Pfister and Stevens (2003) built an individual-based simulation model, but we can do the same in an IPM with individual state $z = (x, q)$ where x is a size measure and q is an individual "quality" variable. Pfister and Stevens (2002, 2003) modeled growth correlation by assuming that the intercept parameter in the linear growth model followed a first-order autoregressive model. So if q_t is an individual's quality at time t, its size at time $t+1$ is Gaussian with mean $a + q_t + bx_t$ and variance σ_X^2. Similarly, its quality at the next time step is Gaussian with mean ρq_t and variance σ_Q^2. If the changes in size and quality are independent, the survival kernel is

$$P(x', q', x, q) = s(x)\varphi(x'; a + bx + q, \sigma_X^2)\varphi(q'; \rho q, \sigma_Q^2) \qquad (6.4.7)$$

where $\varphi(x; \mu, \sigma^2)$ is the Gaussian density with mean μ and variance σ^2, and $s(x)$ is size-dependent survival. In R this is coded as

```
s(x) * dnorm(x1, a + b*x + q, sigmaX) * dnorm(q1, rho*q, sigmaQ)
```

assuming that $s(x)$ and the parameters `a`, `b`, `sigmaX`, `sigmaQ` have been defined.

The hard part with a quality variable is inheritance: how closely do offspring resemble parents? This requires multi-generational data with a known pedigree (e.g., Coulson et al. 2010) or mechanistic assumptions about what quality really is and how it is passed from generation to generation – site quality might be "heritable" (offspring resemble parent) even though it is not under genetic control. If offspring size and quality are independent, and independent of parent size and quality, the fecundity kernel has the form $F(x', q', x, q) = B(x, q)c_x(x')c_q(q')$ where B is state-specific fecundity of parents, and c_x, c_q are the probability densities for size and quality at birth. This might be a reasonable assumption if q reflects site quality, and offspring are dispersed completely at random to sites that differ in quality.

Metcalf et al. (2009a) developed models for 9 species of tropical trees with a discrete quality variable, light environment. Changes in each tree's light environment were assumed to follow a first-order Markov chain, with transition probabilities assumed to be independent of tree size. The transition probabilities for the light environment were estimated separately for each species, because they reflect species properties such as growth rate. Their model distinguished 6 illumination categories, fitting separate demographic functions for each species in each light environment. For simplicity we consider here only two environment states, Low and High light. The Markov chain for light environment is then specified by two parameters, p_{LL} and p_{HH}, where p_{ii} is the probability that a site in environment i at time t is still in environment i at time $t+1$.

Metcalf et al. (2009a) assumed that census in year t is followed by a potential change in light environment, then demographic processes in the (possibly different) time $t+1$ light environment. They also omitted fecundity, so that the model just follows a cohort of existing trees. The form of the model is then

$$n_L(z', t+1) = \int_L^U P_L(z', z) \left[p_{LL} n_L(z, t) + (1 - p_{HH}) n_H(z, t) \right] dz$$

$$n_H(z', t+1) = \int_L^U P_H(z', z) \left[(1 - p_{LL}) n_L(z, t) + p_{HH} n_H(z, t) \right] dz \ . \tag{6.4.8}$$

This can be written schematically in "megamatrix" form

$$\begin{bmatrix} n_L(t+1) \\ n_H(t+1) \end{bmatrix} = \begin{bmatrix} p_{LL} P_L & (1 - p_{HH}) P_L \\ (1 - p_{LL}) P_H & p_{HH} P_H \end{bmatrix} \begin{bmatrix} n_L(t+1) \\ n_H(t+1) \end{bmatrix}, \tag{6.4.9}$$

corresponding to equation (7) in Metcalf et al. (2009a).

Size-quality and other continuous-continuous IPMs lead to large iteration matrices. The iteration matrix for a prototype size-quality model in Section 6.6.2 has 16 million entries. But using methods that we explain in Section 6.6, it's possible to construct the matrix and compute its dominant eigenvalue and eigenvector in just a few seconds.

6.4.5 Stage structure with variable stage durations

de Valpine (2009) used an IPM to model stage-structured populations where individuals vary in the time required to complete each developmental stage. At birth each individual is assigned a vector of stage durations $\mathbf{s} = (s_1, s_2, \cdots, s_m)$. This means that the individual moves from stage 1 to stage 2 when (and if) it reaches age s_1; moves on to stage 3 when it reaches age $s_1 + s_2$; and so on. Birth and death rates are assumed to depend on stage and on age within stage (i.e., how long the individual has been in its current stage). Stage durations can be correlated, with some individuals moving quickly through all stages while others are slower.

Because **s** has dimension m, iterating the population dynamics would require m-dimensional integrals for each age, which is not feasible for m much above 2 using standard numerical integration methods. de Valpine (2009) showed that λ, v, and w can all be calculated by Monte Carlo integration with importance sampling, without ever iterating the population dynamics. Using this method, de Valpine (2009) showed that a model with the observed variance in maturation time substantially increased the estimated population growth rate for Mediterranean fruit fly in California, relative to an otherwise identical model with a constant immature stage duration (de Roos 2008). Early reproduction by rapidly maturing flies more than offsets the delayed reproduction by others (we should expect this to be true very generally because of the nonlinear relationship between r and generation time T in $r \approx \log(R_0)/T$). He also found that in a data-based model for an endangered cactus, vital rate sensitivities were strongly dependent on the assumed distribution of stage durations. Monte Carlo integration is still a very specialized technique for IPMs so we will not give the details here. But it is an important alternative to standard methods, and it may be essential for doing the high-dimensional integrals in models with many individual-level state variables.

6.5 Stable population growth

In previous chapters we have already used the main results about stable population growth, but we outline them here to emphasize that they apply to all of the models described in this chapter. We then outline the assumptions that are required for them to hold in the general IPM described above (Section 6.3). Precise mathematical details are in Section 6.9.

Stable population growth refers to the existence of a unique stable population distribution and long-term growth rate, to which a density-independent population converges from any initial composition:

$$\lim_{t \to \infty} n(z,t)/\lambda^t = Cw(z) \qquad (6.5.1)$$

where λ is the dominant eigenvalue of the kernel and $w(z) > 0$ is the corresponding right eigenvector. The value of the constant C depends on the initial population distribution, but under the assumptions summarized in Section 6.5.1, λ and w do not. The convergence in (6.5.1) is asymptotically exponential, meaning that the error is proportional to $(|\lambda_2|/\lambda)^t$ once initial transients have died out, where $|\lambda_2|$ is the maximum magnitude of nondominant eigenvalues. When $w(z)$ is normalized so that $\int_Z w(z)dz = 1$, it gives the long-run frequency distribution of individual state z within the population. Depending on the model this could be a univariate distribution of individual size; the joint distribution of size and age, size and light environment, or size and individual quality; or the joint size distributions for multiple species within a community.

Corresponding to λ there is also a dominant left eigenvector $v(z) \geq 0$, which is the relative reproductive value. The interpretation of v is exactly the same as in matrix projection models: it measures the relative contribution to future populations, as a function of an individual's current state.

We can also follow population growth on a per-generation basis. As described in Chapter 3, the matrix $R = F(I - P)^{-1}$ projects the population from one generation to the next. That is, if $m(z, k)$ is the distribution of state at first census for individuals in generation k, then Rm is the distribution for individuals in generation $k + 1$. When R satisfies the assumptions for stable population growth, its dominant eigenvalue is the per-generation growth-rate R_0, and its dominant eigenvector is the stable state distribution at birth for all individuals in a generation. It is important to note that the assumptions for stable population growth (Section 6.5.1) have to be checked separately for R, because they don't necessarily hold for R when they hold for K.

Evolutionary analyses of structured models often use the fact that $R_0 - 1$ and $\lambda - 1$ have the same sign. This makes intuitive sense: a population that grows from generation to generation can't be shrinking from year to year, and vice-versa. Under fairly general assumptions, it is also true for IPMs. It's hard to imagine that it could ever *not* be true when there is stable population growth, but should be checked.

6.5.1 Assumptions for stable population growth

Only one assumption is required for stable population growth in matrix population models: the projection matrix A has to be power-positive, meaning that all entries in A^m are positive for some $m > 0$ (Caswell 2001, Section 4.5, calls such a matrix primitive). Note that this is only possible if post-reproductive stages are removed from the model.

The analogous condition in our general IPM, also implying stable population growth, is that the kernel K be power-positive: for some $m > 0$, $K^m(z', z) > 0$ for all $z', z \in \mathbf{Z}$. Here K^m is the mth iterate of the kernel, as defined in Chapter 3: $K^1 = K$ and

$$K^{m+1}(z', z) = \int_{\mathbf{Z}} K(z', y) K^m(y, z) dy. \qquad (6.5.2)$$

When the generation-to-generation kernel $R = F(I - P)^{-1}$ is power-positive, R_0 is a simple, strictly dominant eigenvalue of R, and $R - 1$ and $\lambda - 1$ have the same sign (Ellner and Rees 2006, Appendix B).

Most published IPMs satisfy the power-positivity condition, and in such cases the rest of this section is irrelevant. But sometimes it isn't true, or it's hard to prove that it's true. It isn't true if $\mathbf{Z}$ contains even one state that it is literally impossible for an individual to reach. For example, suppose that the size range is $[0, 1]$, the growth kernel $G(z', z)$ is a beta distribution with mean and variance depending on z, offspring size is another beta distribution, and the shape parameters are all > 1. Then $K(0, z) = K(1, z) = 0$ for all z, and the same is true for any iterate of K. If the endpoints are removed so that $\mathbf{Z} = (0, 1)$ the kernel is now power-positive on $\mathbf{Z}$, but our assumptions are not satisfied because $(0, 1)$ is not a closed interval. This might not seem like a big problem, but it is. It's not hard to construct examples where the kernel is positive and therefore power-positive on $(0,1)$, but the population doesn't converge to a stable structure.[2]

It is therefore very useful that stable population growth also occurs for some kernels that are not power-positive. The mathematical condition goes like this: if there is a probability distribution $u(z)$ on $\mathbf{Z}$ such that for any $n(z,0)$ there are numbers $\alpha, \beta > 0$ depending on $n(z,0)$ such that $\alpha u(z) \leq n(z,m) \leq \beta u(z)$, then the m-step kernel K^m is said to be u-bounded (Krasnosel'skij et al. 1989).[3] Stable population growth occurs in our general IPM if some kernel iterate K^m is u-bounded (Krasnosel'skij et al. 1989; Ellner and Rees 2006).

The important thing about u-boundedness is that it holds under biologically interpretable assumptions. In many IPMs, the kernel is u-bounded as a consequence of *mixing at birth*, which means that the frequency distribution of offspring states is similar for all parents. Technically, writing $K = P + F$, there is mixing at birth if F is not identically 0 and there is a probability distribution c on $\mathbf{Z}$ such that for any population distribution n there are numbers A, B such that

$$Ac(z') \leq Fn(z') \leq Bc(z') \tag{6.5.3}$$

with $A > 0$ whenever $Fn > 0$. The "any population distribution" condition makes (6.5.4) hard to check, but a sufficient (and nearly equivalent) condition is that there are piecewise continuous functions $A(z), B(z)$ such that

$$A(z)c(z') \leq F(z',z) \leq B(z)c(z') \tag{6.5.4}$$

with $A(z) > 0$ whenever a state-z individual has positive fecundity (i.e., $F(z',z) > 0$ for some z').

Many published IPMs have exact mixing at birth (i.e., offspring state distribution is independent of parent state), simply because the data don't identify which parent produced which offspring. However, mixing at birth is likely to really hold in many species. For example, although the size of a parent plant affects the number of seeds produced, there is much less plasticity in seed size or quality. Maternal effects are often small compared to the effect of the environment where a seedling grows (Weiner et al. 1997). So if recruits are censused some time after recruitment, the distribution of recruit sizes will be similar for all parents.

Moreover, it's also sufficient if there is mixing by some age r. That means that instead of $F(z',z)$ in (6.5.4) we have the expected size distribution at age r of offspring born to state-z parents. So the only way mixing at birth can actually fail, is if two individuals who differ enough at birth have no chance of becoming similar over time. And when that's the case, we have every reason

---

[2] A simple example is to have "size" be an unbiased random walk on the real line with Gaussian steps, moved onto $(0,1)$ by the logistic transformation $z \to e^z/(1+e^z)$.

[3] The meaning of "any $n(z,0)$" depends on what kinds of functions we allow as population distributions, but this doesn't matter here because iteration of the kernel produces smooth population distributions; see Section 6.9.

to expect that instead of a unique stable structure, the population has two or more different stable structures depending on the initial structure.[4]

Mixing at birth is not quite enough, but the additional assumptions are also very mild and likely to hold in most data-based models. Using the methods in Chapter 3, P and F define l_a, the survival to age a, and f_a, the mean per-capita fecundity of age-a individuals, for a cohort of individuals with state distribution $c(z)$ at birth. The first additional assumption is that the Leslie matrix describing the cohort up to some age M, with l_{M+1} set to 0, must be power-positive. This will be true, for example, whenever $f_a > 0$ and $f_{a+1} > 0$ for some a. The second is that senescence is not too slow. The precise meaning of "not too slow" is that

$$P^m \leq \delta K^m \quad \text{for some } m > 0, 0 < \delta < 1. \tag{6.5.5}$$

P^m describes how a time-t individual contributes to the time $t + m$ population through their own survival and growth. K^m adds to that the individual's children, grandchildren, etc., at $t + m$. So for $\delta = 1$, equation (6.5.5) is true for all m. The only way it can fail is if the initial difference in state between parents and offspring cannot disappear completely over time. It always holds in an age-structured model with a finite maximum age at reproduction, because post-reproductives have to be removed and therefore $P^m \equiv 0$ for m larger than the maximum age at reproduction. In all models it's true wherever $K^m(z', z) = 0$, because $P^m(z', z)$ must also equal zero. And if there is stable population growth, the dominant eigenvalue of K is larger than that of P (Ellner and Rees 2006, Appendix B), so the difference between K^m and P^m will grow exponentially wherever $K^m(z', z) > 0$ and it will be easy to see numerically that (6.5.5) is true.

In summary, the simplest situation is when K^m is strictly positive for some m, which guarantees stable population growth. When that isn't true, stable population growth still occurs in models with mixing at birth (or mixing at some age r) and senescence that is not too slow. And when none of these are true, it's likely that the model has alternate stable structures.

Stable generation-to-generation growth depends on $R = F(I - P)^{-1}$ having the necessary properties for stable population growth. Again, the simplest situation is when R is power-positive. If not, mixing at birth is still sufficient. Nobody survives from one generation to the next, so R is analogous to the F kernel for an annual with $P = 0$ and the senescence condition (6.5.5) is satisfied. In Section 6.9.1 we show that if equation (6.5.3) holds, then some iterate of R satisfies has mixing at birth, which implies stable population growth. R_0 therefore exists as the dominant eigenvalue of R, and the dominant eigenvector gives the stable birth-state distribution of each new generation.

---

[4] If there is mixing at age $r > 0$ but not at birth, then the additional requirements in stated in the next paragraph must apply to K^{r+1} split up into "reproduction" and "survival" kernels $F_r = P^r F, P_r = K^{r+1} - F_r$.

6.5.2 Alternate stable states

Alternate stable states become possible when offspring state depends strongly on parental state. Evolutionary models are perhaps the clearest example. Suppose that individuals are cross-classified by size and diploid genotype (AA, Aa, or aa). Without mutation, an all-AA population will always be all-AA, and an all-aa population will always be all-aa. And it's not just those extreme cases that cause problems. If heterozygotes are inferior, and size doesn't matter much, a population dominated initially by AA will converge to all-AA, while a population dominated by aa will converge to all-aa. Diploid genetics always has nonlinear frequency dependence and the potential for multiple stable equilibria. In terms of our assumptions, the kernel is not power-positive (no amount of time is sufficient to go from an AA-only state to one with some Aa or aa genotypes), and there is no mixing at birth (an AA parent can't have aa offspring, and vice-versa). Populations that start far apart can follow diverging paths.

The same kind of behavior is harder to arrange with continuous structure. To see why, consider a color-structured population with color z ranging from blue to yellow with green in the middle (represented by the interval $[0, 1]$). Suppose offspring of a color-z parent are uniformly distributed on $[z - 0.1, z + 0.1]$ truncated to lie within $[0, 1]$. There is no mixing at birth, so if fitness is much higher near 0 and 1, you might expect the model to behave like a diploid model with heterozygote inferiority. But 20 steps of size 0.05 between parent and offspring let color go all the way from 0 to 1, so the transition kernel is power-positive unless some colors have exactly zero fitness. The result is a unique, stable color distribution. However, convergence to the stable distribution may be extremely slow, because it takes a long time for an all-blue or all-yellow population to cross the fitness valley and become a mix of blue and yellow individuals.

6.5.3 Time delay models

The time delay model needs some separate attention. Because an individual's lagged state at time $t + 1$ must equal its current state at time t, the delay model is not represented by a smooth kernel. This situation occurs whenever a continuous component of the state changes deterministically; for example, an individual's birth weight, and it's breeding value for a heritable trait, remain constant over an individual's lifetime (see Section 10.4). For the time delay model (6.4.6), the situation is rescued by looking at the second iterate which goes from t to $t + 2$,

$$\begin{bmatrix} P & F \\ I & 0 \end{bmatrix} \begin{bmatrix} P & F \\ I & 0 \end{bmatrix} = \begin{bmatrix} P^2 + F & PF \\ P & F \end{bmatrix}. \tag{6.5.6}$$

The kernels that make up the second iterate are all smooth (i.e., at least piecewise continuous), so long as P and F are. So the general theory in this chapter applies to the second iterate.

However, sensitivity analysis is complicated by the fact that I is not represented by a smooth kernel, so the usual theory doesn't apply to equation (6.4.6).

Sensitivity analysis for models with partially deterministic state dynamics is a gap in current theory. For now, the best approach is to compute sensitivities numerically by making small perturbations in P and F (or to parameters) and computing the effect on λ; see Appendix A of Ellner and Rees (2006) for suggestions on how to do this efficiently.

In addition, because fecundity is a two-step projection (from $t-1$ to $t+1$) while survival is a one-step projection, the total elasticities to F and P do not add up to 1. Instead, the elasticities to F have to be doubled for this to be true:

$$2 \int_L^U \int_L^U e_F(z',z)dz'dz + \int_L^U \int_L^U e_P(z',z)dz'dz = 1$$

(see appendix S5 of Kuss et al. (2008) for details).

6.6 Numerical implementation

We first address the general issue of computing eigenvalues and eigenvectors efficiently. We can't discuss implementation for all possible situations. In this section we consider the size-quality model (illustrating continuous×continuous classification) in this section, and in the next section we use the ungulate case study to show how the same ideas work in the size-age model (illustrating discrete×continuous classification). Between them, these two examples raise all of the main ideas. In this section, and our case studies, we always implement the models using midpoint rule. However, more efficient numerical integration methods may be needed when the state vector is high-dimensional or the kernel is spiky. In Section 6.8 we present several fairly simple methods that can handle many of these situations.

6.6.1 Computing eigenvalues and eigenvectors.

Often λ, w, and v have been obtained using functions that compute the complete set of eigenvalues and eigenvectors. And why not? It's so fast that you might as well. But when the iteration matrix is large it is much more efficient to compute just the dominant eigenvalue and eigenvectors by iteration. That is, choose an arbitrary initial distribution $n(z,0)$, iterate the model to compute $n(z,1), n(z,2), \cdots$, and at each time step rescale n to a total population size of 1. As time goes on, n will converge to w and the integral of Kn will converge to λ (Isaacson and Keller 1966, Section 4.2). If the IPM is represented by one big iteration matrix A, the domEig function defined in Box 6.3 will return λ and a vector proportional to w. domeig(t(A)) will return λ again and a vector proportional to v, because v is the dominant eigenvector of the transposed iteration matrix. This works for any density-independent IPM: iterate, rescale, repeat.

Another option is to use linear algebra routines that can compute just a few leading eigenvalues and eigenvectors. This can be done in R using ARPACK routines (Lehoucq et al. 1998; Dawson 2013) via the igraph library (www.igraph.org):

```
domEig <- function(A,tol=1e-8) {
    qmax <- 10*tol; lam <- 1;
    x <- rep(1, nrow(A))/nrow(A);
    while(qmax > tol) {
        x1 <- A%*%x;
        qmax <- sum(abs(x1-lam*x));
        lam <- sum(x1);
        x <- x1/lam;
    }
    return(list(lambda=lam,w=x/sum(x)))
}
```

Box 6.3: Function to compute the dominant eigenvalue and eigenvector of a matrix by iteration.

```
library(igraph)
matmul <- function(x, A) { A %*% x }
A <- matrix(runif(2500*2500),2500,2500); # random matrix for timings
result <- arpack(matmul, extra=A, options=list(n=nrow(A), nev=1))
lambda <- result$values[1]; w <- Re(results$vectors[,1]); w <- w/sum(w);
```

On the 2500×2500 random matrix above, arpack and domEig took under half a second, while eigen took nearly a minute on a 3.4 GHz desktop. The midpoint rule iteration matrix for a size-quality model, with 50 mesh points for each variable, would be of this size. Increasing to 75 mesh points each (roughly 32 million matrix entries), arpack and domEIg were still under 1 second, but R crashed when we tried to use eigen.

For generation-to-generation growth, we need the dominant eigenvalue and eigenvectors of $R = F(I - P)^{-1}$, so first we need to invert $(I - P)$. If P is too large, solve(I-P) is not feasible. But the inverse can also be calculated by iteration, using $(I - P)^{-1} = I + P + P^2 + P^3 + \cdots$. In either domEig or matmul, the A %*% x step can be replaced with a loop that uses the series expansion of P. After iteration matrices P, F have been built, the matmul below computes Rx approximately without inverting $(I - P)$:

```
PF <- list(P=P, F=F); # this example is in c6/R0withARPACK.R
matmul <- function(x, PF) {
    P <- PF$P; F <- PF$F;
    xtot <- newx <- x;
    for(j in 1:50) {newx <- P%*%newx; xtot <- xtot+newx}
    return(F%*%xtot);
}
result <- arpack(matmul, extra=PF, options=list(n=nrow(P), nev=1))
R0 <- Re(result$values[1]);
```

The j for-loop computes $Ix + Px + P^2x + \cdots P^{50}x$, so the returned vector approximates $F(I - P)^{-1}x$. Why stop at 50? The number of terms needed is actually case-specific, and can be chosen based on an upper bound on the relative error in R_0,

$$\varepsilon_m = (s_{max})^m / (1 - s_{max}) \qquad (6.6.1)$$

where s_{max} is the maximum survival probability. This bound is derived in Section 6.9.1. So if $s_{max} \leq 0.85$, the relative error in R_0 is certain to be less than 2.5% with 35 terms and less than 0.03% with 50 terms, and the actual error is probably much smaller.

6.6.2 Implementing a size-quality model

We assume here that quality is dynamic, with some probability of changing over an individual's lifetime, to illustrate the general approach for cross-classification by two continuous traits. These methods can also be used if quality is static, but the methods for discrete×continuous models may be more efficient.

With continuous size x and quality q, $\mathbf{Z}$ is a rectangle $\mathbf{Z} = [L_x, U_x] \times [L_q, U_q]$, and the IPM iteration is

$$n(x', q', t + 1) = \int_{L_q}^{U_q} \int_{L_x}^{U_x} K(x', q', x, q) \, dx \, dq \qquad (6.6.2)$$

To evaluate the double integral by midpoint rule, mesh points are defined for each variable:

$$x_i = L_x + (i - 0.5)h_x, \ i = 1, 2, \cdots, m_x, \quad h_x = (U_x - L_x)/m_x \quad (6.6.3a)$$
$$q_i = L_q + (i - 0.5)h_q, \ i = 1, 2, \cdots, m_q, \quad h_q = (U_q - L_q)/m_q \quad (6.6.3b)$$

The midpoint rule approximation to (6.6.2) is then

$$n(x_k, q_l, t + 1) = h_x h_q \sum_{i=1}^{m_x} \sum_{j=1}^{m_q} K(x_k, q_l, x_i, q_j) n(x_i, q_j, t) \qquad (6.6.4)$$

It is natural to think of $n(x_i, q_j)$ as the (i, j) entry in a "state matrix" where column j describes the size distribution of individuals with quality q_j. However, matrix languages such as R or Matlab will do the calculations much faster if they are implemented as a large iteration matrix multiplying a large state vector **n**. The state vector is created by stacking the columns of the state matrix into a long vector: column 1 first, column 2 below it, and so on. This is the "megamatrix" layout. The first m_x states in **n** are every size at quality level q_1. The next m_x states are every size at quality q_2, and so on. The projection matrix is structured as an $m_q \times m_q$ matrix of "blocks" (each of size $m_x \times m_x$) in which q is constant. Blocks along the diagonal of the matrix represent transitions in which the q stays the same.

Going from the matrix n to the vector $\mathbf{n}$ is called the "vec" operation: $\mathbf{n} = vec(n)$. In R, *vec* can be done with the `matrix` function,

```
vec <- function(nmat) matrix(nmat,ncol=1);
```

The inverse of vec is also done with `matrix`, in fact it *is* `matrix`. Try

```
a <- matrix(1:12,4,3); matrix(vec(a),nrow=4)-a;
```

Note that `vec` works automatically, but to invert it with `matrix` you have to specify the number of rows of the original matrix or the number of columns.

Next, we need to create the iteration matrix that acts correctly on $\mathbf{n}$. $n(x_i, q_j)$ sits in $\mathbf{n}$ at location

$$\eta(i,j) = (j-1)m_x + i. \tag{6.6.5}$$

So to calculate (6.6.4), we let $\mathbf{A}$ be the matrix whose entry in location $(\eta(k,l), \eta(i,j))$ is $h_x h_q K(x_k, q_l, x_i, q_j)$. The midpoint rule iteration is then

$$\mathbf{n}(t+1) = \mathbf{A}\mathbf{n}(t) \tag{6.6.6}$$

The only tricky part is to construct $\mathbf{A}$ efficiently enough when this matters (e.g., a stochastic model that has a new $\mathbf{A}_t$ each year). Two speedups are illustrated in Box 6.4: compute all η values at once, and vectorize the kernel calculation along its longer axis (x in this case). The kernel is an artificial example in which quality affects growth, size affects survival, and offspring size, and quality is a linear autoregression with autocorrelation $\rho = 1/2$. Despite the large iteration matrix (16 million entries) it only takes ≈ 15 seconds (on a 3.4 GHz desktop) to build the matrix and compute λ, v, and w using `domEig`.

To see w and v as functions of x and q we need to first un-vec them:

```
repro.val=matrix(v,mx,mq); stable.state=matrix(w,mx,mq)
```

This gives us a two-dimensional state matrix, where each column gives the x distribution for one value of q. Then we can plot (see Figure 6.3) the stable size distribution by summing each row (and summing each column gives the stable quality distribution):

```
source("../Utilities/Standard Graphical Pars.R")
set_graph_pars("panel4");
plot(yx,apply(stable.state,1,sum),xlab="Size x",ylab="frequency",type="l")
```

Next, the bivariate state distribution and reproductive value function:

```
matrix.image(stable.state,yq,yx,xlab="Quality q",ylab="Size x",
             do.legend=FALSE)
title(main="Size-quality distribution")
matrix.image(repro.val,yq,yx,xlab="Quality q",ylab="Size x",
             do.legend=FALSE)
title(main="Reproductive value")
```

The drop in the computed reproductive value for very large, high-quality individuals is an artifact of eviction. For anything but this plot it's harmless, because such individuals are never produced by the model.

```
# Demographic process functions
g_x <- function(xp,x,q) dnorm(xp,m=1 + q + 0.7*x,sd=0.3) # Growth
g_q <- function(qp,q) dnorm(qp,m=0.5*q,sd=0.5) # Quality dynamics
s_x <- function(x) exp(x-1)/(1+exp(x-1)) # Survival
c_x <- function(xp,x) dnorm(xp,mean=0.8+x/5,sd=0.35) # Offspring size
c_q <- function(qp) dnorm(qp,mean=0,sd=0.5) # Offspring quality

# Kernel functions
p_xq <- function(xp,qp,x,q) s_x(x)*g_x(xp,x,q)*g_q(qp,q)
f_xq <- function(xp,qp,x,q) 0.75*s_x(x)*c_x(xp,x)*c_q(qp)
k_xq <- function(xp,qp,x,q) p_xq(xp,qp,x,q)+f_xq(xp,qp,x,q)

# Compute meshpoints
mx <- 80; mq <- 50;
Lx <- (-1); Ux<- 7; Lq <- (-2.5); Uq <- (2.5);
hx <- (Ux-Lx)/mx; yx <- Lx + hx*((1:mx)-0.5);
hq <- (Uq-Lq)/mq; yq <- Lq + hq*((1:mq)-0.5);

# Function eta to put kernel values into place in A
eta_ij <- function(i,j,mx) {(j-1)*mx+i}
Eta <- outer(1:mx,1:mq,eta_ij,mx=mx); # Eta[i,j]=eta(i,j)

# Create the A matrix and kernel array
A=matrix(0,mx*mq,mx*mq); Kvals=array(0,c(mx,mq,mx,mq));
for(i in 1:mx){
  for(j in 1:mq){
    for(k in 1:mq){
      kvals=k_xq(yx,yq[k],yx[i],yq[j]) #vector of values
      A[Eta[,k],Eta[i,j]]=kvals
      Kvals[,k,i,j]=kvals
}}}
hxq <- hx*hq; A <- hxq*A;

# Lambda, v and w
out<-domEig(A); out2=domEig(t(A)); lam.stable=out$lambda;
w <- Re(matrix(out$w,mx,mq)); w <- w/(hx*hq*sum(w));
v <- Re(matrix(out2$w,mx,mq)); v <- v/sum(v);
```

Box 6.4: Building the iteration matrix A for a size-quality model.

To compute sensitivities and elasticities we again "forget" that the model is two-dimensional:

```
v.dot.w=sum(hx*hq*stable.state*repro.val)
sens=outer(repro.val,stable.state)/v.dot.w
elas=sens*Kvals/lam.stable;
```

sens and elas are four-dimensional matrices and hard to visualize, so we compute and plot the total elasticity from all transitions from a given (x, q) combination to anywhere else:

```
total.elas=hx*hq*apply(elas,c(3,4),sum);
matrix.image(total.elas,yq,yx,xlab="Quality q",ylab="Size x",
    do.legend=FALSE,main="Total Elasticity");
```

The result (Figure 6.3D) is similar to the stable state distribution, because a transition only matters if somebody makes that transition, but it is pulled a bit towards larger and higher quality individuals.

The method in Box 6.4 assumes that the kernel function k_xq vectorizes in its first argument. The construction is *much* simpler if k_xq vectorizes in all

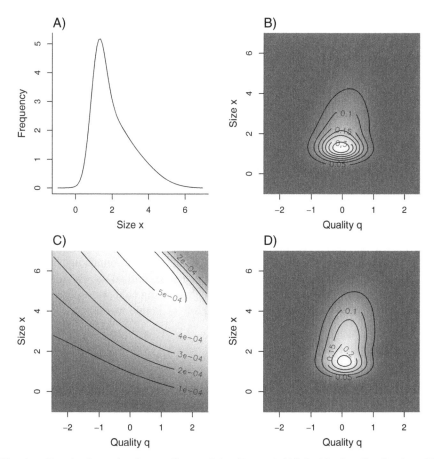

Fig. 6.3 Results from the size-quality model in Box 6.4. (A) Stable size distribution. (B) Stable bivariate distribution of size and quality. (C) Reproductive value function. (D) Total elasticity function. "Total elasticity" is the integral of $e(x', q', x, q)$ over (x', q'), plotted as a function of (x, q). Source file: SizeQualityExample.R

its arguments. This means that if v1,v2,v3,v4 are vectors of equal length then k_xq(v1,v2,v3,v4) returns a vector whose i^{th} entry is

     k_xq(v1[i],v2[i],v3[i],v4[i]).

The key is the expand.grid function, which behaves in exactly the right way to construct IPMs with two continuous state variables. To see this, imagine that the mesh points are yx=c(1,2) for x and yq=c(5,6) for q. Then

Kd <- expand.grid(xp=yx, qp=yq, x=yx, q=yq) produces

```
   xp qp x q
1   1  5 1 5
2   2  5 1 5
3   1  6 1 5
4   2  6 1 5
5   1  5 2 5
6   2  5 2 5
7   1  6 2 5
8   2  6 2 5
9   1  5 1 6
10  2  5 1 6
11  1  6 1 6
12  2  6 1 6
13  1  5 2 6
14  2  5 2 6
15  1  6 2 6
16  2  6 2 6
```

The matrix Kd created by expand.grid has named columns, and its rows are all the possible combinations of mesh points for the final state (x', q') and initial state (x, q). Moreover, the matrix starts with all possible (x', q') combinations paired with the first combination for (x, q) (in the example above, this is the first 4 rows). This is followed by all possible (x', q') combinations paired with the second combination for (x, q) (the next 4 rows), and so on. Better yet, combinations occur in exactly the "megamatrix" ordering that we used to construct the state vector. As a result, the following bit of magic works:

```
A <- with(Kd, k_xq(xp, qp, x, q))
dim(A) <- c(mx * mq, mx * mq)
A <- hx*hq*A
```

That's it: no eta, no loops, no thinking. Try to stay calm. Here's why it works. The first line applies the kernel function to all mesh point combinations. The second line turns A from a vector into a matrix of the right size. Because R "fills in" matrices column-by-column, the first column of the A holds the kernel values for the first (x, q) mesh point combination, and all possible mesh points for (x', q') – exactly what ought to be there. The second column then gets the second (x, q) combination and (x', q'), and so on, exactly as needed. The third line scales the matrix for iteration by midpoint rule. Going back to a 4-dimensional iteration array is again just a matter of re-setting dim(A) <- c(mx,mq,mx,mq).

Many (but not all) IPM kernels in the literature vectorize in the way required for the expand.grid construction. This approach can be much faster than the Box 6.4 method, but it also can be slower or need a prohibitive amount of memory.

6.7 Case Study 2D: Age-size structured ungulate, further calculations

We now return to the ungulate example. We demonstrate how to compute eigenvalues and eigenvectors using the ideas in Section 6.6.1. We then show how to calculate some basic demographic measures and explore the consequences of ignoring age-dependence in the vital rates. We pick things up where we left them at the end of Section 6.2, which means we assume functions such as r_iter are defined and a working IPM model is stored in IPM.sys.

6.7.1 Population growth rate

In Section 6.2.3 we calculated the dominant eigenvalue and stable age-size distribution of the ungulate model by iteration. Using trial and error we found that 100 iterations are sufficient to arrive at reasonably precise estimates of λ and $w_a(z)$. This approach is a little *ad hoc* as we have to determine the number of iterations to use on a case-by-case basis. We can improve on this by constructing a version of the domEig function from Box 6.3 which is adapted to our age-size model: the domEig_r function in Box 6.5. This constructs an initial age-size distribution, then iterates the model, rescales the age-size distribution, and repeats until the state distribution converges to the required tolerance. domEig_r does "right" multiplication using the r_iter function we defined in Box 6.2, returning $w_a(z)$ and λ. We can also construct a similar function to do "left" multiplication (domEig_l, not shown in Box 6.5) to calculate $v_a(z)$ and λ. Both domEig_r and domEig_l can be found in Ungulate Age Calculations.R.

Alternatively, we can use the linear algebra routines introduced earlier to compute the leading eigenvalue and eigenvectors (right or left) of an age-size IPM. Sticking with the ARPACK routines, we need to define an R function to iterate the model. We have already done this with r_iter. However, the arpack function expects the age-specific size distributions to be stacked into a single long vector, so we have to wrap r_iter in a function that converts a stacked state vector to the list form our model uses, iterates the model one step, and then converts the new list of age-specific size distributions to the stacked form:

```
matmul.r <- function(x.vec, IPM.sys) {
  with(IPM.sys, {
    x.list <- list()
    for (ia in seq_len(na)) x.list[[ia]] <- x[seq.int(ia*m-m+1, ia*m)]
    unlist(r_iter(x.list, na, F, P))
  })
}
```

We then pass this function to arpack along with IPM.sys to compute the eigenvalue and eigenvector:

```
domEig_r <- function(IPM.sys, tol) {
  with(IPM.sys, {
    # initial state
    wt <- lapply(seq_len(na), function(ia) rep(0, m))
    wt[[1]] <- rep(1 / m, m)
    # iterate until converges in lambda
    lam <- 1; qmax <- 10*tol
    while (qmax > tol) {
      wt1 <- r_iter(wt, na, F, P)
      qmax <- sum(abs(unlist(wt1) - lam * unlist(wt)))
      lam <- sum(unlist(wt1));
      wt <- lapply(wt1, function(x) x/lam)
    }
    list(lambda = lam, wz = wt)
  })
}
```

Box 6.5: R function to compute the dominant eigenvalue and eigen-vector of an age-size structured IPM using iteration.

```
eigs.r <- arpack(matmul.r, extra=IPM.sys,
                 complex=FALSE, options=list(n=with(IPM.sys, m*na), nev=1))
Re(eigs.r$val[1])
```

So which approach is best, iteration or ARPACK? In this example arpack takes about twice as long as domEig_r with 100 mesh points and 22 age classes, so there isn't really much between them. They are both fast, taking only a fraction of a second to run on a 4 Ghz workstation.

Are there better alternatives? In an early application Childs et al. (2003) implemented their age-size IPM using a single block matrix, where the $F_a(z', z)$ and $P_a(z', z)$ were stored as submatrices ('blocks') arranged so that they resembled a Leslie matrix, with the remaining blocks containing zero elements. This implementation is analogous to the age-stage matrix model originally proposed by Goodman (1969). The matrix was stored as standard matrix object and eigen was used to compute the eigenvalues and eigenvectors. This implementation of the ungulate age-size model is performed by the IPM_list_to_mat function in Ungulate Age Calculations.R. The take home message is that the block matrix approach is very inefficient: the time taken to calculate the leading eigenvalue is nearly 3 orders of magnitude greater than using iteration with the "list form" version of the same model. This difference can be cut to about 1–2 orders of magnitude by switching to ARPACK. The remaining bottleneck occurs because the matrix has approximately 5 million elements. Since most of these are zero, a further speedup may be achieved by using sparse matrix routines, though we have not tested this assertion.

In general, when working with age-size IPMs we recommend storing the size distribution associated with each discrete state in a list, and then writing functions to handle the one-step iteration required to compute eigenvalues and eigenvectors. It is hard to give concrete advice for general continuous-discrete classified IPMs. If the number of discrete states is large but only a small number of transitions among states are possible in a single step, then using the "list form" is probably best. Otherwise, storing all the kernels in one big matrix is probably simpler and will not incur significant efficiency losses. In either case, eigenvalues and eigenvectors should be calculated using iteration or efficient linear algebra routines such as ARPACK.

6.7.2 Other demographic measures

The long-run, generation-to-generation growth rate R_0 for an age-size IPM is calculated in essentially the same way as for a simple IPM. We need to add up the expected offspring production in each year of life from a cohort with size distribution $n_0(z)$ at birth. If there is a maximum age M beyond which no individuals survive, this is given by

$$(F_0 + F_1 P_0 + F_2 P_1 P_0 + F_3 P_2 P_1 P_0 + \cdots + F_M P_{M-1} \cdots P_0) n_0$$
$$= \left(F_0 + \sum_{a=1}^{M} F_a \prod_{m=0}^{a-1} P_m \right) n_0 = R n_0. \tag{6.7.1}$$

The R in this expression is thus the next generation kernel for an age-size IPM with a maximum age. If instead of a hard limit on age individuals beyond age M are lumped into a single "$M+1$ or older" class, then we need to include an additional term in the expression for R to account for the contributions from this class. This is

$$F_{M+1}(I + P_{M+1} + P_{M+1}^2 + \cdots) \prod_{m=0}^{M} P_m = F_{M+1}(I - P_{M+1})^{-1} \prod_{m=0}^{M} P_m. \tag{6.7.2}$$

R_0 is the dominant eigenvalue of the R kernel, and as usual we can calculate this in various ways. Box 6.6 contains the function matmulR that we can use with the age-size ungulate model and arpack. It takes an initial cohort size vector and iterates it one generation. Rather than inverting $(I - P_{M+1})$, we have approximated the sum on the left-hand side of equation (6.7.2) by summing the first 100 terms. R_0 is calculated with:

```
R0.arpack <- arpack(matmulR, extra=IPM.sys, complex=FALSE,
            options=list(n=IPM.sys$m, nev=1))$val[1]
```

Running this code we find that $R_0 = 1.084$; the population increases by about 8% each generation.

```
matmulR <- function(x, IPM.sys, n.iter=100) {
  with(IPM.sys, {
    xtot <- rep(0, length(x))
    for (ia in seq_len(n.iter)) {
      if (ia < na) {Pnow <- P[[ia]]; Fnow <- F[[ia]]}
      else         {Pnow <- P[[na]]; Fnow <- F[[na]]}
      xtot <- Fnow %*% x + xtot; x <- Pnow %*% x
    }
    xtot
  })
}
```

Box 6.6: R function to iterate the next generation matrix of an age-size structured IPM.

It may be tempting to calculate R_0 by extracting the stable size distribution of age 0 individuals born in a given year, and using the matmulR function to iterate this cohort one generation ahead. The following R code does this:

```
c.init <- wz[[1]] / sum(wz[[1]])
sum(matmulR(c.init, IPM.sys))
```

This returns a value of 1.078, which is not exactly the same as the estimate we calculated above. The second estimate is in fact wrong. c.init above is the distribution of size at age 0 among all individuals born in a given year during stable population growth. The calculation above assumes that c.init is the dominant eigenvector of R, but the dominant eigenvalue of R is the distribution of size at age 0 among all individuals born in a given generation during stable population growth. These will only be the same as c.init if either $\lambda = 1$ or if offspring size is independent of parent size. Neither of these is true for the age-size ungulate model.

6.7.3 Consequences of age-structure

We generally recommend including a state variable in an IPM if the statistical analysis of the data suggest it influences vital rates. After all, why throw away information if you do not have to? However, including an extra state variable or two obviously increases the amount of effort we have to expend to construct and analyze the model. The obvious question is then, does it matter if we ignore a feature like age-dependence or size-dependence in the vital rates? A simple way to address this question is to construct two different models – one that includes the additional state variable and another that ignores it – and then compare their predictions.

We can use this strategy to explore the consequence of ignoring either age-structure or size-structure in the Soay case study. In Ungulate Age Calculations.R we refit the vital rate functions without including age as a covariate, and then

construct and analyze the resulting size-structured IPM using the appropriate
tools from Chapters 2 and 3. When we do this we find that the size-only IPM
predicts λ and R_0 to be 1.025 and 1.337, respectively (for the particular sample
in the simulations), whereas the age-size IPM gives 1.013 and 1.083, respectively. The λ values are very similar - the difference is well within the range of
sampling variability - but R_0 is estimated poorly by the size-only model. This
differences arise because the average lifespan is over-estimated if we ignore the
age-dependent decline in survival. We can see this difference very clearly if we
compute the mean age-specific survivorship ($\tilde{l}_a$) of each model using the calculations in Chapter 3 (Figure 6.4A). The l_a predicted by the size-only model
converges quickly to an age-independent asymptote, reflecting the convergence
of survivors to a stable size distribution (the dominant right eigenvector of P),
while the age-size model's l_a starts to decline steeply after age 5. So although
4- and 8-year-old females have very similar size distributions, the population
projections starting from hundred 4-year-olds and hundred 8-year-olds are very
different (Figure 6.4B). As a result, the size-only model badly mis-predicts generation time (Figure 6.4C) and likely other life cycle properties that depend on
age-specific vital rates.

There are also large effects of ignoring the fact that similar-age individuals
differ in size. For example, the 5th and 95th percentiles of the stable size distribution for 4-year-old prime age females are roughly 2.9 and 3.3. This difference
in size makes the difference between between 74% and 98% predicted survival
to age 5, using our "true" parameter values that were estimated from the actual
Soay data, a roughly 20-fold difference in mortality rates. This again affects life
cycle properties (Figure 6.4D), although the error is much smaller than ignoring
age.

6.8 Other ways to compute integrals

The models in this chapter sometimes lead to very large iteration matrices if
our default method, midpoint rule, is used. We generally try the midpoint rule
first because it is simple, robust, and usually efficient enough. Need to invert
a 250×250 matrix (for example, to compute R_0)? In R that takes under 0.1
seconds. Typing solve(A) takes a lot longer.

But if you take a size-structured IPM and add a second continuous variable
with 50 mesh points, the time needed to invert the 50×larger iteration matrix
is about $50^3 = 125{,}000$ times longer in R.[5] So sometimes it is essential to
implement an IPM in a different way that can get accurate results with a smaller
iteration matrix. The same issue arises in a one-dimensional IPM when there is
small variance in growth or offspring state, because the kernel then has sharp
spikes or ridges and midpoint rule needs a prohibitive number of mesh points.
If midpoint rule meets your IPM needs, you can skim this section (ignoring the
R code but reading the last two paragraphs), and come back when you need to.

Which integration method will give the best results depends on the situation.
Dawson and Emmett (see Dawson 2013) developed recommendations based on a

---

[5] Matrix inversion by solve.default uses DGESV from LAPACK, which uses $O(N^3)$
operations to invert an $N \times N$ matrix. Numerical experiments show that the run time
for qr.solve is also $O(N^3)$.

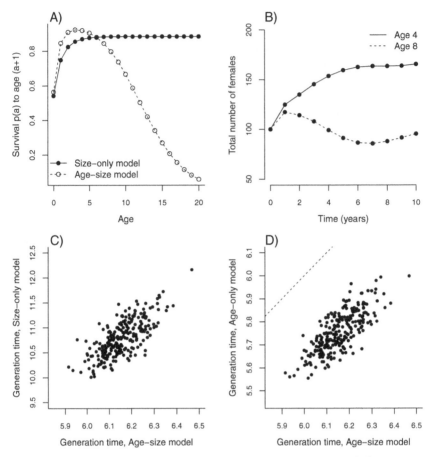

Fig. 6.4 Comparison of age-size and size-only ungulate models. (A) Survivorship l_a curves calculated by the methods in Chapter 3. (B) Population projections starting from hundred 4-year-old and hundred 8-year-old females. In both cases the initial populations had the stable size distribution conditional on age. (C, D) Comparison of estimated generation times T between the age-size model and the size-only (C) and age-only (D) models. The plotted points represent estimates from 250 samples of size 3000 from simulation of the age-size IBM. Source files Ungulate Age calculations.R and UngulateAgeSizeComps.R

suite of test kernels representing different demographic properties. If the kernel is smooth, there can be substantial benefits to using a higher-order integration method such as Gauss-Legendre quadrature ("quadrature" means numerical integration). Gauss-Legendre uses unevenly spaced mesh points, say z_i, and unequal weights W_i, chosen so that the N-point approximate integration

$$\int_a^b f(x)\,dx \approx \sum_{i=1}^{N} W_i f(z_i) \qquad (6.8.1)$$

is exactly correct whenever f is a polynomial of degree $2N + 1$ or lower.

Gauss-Legendre mesh points are densest near the endpoints of the range of integration, but that might not be where the IPM kernel is most nonlinear and hardest to integrate accurately. To get around this, *sub-interval methods* divide the range of integration into sub-intervals, and use medium-order quadrature on each sub-interval (Dawson and Emmett suggest 9-point Gauss-Legendre). The approximate integral is a sum like (6.8.1) but with more evenly spaced mesh points. Gauss-Legendre and sub-interval Gauss-Legendre can be orders of magnitude more accurate than the midpoint rule when the function being integrated is smooth and non-spiky (Figure 6.5A, B). In that example the sub-interval method looks like a loser, but it wins when the kernel has occasional jumps, and sub-intervals are placed so that all jumps are at sub-interval endpoints (Dawson 2013).

In the one-dimensional case, the IPM iteration is approximated as

$$n(z_i, t+1) = \sum_{j=1}^{N} W_j K(z_i, z_j) n(z_j, t). \qquad (6.8.2)$$

The right-hand side uses weights W_i because it represents the integral of $K(z', z)n(z, t)$ with respect to z. The iteration matrix $\mathbf{A}$ therefore has entries $A_{ij} = K(z_i, z_j)W_j$. As with midpoint rule, λ, v, w are the dominant eigenvalue and eigenvectors of the iteration matrix. Similarly, P and F are represented by matrices whose entries are $P_{ij} = P(z_i, z_j)W_j, F_{ij} = F(z_i, z_j)W_j$.

However, any formula involving an integral has to use the weights W_i. For example, for sensitivity analysis the inner product $\langle v, w \rangle$ is approximated as $\sum_{i=1}^{N} W_i v_i w_i$. So the calculations for a multivariate IPM, such as a size-quality model, have to get the W's in the right place. And this needs to be done in such a way that integrals can be computed without 4-dimensional loops (on current desktops a 3-dimensional loop like in Box 6.4 takes a few seconds, but 4-dimensional loops are painfully slow). Box 6.7, excerpted from GaussQuadSub.R and SizeQualityGaussQuad.R, illustrates Gauss-Legendre implementation of a size-quality model. Notice how weight matrices outer(Wx,Wq) for two-dimensional integrals, and W for four dimensions, are structured so that the sums approximating integrals are done without loops.

A spiky kernel presents a different challenge. Figure 6.5C, D illustrate that higher-order quadrature is only effective once mesh points are numerous and closely spaced. Until then, midpoint rule is more accurate - but comparing against Figure 6.5A, B (note the difference in x-axis scale), that's not saying much. For this situation, to get better accuracy with a small matrix, Dawson and Emmett recommend "bin-to-bin" methods. Bin-to-bin methods are like midpoint rule except that each kernel value is replaced by its average over the corresponding grid rectangle. In a basic size-structured IPM, this means that the (i, j) entry of the iteration matrix is

```
require(statmod); # for gauss.quad function
# Gauss-Legendre quadrature nodes and weights on (L,U)
gaussQuadInt <- function(L,U,order=7) {
    out <- gauss.quad(order); # nodes & weights on [-1,1]
    w <- out$weights; x <- out$nodes;
    return(list(weights=0.5*(U-L)*w, nodes=0.5*(U+L) + 0.5*(U-L)*x));
}


# Compute meshpoints and weights
mx <- 50; mq <- 30; Lx <- (-1); Ux<- 7; Lq <- (-2.5); Uq <- (2.5);
out=gaussQuadInt(Lx,Ux,mx); yx=out$nodes; Wx=out$weights;
out=gaussQuadInt(Lq,Uq,mq); yq=out$nodes; Wq=out$weights;


A=matrix(0,mx*mq,mx*mq); W=Kvals=array(0,c(mx,mq,mx,mq));
for(i in 1:mx){
    for(j in 1:mq){
      for(k in 1:mq){
         kvals=k_xq(yx,yq[k],yx[i],yq[j]);
         A[Eta[,k],Eta[i,j]]=kvals*Wx[i]*Wq[j]
         W[,k,i,j]=Wx*Wq[k]*Wx[i]*Wq[j] # weight for 4D integration
}}}
out<-domEig(A); out2<-domEig(t(A)); lam.stable=out$lambda;
w <- Re(matrix(out$w,mx,mq)); w <- w/sum(outer(Wx,Wq)*w);
v <- Re(matrix(out2$w,mx,mq)); v <- v/sum(v);


# Compute elasticity matrix
repro.val=matrix(v,mx,mq); stable.state=w;
v.dot.w=sum(outer(Wx,Wq)*stable.state*repro.val)
sens=outer(repro.val,stable.state)/v.dot.w
elas=sens*Kvals/lam.stable;


# One check that everything went where it belonged
cat("Integrated elasticity =",sum(W*elas)," and it should = 1","\n");


# Compute matrix of total(=integrated) elasticities
#  for all transitions (x_i,q_j) -> anywhere
total.elas=matrix(0,mx,mq);
for(i in 1:mx) {
for(j in 1:mq) {
    total.elas[i,j]=sum(elas[,,i,j]*outer(Wx,Wq))
}}
```

Box 6.7: Implementing a size-quality model with Gauss-Legendre quadrature.

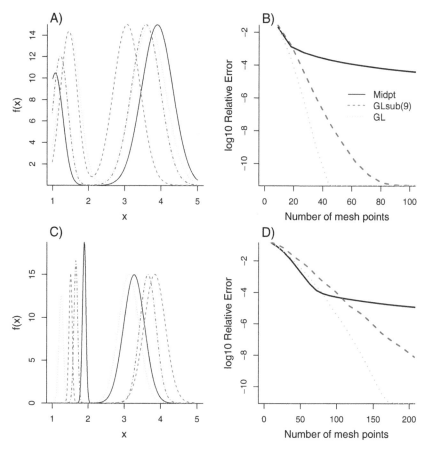

Fig. 6.5 Comparing the accuracy of quadrature methods. A) Four examples of randomly generated smooth functions with relatively broad peaks. B) Relative error as a function of the number of mesh points for numerical integration of the functions by midpoint rule (black solid), Gauss-Legendre (green dotted), and 9th order Gauss-Legendre on equal-size sub-intervals (red dash). The curves show the average relative error over 250 randomly generated functions. Panels C) and D) show the same for functions with sharper peaks. Source files: GaussQuadTest.R, GaussQuadSub.R

$$h^{-1} \int\limits_{z_j-h/2}^{z_j+h/2} \left(\int\limits_{z_i-h/2}^{z_i+h/2} K(z',z)dz' \right) dz \qquad (6.8.3)$$

instead of $hK(z_i, z_j)$. The factor h^{-1} is the result of two steps: dividing the double integral by h^2 gives the kernel average (over a square of area h^2), and this average is multiplied by h (exactly as in midpoint rule) to get iteration matrix entries. Similarly, in size-quality model the (i, j, k, l) iteration matrix entry is

$$(h_x h_q)^{-1} \underset{B_{ijkl}}{\iiiint} K(x', q', x, q) dx' \, dq' \, dx \, dq \qquad (6.8.4)$$

where B_{ijkl} is the 4-dimensional rectangle with sides (h_x, h_q, h_x, h_q) centered at (x_i, q_j, x_k, q_l). The integrals need to be evaluated by some fairly fast method, such as low-order Gauss-Legendre. Once the iteration matrix is constructed, it can be used exactly as if it had been constructed using midpoint rule.

To illustrate bin-to-bin we use a simplified version of a kernel parameterized by Zuidema et al. (2010) for the subtropical forest tree *Parashorea chinensis* in Vietnam. Our simplified kernel skips the "seedling" stage (dbh < 1 cm, for which Zuidema et al. (2010) used a four-stage matrix model). Instead, offspring are given the size distribution of new "saplings" (mean dbh=1.1, sd=0.4); we also rounded some parameter values.

$$\text{Growth: } z' \sim Norm\left(\mu = z + \frac{325 z^{1.26}}{(144 + z^{2.26}/42)^2}, \sigma = 0.1\right) \qquad (6.8.5a)$$

$$\text{Survival: } s(z) = 0.98 \qquad (6.8.5b)$$

$$\text{Fecundity: } b(z) = \frac{0.83 e^{-3.2+0.146z}}{1 + e^{-3.2+0.146z}} \text{ if } z > 10, \text{otherwise } 0 \qquad (6.8.5c)$$

$$\text{Offspring size: } c(z) \sim Norm(\mu = 1.1, \sigma = 0.4) \qquad (6.8.5d)$$

The problem with this kernel is that growth is very predictable ($\sigma = 0.1$) and very slow (< 1 cm/year) relative to the range of tree sizes (up to > 100 cm dbh).

Figure 6.6A shows the dominant eigenvalue λ as a function of iteration matrix size using midpoint and bin-to-bin. At the smallest size (100 mesh points) the dominant eigenvector of the midpoint rule iteration matrix is $(1, 0, 0, \cdots, 0)$. Trees in the smallest grid cell $[0, h]$ don't grow enough to leave that cell, but the sharp peak in the growth kernel produces a spurious increase in their numbers each year. This problem goes away with enough mesh points. Here that's feasible, but 600 mesh points for size would be unworkable in a size-quality model.

Bin-to-bin gives accurate results with a much smaller iteration matrix (Figure 6.6A). These calculations used 3-point Gauss-Legendre for integrating over initial size z, and 7-point Gauss-Legendre for integrating over final size z'. Initial and final size are treated differently because the main problem with spiky kernels is analogous to eviction. If $K(z', z_k)$ has a tall, narrow spike as a function of z', but there are no mesh points near the top of the spike, then the size-transitions or births that the spike represents are lost. And if there happens to be a mesh point near the top of the spike, the peak transition rate is applied to the whole cell containing the spike, giving a survival rate > 1. To avoid these errors, it pays to do the final size integral more accurately.

Functions are often available that make it easy to compute the "cumulative kernel" $\int\limits_{-\infty}^{b} K(z', z) dz'$. The integral over z' in (6.8.3) can then be computed quickly and accurately, using the fact that $\int_a^b f(z') dz' = \int_c^b f(z') dz' - \int_c^a f(z') dz'$

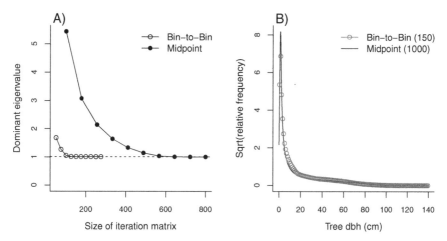

Fig. 6.6 Comparison of midpoint rule with bin-to-bin for the *Parashorea chinensis* IPM. The bin-to-bin matrix was constructed using 7-point Gauss-Legendre to average over destination cells (z'), and 3-point Gauss-Legendre to average over starting cells (z). A) Dominant eigenvalue of the iteration matrices as a function of the matrix size (number of rows, equal to the number of mesh points). B) Stable size distributions, from midpoint rule with 1000 mesh points and bin-to-bin with 150 mesh points. Note, that the distributions are plotted on square-root scale to emphasize the tail of larger individuals. Source file: TreeKernelB2B.R

for any a, b, c. The *P. chinensis* kernel is a good example. The growth distribution can be coded as

```
g_z1z <- function(z1,z) {
    mu <- z + 325*(z^1.3)/((144+(z^2.3)/42)^2);
    return(dnorm(z1,mean=mu,sd=0.1))
}
```

and the cumulative growth distribution is identical except that pnorm is substituted for dnorm:

```
intg_z1z <- function(z1,z) {
    mu <- z + 325*(z^1.3)/((144+(xz2.3)/42)^2);
    return(pnorm(z1,mean=mu,sd=0.1))
}
```

In Box 6.8 we show how to compute the cumulative P and F kernels for *P. chinensis* and then construct an iteration matrix by combining those with Gauss-Legendre integration over initial size. The script TreeKernel-intB2B.R redraws Figure 6.6 using the cumulative kernel instead of Gauss-Legendre for final size. With the cumulative kernel method, a 50×50 iteration matrix is already big enough to estimate λ with only about 2% relative error. The script SizeQuality-intB2B.R does the same for a size-quality model.

Figure 6.7 compares midpoint, Gauss-Legendre, and cumulative kernel bin-to-bin implementation of the *P. chinensis* kernel, as a function of the number of

```
require(statmod); source("domEig.R"); source("GaussQuadSub.R");
### Integral of Growth
intg_x1x <- function(x1,x) {
   mu <-  x + 325*(x^1.3)/((144+(x^2.3)/42)^2);
   return(pnorm(x1,mean=mu,sd=0.5))
}
### Survival
s_x <- function(x) 0.98
### Fecundity
b_x <- function(x){
    u = -3+ 0.15*x;
    return(ifelse(x>10, 0.83*exp(u)/(1+exp(u)), 0))
}
### Integral of offspring size
intc_x1 <- function(x1) pnorm(x1,mean=1.1,sd=0.35)

### Cumulative kernel functions
intp_x <- function(x1,x) s_x(x)*intg_x1x(x1,x)
intf_x <- function(x1,x) b_x(x)*s_x(x)*intc_x1(x1)
intk_x <- function(x1,x) intp_x(x1,x)+intf_x(x1,x)

###### Make iteration matrix using GL(order) for initial size
mk_K_int <- function(m,L,U,order) {
    h <- (U - L)/m; meshpts <- L + ((1:m) - 1/2) * h
    out <- gaussQuadInt(-h/2,h/2,order); nodes <- out$nodes;
        weights <- out$weights; F <- P <- matrix(0,m,m);
    for(i in 1:m){
    for(j in 1:m){
       fvals1=intf_x(meshpts[i]-h/2,meshpts[j]+out$nodes);
       fvals2=intf_x(meshpts[i]+h/2,meshpts[j]+out$nodes);
       F[i,j]=sum(weights*(fvals2-fvals1))
       pvals1=intp_x(meshpts[i]-h/2,meshpts[j]+out$nodes);
       pvals2=intp_x(meshpts[i]+h/2,meshpts[j]+out$nodes);
       P[i,j]=sum(weights*(pvals2-pvals1))
    }}
    P<- P/h; F<-F/h; K=P+F;
    return(list(K = K, meshpts = meshpts, P = P, F = F))
}
KB2B <- mk_K_int(150,L=0,U=140,order=3);
```

**Box 6.8: Constructing a cumulative bin-to-bin iteration matrix for *P.
chinensis* using Gauss-Legendre for initial size**, from TreeKernel-intB2B.R.
See SizeQuality-intB2B.R for analogous code for the size-quality model.

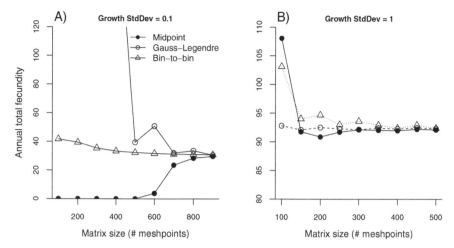

Fig. 6.7 Total fecundity of 1000 individuals at the stable size distribution for the simplified *P. chinensis* IPM. The curves show the values from three ways of implementing the model - midpoint rule, Gauss-Legendre quadrature, and bin-to-bin with cumulative kernels - as a function of the number of mesh points. In A) the conditional distribution of z' given z has standard deviation 0.1, as estimated by Zuidema et al. (2010); in B) the standard deviation is increased to 1. Source file: `ThreeWayTotalTreeF.R`

mesh points. For variety, the output is the estimated annual fecundity of 1000 individuals at the stable size distribution,

$$\int_0^{140} b(z)w(z)\,dz, \quad \text{with } Kw = \lambda w, \int_0^{140} w(z)\,dz = 1000.$$

When the growth distribution has `sd=0.1` as in the code above (Figure 6.7A), midpoint rule and Gauss-Legendre are very far off (in opposite directions) unless the iteration matrix is 700×700 or larger, but bin-to-bin does reasonably well at 200×200. When the kernel is less spiky (Figure 6.7B), Gauss-Legendre is the best with few mesh points, but midpoint is nearly as good. The variability in growth is still quite small compared to the population's size range, but it's enough that there is no payoff for bin-to-bin.

The loopless construction using `expand.grid` (Section 6.6.2) can also be used with bin-to-bin and cumulative bin-to-bin. The key idea is to add the quadrature nodes to the `expand.grid` call. Evaluating the kernel function on the matrix from `expand.grid` then gives all of the kernel values needed to compute the iteration matrix. This is illustrated in `SizeQualityB2B.R` for bin-to-bin and in `TreeKernel-intB2B.R` for cumulative bin-to-bin. Loopless is often much faster for bin-to-bin, but it uses more memory ($\approx$32 Gb for the size-quality example) and may be very slow on computers without enough memory.

In summary (echoing Dawson 2013), we recommend midpoint rule as the default. A manageable number of mesh points may be enough if you only compute the dominant eigenvalue/vector instead of all of them. If iteration matrix size is an issue, Gauss-Legendre or sub-interval Gauss-Legendre may help if the kernel is smooth or smooth with jumps. If the kernel has sharp spikes, we recommend bin-to-bin using Gauss-Legendre for the initial state, and cumulative kernel for final state whenever possible. Bin-to-bin requires more kernel evaluations, so midpoint rule or Gauss-Legendre are usually more accurate for equal run time on a smooth kernel (Dawson 2013). The payoff from bin-to-bin is that much smaller iteration matrices can be used for spiky kernels. All of these methods slow down exponentially as the dimension of the model increases; in Section 10.2 we describe some potential methods for high-dimensional models.

So overall we stand by "keep it simple": try midpoint, then bin-to-bin, and others as a last resort. Midpoint and bin-to-bin give an iteration matrix that can be used as if it were a big matrix model, with no quadrature weights to worry about. That may be worth a lot more than the computational speedup from higher-order methods.

6.9 Appendix: the details

The goal of this section is to state the precise assumptions and main results of the theory for general deterministic IPMs, and give an indication of the proofs. This requires more mathematics background than the rest of the book, specifically some familiarity with functional analysis. Dunford and Schwartz (1958) is the source for assertions below that are stated without proof.

We assume that $\mathbf{Z}$ is a compact metric space, hence it is a complete and separable Hausdorff space, Borel measure is defined, and the spaces $L_p(\mathbf{Z}), 1 \leq p < \infty$ are Banach spaces. We make $\mathbf{Z}^2 = \mathbf{Z} \times \mathbf{Z}$ a metric space using the product metric

$$d\left((z', z), (x', x)\right) = \sqrt{d(z', x')^2 + d(z, x)^2}.$$

The topology induced by this metric is the same as the product topology, so $\mathbf{Z} \times \mathbf{Z}$ is also a compact metric space.

The IPM is defined relative to a measure space $(\mathbf{Z}, \mathcal{B}, \mu)$ where $\mathcal{B}$ is the Borel σ-field on $\mathbf{Z}$ and μ is a finite measure on $(\mathbf{Z}, \mathcal{B})$. The state of the population at time t is given by a nonnegative function $n(z, t)$ with the interpretation that $\int_S n(z, t) d\mu(z)$ is the number of individuals whose state at time t is in the set S. The population dynamics are defined by piecewise continuous functions $K(z', z) = F(z', z) + P(z', z)$ on $\mathbf{Z} \times \mathbf{Z}$,

$$n(z', t+1) = \int_{\mathbf{Z}} K(z', z) n(z, t) d\mu(z). \tag{6.9.1}$$

For simplicity we use K to denote both the kernel and the linear operator (6.9.1). *Piecewise continuous* means that

- There is a partition of $\mathbf{Z}^2$ into a finite number of disjoint open sets $\mathcal{U}_k$ whose boundaries are a finite set of continuous curves, such that $\bigcup_k \bar{\mathcal{U}}_k = \mathbf{Z}^2$ where $\bar{\mathcal{U}}_k$ is the closure of $\mathcal{U}_k$.
- F and P are continuous on each $\mathcal{U}_k$ and can be extended as continuous functions on $\bar{\mathcal{U}}_k$.

When we say that F, P, or K has some pointwise property (e.g., F is positive) we mean that the property is true pointwise in all of the continuous extensions. When we say that a property holds at some point, this includes the possibility that it holds on a point of a boundary curve in one of the continuous extensions.

Properties of K depend on the domain of functions n on which it acts. It is most natural to think of K as an operator on $L_1(\mathbf{Z})$ (state distributions with finite total population) but we can also think of it as operating on $L_2(\mathbf{Z})$ and on $C(\mathbf{Z})$. Because $\mathbf{Z}$ has finite measure, $L_2(\mathbf{Z}) \subset L_1(\mathbf{Z})$ (proof: $|n| \leq 1 + |n|^2$, so if $n \in L_2$ then $|n|$ has finite integral). And because $\mathbf{Z}$ is compact any function in $C(\mathbf{Z})$ is bounded and therefore in L_1 and L_2. Thus $C \subset L_2 \subset L_1$. Moreover, K maps L_1 into C (proof: let $n \in L_1$ and $z_n \to z'$. K is almost everywhere continuous and bounded, so the functions $f_n(z) = K(z_n, z)n(z)$ are bounded above by the L_1 function $\sup(K)n(z)$ and converge almost everywhere to $f(z) = K(z', z)n(z)$. The dominated convergence theorem therefore implies that $Kn(z_n) \to Kn(z'))$.

In short, under our assumptions K maps $L_1 \cup L_2 \cup C$ into $L_1 \cap L_2 \cap C$. As a result, an eigenvector of K in any of these spaces is in C (because an eigenvector w is a multiple of Kw which is in C). This has three important implications. First, the eigenvectors of K really are well-defined pointwise, as we have assumed in our sensitivity analysis formulas. Second, eigenvectors are in L_1 and therefore represent finite populations. Third, the eigenvalues and eigenvectors of K are the same on all three spaces, so we can talk about eigenvalues without specifying the space.

The next issue is compactness. An operator is called *compact* if it maps the closed unit sphere $\{n : ||n|| \leq 1\}$ into a set with compact closure. Compactness is important because the dominant eigenvalue of a compact operator is always isolated (and the spectrum consists only of the eigenvalues and possibly 0). The dominant eigenvalue λ of any u-bounded positive operator is strictly larger in absolute value than any other eigenvalue. But that's not enough, because there could still be an infinite sequence of eigenvalues whose absolute values converge to λ, in which case there would not be exponentially fast convergence to the stationary distribution. If that were true, the eigenvalues would have a point of accumulation on the sphere $|\zeta| = \lambda$. However, only 0 can be an accumulation point for the eigenvalues of a compact operator. Thus, if K is compact, there exists $0 \leq \lambda_2 < \lambda$ such that all sub-dominant eigenvalues have absolute value $\leq \lambda_2$.

Because the kernel K is bounded and therefore square-integrable on $\mathbf{Z} \times \mathbf{Z}$, the operator K is compact on L_2 (i.e., as an operator mapping L_2 into itself; Jörgens (1982), Theorem 11.6). In Section 6.9.1 we show that K is compact on L_1. If K is continuous rather than just piecewise continuous, then K is compact on C (this is also proved in Section 6.9.1). It is an open question (to us) whether K is necessarily compact on C if it is merely piecewise continuous.

We have stated two different positivity assumptions about K: power positivity and u-boundedness. Power-positivity is the stronger assumption because if K^m is positive, it is u-bounded for $u(z) \equiv 1$. So under either assumption, some kernel iterate K^m is compact and u-bounded. K^m therefore satisfies the assumptions of Theorem 11.5 in Krasnosel'skij et al. (1989) in the spaces $L_1(\mathbf{Z})$ and $L_2(\mathbf{Z})$, implying that

(a) K^m has an eigenvalue equal to its spectral radius λ^m, where λ is the spectral radius of K.

(b) λ^m is a simple eigenvalue, and the corresponding eigenvector w is the unique (up to normalization) positive eigenvector of K^m (positive here means that w is nonnegative and nonzero).

(c) All other eigenvalues of K^m are have absolute value $\leq q\lambda^m$, for some $q < 1$.

Easterling (1998) showed that the three properties above also hold for K itself, with λ as the dominant eigenvalue. Stable population growth, exponentially fast convergence to stable distribution w, and the rest of stable population theory therefore follow directly from the Riesz-Schauder theory for compact operators (e.g., Zabreyko et al. (1975, p. 117) or Jörgens (1982, Section 5.7)). See Ellner and Easterling (2006) for the details, which are very similar to the finite-dimensional case.

We also want to know about v, the dominant "left eigenvector" (i.e., eigenvector of the adjoint operator K^*). K^* has the same spectrum as K (Dunford and Schwartz 1958, VII.3.7) so λ is also the strictly dominant eigenvalue of K^* and equal to its spectral radius. Because K is compact, the eigenspace decomposition for K^* has the same structure same as that for K (Jörgens 1982, Section 5.7), so there is simple eigenvector v of K^* corresponding to λ. K^* is compact and the cone of nonnegative functions is reproducing in L_1; under these conditions, Theorem 9.2 of Krasnosel'skij et al. (1989) says that there is an eigenvector corresponding to the spectral radius which lies in the cone. Since v is the unique eigenvector corresponding to the spectral radius λ, this tells us that $v \geq 0$. Hence v is strictly positive whenever K and therefore K^* are power-positive. If x is any other eigenvector (actually, any vector orthogonal to v) then

$$0 = \lim_{n \to \infty} \langle \lambda^{-n}(K^*)^n x, w \rangle = \langle x, \lambda^{-n} K w \rangle = \langle x, w \rangle$$

so v is (up to constant multiples) the unique eigenvector of K^* with $\langle v, w \rangle > 0$ (we can't also have $\langle v, w \rangle = 0$, because in that case w would be orthogonal to everything and therefore 0).

The results above can also be applied to the next-generation operator $R = F(I-P)^{-1}$, to ensure existence of R_0 as the dominant eigenvalue of R. However, the intuitive result that $R_0 - 1$ and $\lambda - 1$ have the same sign has not been proved, to our knowledge, without additional assumptions. It is true in our general IPM when K and $Q = (I - P)^{-1}F$ are both power-positive (Ellner and Rees (2006, Appendix B)), and see below). Smith and Thieme (2013, Theorem 2.4) showed that it is true when there is exact mixing at birth, $F(z', z) = c(z')b(z)$. It seems that it ought to be true whenever K and R are both compact and u-bounded, but so far this has not been proved.

6.9.1 Derivations

1. Derivation of (6.6.1). We work here in $L_1(\mathbf{Z})$ using L_1 norm for vectors and the L_1 operator norm for operators. We assume that R satisfies our assumptions for stable population growth, hence has a positive dominant eigenvalue that equals its spectral radius $\rho(R)$. Let $R_m = F(I + P + P^2 + \cdots P^m)$ and $E_m = R - R_m \geq 0$. We use Gelfand's formula $\rho(A) = \lim\limits_{k \to \infty} ||A^k||^{1/k}$ to bound $\rho(R_m)$. First, because $0 \leq R_m \leq R$, $||R_m^k|| \leq ||R^k||$ for all k and therefore $\rho(R_m) \leq \rho(R) = R_0$. Gelfand's formula implies that if there exist $\lambda \geq 0, x \geq 0$ such that $Ax \geq \lambda x$ then $\rho(A) \geq \lambda$. We use this to get a lower bound on $\rho(R_m)$. Let $x \geq 0$ be the dominant right eigenvector of R, so

$$Rx = R_0 x, \quad R_m x = R_0 x - E_m x. \tag{6.9.2}$$

From the first equality in (6.9.2) we have $Fx \leq F(I - P)^{-1}x = R_0 x$, therefore

$$E_m x = F(P^{m+1}x + P^{m+2}x + \cdots) \leq F(\varepsilon_m x) \leq \varepsilon_m R_0 x.$$

So from the second equality in (6.9.2) $R_m x \geq (1 - \varepsilon_m)R_0 x$ and therefore $\rho(R_m) \geq (1 - \varepsilon_m)R_0$. Hence $\rho(R_m)$ differs from R_0 by at most $\varepsilon_m R_0$. $\quad\square$

2. Proof that K is compact as an operator from $L_1(\mathbf{Z})$ to $L_1(\mathbf{Z})$. Theorem 7.1 in Luxemburg and Zaanen (1963) gives a condition for compactness of integral operators from one Banach function space to another, including operators from $L_p(\mathbf{Z})$ to itself under our assumptions: T is compact if and only if the set $\{Tn : ||n|| \leq 1\}$ is of uniformly absolutely continuous norm. Here $|| \bullet ||$ is the norm on the space where T acts, and a set of functions S is of *uniformly absolutely continuous norm* if for any sequence of measurable sets $E_m \downarrow \emptyset$ and any $\varepsilon > 0$, there is an number $N > 0$ such that $||f\chi_{E_m}|| < \varepsilon$ for all $m > N$ and for all functions $f \in S$ where χ_A is the indicator function of A ($\chi_A(z) = 1$ for $z \in A$, 0 for $z \notin A$).
Choose $n \in L_1(\mathbf{Z})$ with $||n||_1 \leq 1$. Then

$$0 \leq Kn(z') = \int_{\mathbf{Z}} K(z', z)n(z)d\mu(z) \leq sup(K)||n||_1 \leq sup(K).$$

Let $\epsilon > 0$ be given, and $E_m \downarrow \emptyset$ a sequence of measurable sets in $\mathbf{Z}$. Then

$$||Kn\chi_{E_m}||_1 = \int_{E_m} |Kn(z')|d\mu(z') \leq sup(K)\mu(E_m). \tag{6.9.3}$$

Since $E_m \downarrow \emptyset$, we have $\lim\limits_{m \to \infty} \mu(E_m) = \mu\left(\bigcap\limits_{i=1}^{\infty} E_i\right) = 0$. So given $\epsilon > 0$, choose N such that $\mu(E_m) < \varepsilon/sup(K)$ for $m > N$. $\square$

3. Proof that if the kernel K is continuous, K is compact as an operator on $C(\mathbf{Z})$. Because $\mathbf{Z}$ is compact any continuous K is uniformly continuous. Hence for any $\epsilon > 0$ there is a $\delta > 0$ such that $|K(z_1, z) - K(z_2, z)| < \epsilon$ whenever $d(z_1, z_2) < \delta$. Let n be any function in the unit sphere of C, i.e., a continuous function with $|n| \leq 1$ everywhere. Then if $d(z_1, z_2) < \delta$,

$$|Kn(z_1) - Kn(z_2)| \leq \int_{\mathbf{Z}} |K(z_1, z) - K(z_2, z)| \, |n(z)| \, d\mu(z) < \epsilon\mu(\mathbf{Z}).$$

Clearly $|Kn| \leq \sup(K)\mu(\mathbf{Z})$. The image of the unit sphere is therefore equicontinuous and uniformly bounded, so by the Arzela-Ascoli Theorem it has compact closure in C, hence K is compact. $\square$

4. Proof that (6.5.4) implies mixing at birth for R. If F satisfies (6.5.4), then because $Rn = F\left((I - P)^{-1}n\right)$, we have

$$A\left((I - P)^{-1}n\right) c(z') \leq Rn \leq B\left((I - P)^{-1}n\right) c(z') \tag{6.9.4}$$

as desired. And if $Rn \neq 0$ it must be that $(I-P)^{-1}n \neq 0$, so $A\left((I - P)^{-1}n\right) > 0$ whenever $Rn \neq 0$. $\square$

5. Proof that $R_0 - 1$ and $\lambda - 1$ have the same sign in our general IPM when $Q = (I - P)^{-1}F$ *and K are power-positive.* Q and R have the same eigenvalues (this is equivalent to the familiar result for matrices that A and BAB^{-1} have the same eigenvalues for any invertible matrix B). If $\lambda = 1$ then $(P+F)w = w$, so $Fw = (I - P)w$ and therefore $Qw = w$, so 1 is an eigenvalue of Q. Since w is a positive eigenvector of Q it must be the dominant eigenvector, hence 1 is the dominant eigenvalue of Q and therefore of R, i.e., $R_0 = 1$. Reversing this argument shows that $\lambda = 1$ whenever $R_0 = 1$.

Suppose that $\lambda > 1$. Then $(P + F)w = \lambda w$ and therefore $Fw = \lambda w - Pw = (\lambda - 1)w + (I - P)w$, so

$$Qw = (I - P)^{-1}Fw = (\lambda - 1)(I - P)^{-1}w + w \geq (\lambda - 1)w + w = \lambda w.$$

By theorem 9.4 of Krasnosel'skij et al. (1989) this implies that some eigenvalue of Q, and hence the dominant eigenvalue has absolute value of λ or larger, hence $R_0 > 1$.

Suppose that $R_0 > 1$. Then consider the family of operators $K_a = P + aF, 0 < a \leq 1$. It is easy to see that K_a and the corresponding R_a are power-positive, so there exist $\lambda(a), R_0(a)$, and so on. Because $Q_a = aQ(1)$, we have $R_0(a) = aR_0(1) > 1$ so there is an $a < 1$ such that $R_0(a) = 1$ and therefore $\lambda(a) = 1$. When K is power-positive, so is $aK \leq K_a$ hence K_a is power-positive. Choose m so that K_a^m is strictly positive. Then

$$K^{m+1}w_a - w_a = K^{m+1}w_a - K_a^{m+1}w_a \geq K_a^m(Kw_a - K_aw_a) = (1 - a)K_a^m Fw_a$$

Since $R_0 > 0$ we cannot have $F = 0$, and w_a is strictly positive so Fw_a is nonzero, hence $(1 - a)K_a^m Fw_a$ is strictly positive. It is also uniformly continuous, therefore bounded below by some positive number and therefore greater than some positive multiple of w_a. Consequently there is an $\epsilon > 0$ such that $K^{m+1}w_a \geq (1+\epsilon)^{m+1}w_a$, hence the spectral radius of K^{m+1} is greater than $(1+\epsilon)^{m+1}$. Therefore the spectral radius of K, which is λ, must be greater than $1 + \epsilon$.

We have shown that $R_0 = 1 \Leftrightarrow \lambda = 1$ and $R_0 > 1 \Leftrightarrow \lambda > 1$ so we must also have $R_0 < 1 \Leftrightarrow \lambda < 1$. $\square$

Chapter 7
Environmental Stochasticity

Abstract Environmental stochasticity occurs when the parameters that determine individual fates and population growth vary from year to year. This is a ubiquitous feature of natural ecosystems. To illustrate the process of incorporating environmental stochasticity into an IPM, we use an extended case study based on a well-studied field system. Methods for characterizing between-year variation in model parameters using both fixed and mixed effects models are presented. Methods for construction and implementation of density-independent IPMs are then presented. The key properties of the stochastic growth rate (λ_S) are summarized, along with methods for its estimation and appropriate perturbation theory. Methods for implementing Life Table Response Experiments (LTREs) analyses are discussed.

7.1 Why environmental stochasticity matters

Even the most cursory observation of natural systems suggests they vary from year to year. Understanding how these changes in the environment impact populations is of course difficult, especially when mediated by an individual's state. For example in Soay sheep, survival of lambs and males was determined by weather throughout winter, whereas in females survival was most strongly influenced by rainfall at the end of winter. Density and weather interacted, with bad weather depressing survival at high density, especially in young and old animals (Coulson et al. 2001). This is one example out of many.

Multi-annual demographic studies (reviewed by Hairston et al. 1993) have documented enormous interannual variation in per capita reproductive success, by 1 to 4 orders of magnitude, in a wide range of taxa and habitats (plants in forest, grassland, and deserts; birds, fish, and other vertebrates; insects and marine invertebrates).

This variation, where vital rates fluctuate from one year to the next, is called *environmental stochasticity*. This differs from *demographic stochasticity* which occurs as a result of applying fixed vital rates to a finite population, Section 10.3. Our aim will be to review the complete process from parameter estimation to

© Springer International Publishing Switzerland 2016   187
S.P. Ellner et al., *Data-driven Modelling of Structured Populations*,
Lecture Notes on Mathematical Modelling in the Life Sciences,
DOI 10.1007/978-3-319-28893-2_7

IPM construction and analysis. As in Chapter 2 we will use an individual-based model to generate "data" and use this for constructing the IPM. In terms of analysis we will cover sensitivities and elasticities, at the kernel, demographic function and underlying model parameter levels, and also Life Table Response Experiments (LTREs). When parameters vary between years there is a proliferation of things we can perturb, for example the mean parameter or its variance. This leads to an impressive variety of perturbation formulae, but luckily the basic methodology is reasonably straightforward. For parameter estimation we cover the main current approaches, but some newer ones (e.g., INLA and Stan) are not covered. A recent simulation study suggests that the choice of statistical method is relatively unimportant, providing covariation between different vital rates is accurately modeled (Metcalf et al. 2015). Overall Metcalf et al. (2015) found that kernel selection (see below) using fixed-effects demographic models was one of the best performing methods. But it always takes a lot of data to estimate environmental variability accurately, so it is especially important to quantify your uncertainty.

7.1.1 Kernel selection versus parameter selection

There are two basic approaches to building an IPM with environmental stochasticity:

1. *Kernel Selection.* For each year in the study, a fitted fixed-effects model gives us a vector of year-specific values for each parameter, and a fitted mixed-effects model gives us a vector of fitted values or posterior modes for each parameter. In either case, we construct a set of kernels using each of the year-specific sets of parameters, and then simulate the model by selecting from the set of kernels at random. Fixed-effects parameter estimates are biased towards over-estimating the between-year variation because sampling error and true between-year variation are confounded, so when possible the results should be compared with an IPM constructed using a mixed-effects model. However, in practice the difference is often small compared to other sources of error.

2. *Parameter Selection.* If you have characterized the between-year variation in the parameters using mixed-effects models, then you can sample from the fitted distributions (see Section 7.2.1) and build a unique kernel for each year that you simulate.

Kernel selection has advantages and disadvantages. An IPM using kernel selection runs much faster than one using parameter selection, as all the kernels can be constructed before iterating the model. Parameter estimation is easier for kernel selection as you don't have to worry about correlations between different demographic processes, as in the *Carlina* data set. These correlations are already "built in" to the year-specific parameter estimates. Kernel selection is also less sensitive to what you assume about the distribution of parameter variation; with only a few years of data it may be unclear if (for example) a normal or lognormal distribution is more reasonable. However, it is difficult to vary the correlations between different processes or parameters. Some model predictions (e.g., quasi-extinction risk) can be very sensitive to correlations between vital rates, so the sensitivity to correlations is then an important part of assessing the uncertainty in model predictions (Fieberg and Ellner 2001). With parameter selection, you're not limited to the parameter values estimated for each year in the data set. Providing you're happy with the assumed between-year distributions of parameters, the model can then be more fully explored.

Models in which demographic rates are functions of environmental covariates are an important example of parameter selection; Dalgleish et al. (2011) and Simmonds and Coulson (2015) are two recent examples. Dalgleish et al. (2011) modeled effects of precipitation and temperature on all vital rates of three sagebrush steppe plants, including effects of climate in previous years; Simmonds and Coulson (2015) linked vital rates of Soay sheep on St. Kilda to the NAO (North Atlantic Oscillation) index, based on previous studies of NAO effects. This kind of model is important because projections and analyses can take advantage of long-term climate data. As we will see below, forecast precision is seriously limited by the duration of the demographic study, because 10 years (for example) often provides only a rough picture of demographic variability. One way past this limit is to link (when possible) most of the demographic variability to measured environmental covariates for which long-term data are available, based on observational data or experiments that manipulate environmental conditions and observe the response (Fieberg and Ellner 2001).

7.2 Case Study 1C: Another monocarpic perennial

Our case study will be a monocarpic perennial like *Oenothera* introduced in Chapter 2. However, we will use the *Carlina vulgaris* dataset which we have previously used in several papers (Rose et al. 2002; Childs et al. 2003, 2004; Rees et al. 2006; Metcalf et al. 2008; Rees and Ellner 2009). All data were collected from a permanent quadrat of $100\,\text{m} \times 0.5\,\text{m}$ in chalk grassland in Sussex, England. The site had a continuous record of winter grazing by cattle and increasing incidence of year-round grazing by rabbits. *Carlina* had one of the lowest mean densities of all the turf-compatible short-lived species in the quadrat (Grubb 1986). The population was censused in early August, for every year between 1979 and 1997. All emerging plants were mapped, so it was possible to follow individuals throughout their entire life span. Thus, apart from the plants already present in 1979, all individuals could be accurately aged. For more details of the data collection, see Rose et al. (2002).

During each census, individuals were recorded as dead or alive, whether they had flowered, and how large they were, measured as length of the longest leaf (mm). In other studies of size-related flowering in plants, rosette diameter is commonly used to measure size (e.g., Kachi and Hirose 1985; Wesselingh et al. 1997; Rees et al. 1999). However, in *Carlina* rosette diameter would not have correlated well with leaf area per plant, because leaves (especially of larger individuals) were often semi-erect in the turf that had grown around the rosette. Therefore length of the longest leaf was used to measure plant size.

The events in the life cycle are as follows: plants are censused in August, we then assume mortality occurs over winter, and those that survive flower with

some probability depending on their size (flowering is fatal), those that don't
die or flower, grow, and their size is measured at the next census. Flowering
occurs between June and August, and the seeds are retained in the flower heads
until they are dispersed during dry sunny days in late autumn, winter, or spring
(P. J. Grubb, *unpublished data*). Seeds germinate from April to June, and there
is little evidence of a persistent seed bank. It was not possible to estimate the
relationship between plant size and number of seeds produced from the data
available. However, other studies of monocarpic species similar to *Carlina* have
estimated fecundity as a function of longest leaf length or rosette diameter (i.e.,
2×longest leaf length) and based on these studies we used the relationship

$$\text{seeds} = \exp(A + Bz) \qquad (7.2.1)$$

where z is log longest leaf length in year t. Several studies (Metcalf et al. 2003)
have shown that seed production is proportional to rosette diameter squared,
and so the value of B was fixed at 2.0. We set $A = 1$ in all simulations, and the
probability of recruitment, p_r was set at 0.00095 so the population increase was
similar to that observed in the field.[1]

Rees and Ellner (2009) used mixed models to describe the effects of temporal variation in *Carlina* demography; the models used are very similar to the
Oenothera models. The probabilities of survival and flowering were described
by logistic models but with intercepts varying between years, and for survival
the slope also varying between years. Seed production as a function of size was
assumed to not vary between years. Growth and recruit size also vary between
years and in this case the intercepts are positively correlated ($r = 0.77$), so
in good years for growth, recruits are large. For growth the slope of the linear
model also varied among years. A summary of the fitted models is given in
Table 7.1.

As in the *Oenothera* example we will use these parameter estimates to build
an individual-based model so we can explore the process of building models. The
mean parameter estimates are stored in m.par.true, the standard deviations are
in m.par.sd.true, and the variance-covariance matrix for growth and recruitment
intercepts is VarCovar.grow.rec in the script file Carlina Demog Funs DI.R. The
code for the IBM (in Carlina DI IBM.R) is very similar to the *Oenothera* IBM
except that we generate year-specific parameter values to describe the demography in each year. We use the parameter selection approach, meaning that the
values for each year are obtained by sampling from the estimated parameter
distributions given in Table 7.1. As all the estimated distributions are Gaussian we need to bound the range of parameters generated, so we truncate the
distributions at the 0.001 and 0.999 percentiles. Twelve parameters define the
model and in R we can generate truncated, standard Gaussian distributions
($\mu = 0, \sigma^2 = 1$) using the following trick,

```
qnorm(runif(12,0.001,0.999))
```

---

[1] Only the product $p_r \exp(A)$ affects the number of recruits per parent, so A can be set
to any value so long as p_r is adjusted to compensate.

Table 7.1 Estimated demographic functions in the mixed-models analysis for *Carlina*. In all cases z is size in year t (log transformed) and z' the size the following year, z_R is recruit size, $N(\mu, \sigma)$ is a normal distribution with mean μ and standard deviation σ, and $MVN(\mu, \Sigma)$ is a multivariate normal distribution with mean vector μ and variance-covariance matrix Σ.

Demographic process	Model	Parameter estimates
Size dynamics: rosette growth and recruit size	$z' = a_0 + b_z z + \epsilon$ $z_R = a_R + \omega$	$b_z \sim N(0.74, 0.13)$ $\epsilon \sim N(0, 0.29)$ $\omega \sim N(0, 0.50)$ $a_0, a_R \sim MVN(\mu, \Sigma)$ $\mu = (1.14, 3.16)$ $\Sigma = \begin{pmatrix} 0.037 & 0.041 \\ 0.041 & 0.075 \end{pmatrix}$
Probability of survival	$\text{logit}(s(z)) = m_0 + m_z z$	$m_0 \sim N(-2.28, 1.16)$ $m_z \sim N(0.90, 0.41)$
Probability of flowering	$\text{logit}(p_b(z)) = \beta_0 + \beta_z z$	$\beta_0 \sim N(-16.19, 1.03)$ $\beta_z = 3.88$
Seed production	$b(z) = \exp(A + Bz)$	$A = 1$, $B = 2$

`runif(12,0.001,0.999)` generates 12 uniform random deviates between 0.001 and 0.999. `qnorm` is the Gaussian quantile function, which takes a quantile, in this case given by the uniform deviates, and returns the corresponding value from a Gaussian distribution with mean 0 and variance 1. We then scale these Gaussian deviates, d, so they have mean μ and variance σ^2 using $\mu + \sigma d$. For correlated parameters the calculation is slightly more complicated involving: 1) Cholesky factorization of the variance-covariance matrix, 2) generating independent, truncated Gaussian deviates, then 3) transforming these using the Cholesky factorization.

Output from 30 simulated populations is shown in Figure 7.1A which should be compared with the *Oenothera* IBM, Figure 2.5A. The *Oenothera* IBM only includes the effects of demographic stochasticity and so fluctuations become smaller as population size increases. The *Carlina* IBM includes both demographic and environmental stochasticity, and so several populations go extinct despite starting with 100 individuals, and the effects of stochasticity do not become smaller as population size increases (Figure 7.1A).

7.2.1 Building an IPM

The *Carlina* IPM includes the effects of environmental stochasticity but ignores demographic stochasticity (see Chapter 10 for a discussion of models with demographic stochasticity). The general form of the model is

$$n(z', t+1) = \int_L^U K(z', z; \theta(t)) n(z, t) dz \qquad (7.2.2)$$

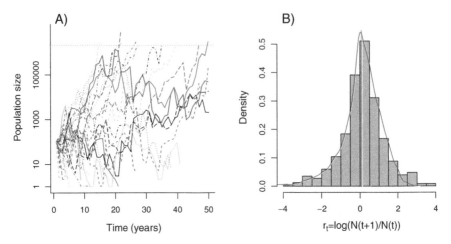

Fig. 7.1 Simulation of the density-independent *Carlina* IBM. A) Population trajectories through time, for 30 replicate populations, the dotted red line indicates the upper bound on the simulated population sizes; simulations were terminated if the population exceeded this, and B) the distribution of one-step population growth rates, $r_t = \log(N(t+1)/N(t))$ from the IBM. The red line is the corresponding density estimate from the stochastic IPM, and the turquoise line the stochastic growth rate from the IPM. Source file: `Carlina simulate stoch DI IBM.R`.

where $\theta(t)$ is a vector containing the yearly parameter estimates. At this point before constructing the kernel we need to draw a life cycle diagram. Luckily we can use Figure 2.2B which corresponds closely to the census times and life cycle of *Carlina*, the main difference being that reproduction if fatal - this leads to a kernel of the form,

$$K(z', z) = s(z)(1 - p_b(z))G(z', z) + s(z)p_b(z)b(z)p_r c_0(z'). \qquad (7.2.3)$$

Reading this aloud we interpret this as "Plants are censused in August, mortality occurs over winter, and those that survive flower with probability $p_b(z)$. Plants that survive and don't flower grow according to $G(z', z)$. Flowering plants produce $b(z)$ seeds, these recruit with probability p_r, and recruit size is independent of parental size, $c_0(z')$." This is what we assume happens in the field.

Implementing the IPM is straightforward using the functions defined in Chapter 2, the main difference being that, as in the IBM, we need a set of parameters to define the demographic functions in each year. The code (in `Carlina simulate stoch DI IBM.R`) can be summarized as follows

1. Generate yearly parameters and store them in `params.t`.
2. Construct the kernel using

   ```
   year.K <- mk_K(nBigMatrix,params.t,minsize,maxsize)
   ```

   where the `mk_K` function constructs the kernel in year t, this is the same function we used in Chapter 2. The arguments are the number of mesh points, demographic parameters, and minimum and maximum sizes.

3. Project the population forward using

```
nt1 <- year.K$K %*% nt
```

and then normalize the distribution of nt1 so it sums to one - this prevents the population from increasing without bound, remember this model has no density dependence.
4. Calculate summary statistics.

We wrap this up in a function iterate_model whose arguments are the mean and standard deviation of the demographic parameters, number of iterates of the model, and whether to store the yearly parameters.

7.2.2 Basic analyses by projection

Before considering model fitting, let's use the actual IBM parameters in the IPM. We run the IPM using

```
iter <- iterate_model(m.par.true ,m.par.sd.true, n.est, store.env=F).
```

The mean size each year from the IBM is stored in mean.z.t and we can calculate the mean from the IPM using equation (2.5.2), the only difference being that $w(z)$ now varies from year to year. There are several ways of doing the calculation:

$$
\bar{z} = \int \phi(w_t) \int w_t(z) z \, dz \, dw_t
$$

$$
= \int z \int \phi(w_t) w_t(z) \, dw_t \, dz = \int \overline{w(z)} z \, dz, \qquad (7.2.4)
$$

where ϕ is the probability density function of w_t. So we can average the mean sizes in each year with respect to $\phi(w_t)$ or average $w_t(z)$ with respect to $\phi(w_t)$ and then use this to average z. Implementing these calculations we find excellent agreement between the approaches.

```
> # Mean plant size from IBM (log scale)
> mean(store.mean.z.t)
[1] 3.281639
> # from IPM (log scale)
> sum(iter$size.dist * iter$meshpts)
[1] 3.282155
```

We can calculate a wide range of other statistics, for example, the mean arithmetic flowering size. From the IBM we average the mean flowering size each year, whereas in the IPM we calculate the average distribution of flowering sizes $\overline{w_f}$, by averaging $w_t(z)s(z;\theta_t)p_b(z;\theta_t)$ over the course of the simulation, and then calculating

$$
\int \overline{w_f}(z) \exp(z) \, dz \qquad (7.2.5)
$$

Again there is good agreement between the approaches

```
> # Mean flowering size from IBM
> mean(store.mean.fl.z.t,na.rm=TRUE)
[1] 61.44376
> # from IPM
> sum(iter$size.dist.fl * exp(iter$meshpts))
[1] 64.34467
```

In our final comparison between the IBM and IPM we will look at population growth. For stochastic models the key parameter is the stochastic growth rate λ_S, which we discuss extensively in Section 7.4. For now, it's enough to know that λ_S is the analog of λ in a deterministic IPM, and that we can estimate $\log \lambda_S$ as the average of the one-step log population growth rates. Let $r(t) = \log(N(t+1)/N(t))$, then the estimate is

$$\widehat{\log(\lambda_S)} = \frac{1}{T} \sum_{t=0}^{T-1} r(t). \tag{7.2.6}$$

The distributions of $r(t)$ from the IBM and IPM are in good agreement, Figure 7.1B. As $\widehat{\log(\lambda_S)}$ is an average we can calculate its approximate standard error as $\sqrt{V(r(t))/T}$ where $V(r(t))$ is the variance in the one-step growth rates. Comparison of the IBM and IPM

```
> cat("Log LambdaS IPM =",stoch.lambda,"from IBM =",stoch.lambda.IBM,"\n")
Log LambdaS IPM = 0.1151448 from IBM = 0.1639114
> cat("Approximate 95% CI for log LambdaS from IBM ",
+       stoch.lambda.IBM - 2*SE.lambda.IBM, " to ",
+       stoch.lambda.IBM + 2*SE.lambda.IBM,"\n")
Approximate 95% CI log LambdaS from IBM 0.06434863 to 0.2634742
```

shows there are large differences between the two methods, $\widehat{\log(\lambda_S)} = 0.12$ *vs* 0.16. This arises because despite having a sample size of ≈ 500 one-step growth rates, the confidence intervals on $\log \lambda_S$ from the IBM are substantial (0.06 to 0.26).

These confidence intervals only take into account the uncertainty due to finite simulation length, not the uncertainty associated with parameter estimation. With large variation among years, parameters will be hard to estimate accurately. When one year can be very different from another, precision is limited by the number of years for which you have data, regardless of how much data you have for each year or how build your model (Metcalf et al. 2015). As a result, *precise estimation of λ_S is not easy.*

To emphasize this point, we generated a large number of 20-year parameter sets using the distributions from Table 7.1, and then calculated $\log(\lambda_S)$ for each of these using kernel selection (Section 7.1.1). This is equivalent to observing an infinite population perfectly for 21 years and knowing all the demographic

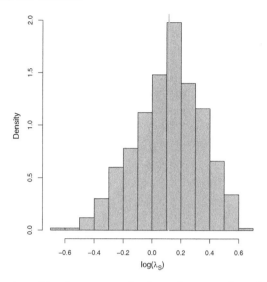

Fig. 7.2 Distribution of $\log \lambda_S$ generated using a large number of 20-year parameter sets and iterating the resulting density-independent *Carlina* IPM for 10000 years. The turquoise line is the "true" value. Source file: `Carlina simulate stoch DI IBM.R`.

functions exactly. The result is rather sobering (Figure 7.2), as in about 30% of cases we would get the sign wrong! An exercise like this is essential, to see how much you really know about λ_S.

7.3 Modeling temporal variation

In the following sections we will use data from our monocarp IBM to illustrate various statistical approaches that can be used to parameterize an IPM with environmental stochasticity. We will use a large dataset so comparisons with the underlying processes generating the data (Table 7.1) are simpler. Appropriate statistical background reading would be Pinheiro and Bates (2000) for information on `lme` and a wealth of ideas related to mixed modeling, Bates (`lme4.r-forge.r-project.org/lMMwR/lrgprt.pdf`) for `lmer`, Hadfield (2010) for `MCMCglmm` (see also

   `cran.r-project.org/web/packages/MCMCglmm/vignettes/CourseNotes.pdf`

for a detailed vignette), and Plummer (2003) for JAGS. There are also numerous resources on the web.

7.3.1 Fixed versus random effects

There are two ways of incorporating temporal variation into our demographic models. The first is to fit year as an unordered factor (fixed effect), so for survival we might fit a series of models

```
mod.Surv <- glm(Surv ~ Yeart, family = binomial, data = sim.data)
mod.Surv.1 <- glm(Surv ~ Yeart + z, family = binomial, data = sim.data)
mod.Surv.2 <- glm(Surv ~ Yeart * z, family = binomial, data = sim.data)
```

The first of these models assumes survival is independent of size but varies between years, the second allows survival to vary between years but assumes a common size slope, while the last allows the slopes and intercepts to vary between years. We can compare the models using the anova command, which produces an analysis of deviance table.

```
> anova(mod.Surv,mod.Surv.1,mod.Surv.2,test="Chisq")
Analysis of Deviance Table

Model 1: Surv ~ Yeart
Model 2: Surv ~ Yeart + z
Model 3: Surv ~ Yeart * z
  Resid. Df Resid. Dev Df Deviance Pr(>Chi)
1     18428      18617
2     18427      18055  1   562.36 < 2.2e-16 ***
3     18408      17927 19   128.16 < 2.2e-16 ***
```

Each model is compared with the preceding one and so we conclude that a model requiring separate slopes and intercepts in each year is required - the simplification going from slopes varying between years to a common slope is highly significant. This is reassuring as we know both the intercepts and slopes of the survival-size relationship vary between years, Table 7.1. To simplify extracting the model parameters we re-fit the model as follows

```
mod.Surv <- glm(Surv ~ Yeart/ z -1 , family = binomial, data = sim.data)
```

which means that coef(mod.Surv) then gives the slopes and intercepts in each year, rather than contrasts. For the other demographic processes we use essentially the same approach; see Carlina Fixed Effects Demog.R for the code. We can also fit survival models in a mixed model framework using the lme4 package. As before we first fit a series of models of increasing complexity and then compare them with the anova command.

```
mod.Surv.glmer  <- glmer(Surv ~ 1 + (1|Yeart),
                        family = binomial, data = sim.data)
mod.Surv.glmer.1 <- glmer(Surv ~ z + (1|Yeart),
                        family = binomial, data = sim.data)
mod.Surv.glmer.2 <- glmer(Surv ~ z + (1|Yeart) + (0 + z|Yeart),
                        family = binomial, data = sim.data)
mod.Surv.glmer.3 <- glmer(Surv ~ z + (z|Yeart),
                        family = binomial, data = sim.data)
```

The first model assumes survival varies between years, the second adds a common size effect, the third has intercepts and slopes varying between years but assumes they are uncorrelated, and finally the forth model has correlated year-specific intercepts and slopes. Model selection for mixed models is complicated even for linear mixed models (Mueller et al. 2013) and we need generalized linear mixed models (GLMMs) - on the Wiki dealing with GLMMs they state "(G)LMMs are hard - harder than you may think," so you have been warned. Here we suggest a pragmatic approach of screening through possible models using AIC or BIC, selecting a range (2 or 3) plausible models, and using these to test if the resulting IPM predictions are robust. Using anova command we find

```
> anova(mod.Surv.glmer,mod.Surv.glmer.1,mod.Surv.glmer.2,mod.Surv.glmer.3)
```

```
Data: sim.data
Models:
mod.Surv.glmer: Surv ~ 1 + (1 | Yeart)
mod.Surv.glmer.1: Surv ~ z + (1 | Yeart)
mod.Surv.glmer.2: Surv ~ z + (1 | Yeart) + (0 + z | Yeart)
mod.Surv.glmer.3: Surv ~ z + (z | Yeart)
                   Df    AIC    BIC  logLik deviance   Chisq Chi Df Pr(>Chisq)
mod.Surv.glmer      2  18756  18772 -9376.1    18752
mod.Surv.glmer.1    3  18197  18221 -9095.6    18191  561.0758     1   <2e-16***
mod.Surv.glmer.2    4  18123  18154 -9057.4    18115   76.3497     1   <2e-16***
mod.Surv.glmer.3    5  18124  18163 -9056.8    18114    1.2684     1     0.2601
```

So AIC and BIC both suggest mod.Surv.glmer.2 is best, and the likelihood ratio tests support this conclusion. Likelihood ratio tests when the parameter is on the boundary of parameter space (variances can't be negative, and we want to test if $\sigma^2 = 0$) are typically conservative (p values are too big), so nonsignificant terms need to be treated with caution. In this case mod.Surv.glmer.2 happens to be the right model (we know that) but in a real application we might consider both mod.Surv.glmer.2 and mod.Surv.glmer.3 as these have similar AICs although their BICs differ by 9 suggesting mod.Surv.glmer.2 is better.

BIC penalizes complex models more harshly than AIC, so typically leads to simpler models. BIC is based on the idea of identifying the "true model," whereas AIC targets out-of-sample predictive accuracy. These are not the same thing. When sample size is large, the damage from leaving out a real term is bigger than the damage from including a spurious one (because its impact on predictions will be small). That asymmetry means that the AIC-best model may include a term even if the evidence for its presence is equivocal. Another caveat is that the parameter count (Df) used in lme4 is appropriate for totally out-of-sample prediction, meaning a year that is not in your data set. The appropriate parameter count for prediction within observed years (a "trace of the hat matrix" kind of Df) is typically higher.

We can also fit GLMMs using a Bayesian approach using the MCMCglmm package. For example, if we want to fit the same set of survival models as we fitted with glmer, we would use

```
d<-1
prior=list(R=list(V=1, fix=1), G=list(G1=list(V=diag(d), nu=d,
           alpha.mu=rep(0,d), alpha.V=diag(d)*1000)))
mod.Surv.MCMC <- MCMCglmm(Surv ~ 1, random = ~ Yeart,
                    family = "categorical", data = sim.data,
                    slice=TRUE, pr=TRUE, prior=prior)
mod.Surv.MCMC.1 <- MCMCglmm(Surv ~ z, random = ~ Yeart,
                    family = "categorical", data = sim.data,
                    slice = TRUE, pr = TRUE, prior = prior)
d<-2
prior=list(R=list(V=1, fix=1),
         G=list(G1=list(V=diag(d), nu=d, alpha.mu=rep(0,d),
         alpha.V=diag(d)*1000)))
mod.Surv.MCMC.2 <- MCMCglmm(Surv ~ z, random= ~ idh(1+z):Yeart,
                    family = "categorical", data = sim.data,
                    slice = TRUE, pr = TRUE, prior = prior)
mod.Surv.MCMC.3 <- MCMCglmm(Surv ~ z, random= ~ us(1+z):Yeart,
                    family = "categorical", data = sim.data,
                    slice = TRUE, pr = TRUE, prior = prior)
```

Using DIC to select between models, we again choose the final two models as DIC for these only differs by 2 units. For the more complex final model the estimated covariance is -0.2498 with a 95% confidence interval of [-0.9035, 0.3496] suggesting that there really isn't much evidence for covariance between the intercepts and slopes - which we know is true. If we need yearly parameters we can extract these with

```
post.modes <- posterior.mode(mod.Surv.MCMC.2$Sol)
```

Box 7.1: Fitting mixed models using MCMCglmm

The other demographic models for growth, recruit size, and the probability of flowering are fitted using very similar approaches and methods for model simplification; see Carlina Fixed Effects Demog.R for the code. In each of these cases AIC and BIC get the answer right but then again we have a lot of data, and so in real examples a range of models should be explored. We can also fit models using a Bayesian approach using the MCMCglmm package. This requires a bit of extra work but not much; see Box 7.1.

For kernel selection, where we construct a kernel for each year in the study (i.e., year 1 to 2), these are all the demographic models we need. For each year in the study a fitted fixed-effects model gives us a vector of year-specific values for each parameter, and a fitted mixed-effects model gives us a vector of posterior modes for each parameter. In either case, we simulate the model by using the complete parameter vectors to construct yearly kernels, and then sampling at random from these.

For parameter selection, where we model the distribution of model parameters and sample from this to construct yearly kernels (as in Section 7.2.1), mixed-effects models like the ones we have just fitted give us all the information we need if there are no correlations between different demographic processes. However, things are more complicated when such correlations exist, as in *Carlina* where the intercept of the growth function is positively correlated with average recruit size. When a separate model is fitted for each demographic process, such correlations cannot be included. The previous health warnings for model selection (and fitting) are writ large for multivariate mixed models, and again we suggest a simple, pragmatic approach. Following Rees and Ellner (2009) we recommend:

1. Fit demographic models with year as a factor for each demographic processes with between-year variability, and store the yearly parameter estimates in a matrix m.par.est, that is

   ```
   m.par.est <- matrix(NA,nrow=12,ncol=20)
   m.par.est[1,] <- coef(mod.Surv)[1:20]
   m.par.est[2,] <- coef(mod.Surv)[21:40]
   ```

   so each row corresponds to a parameter and the columns to specific years. The estimates could be year-specific values from a fixed-effects model with year as a factor, or fitted values (BLUPS) from a mixed-effects model with year as a random effect.

2. Compute the correlation between yearly parameter estimates of the different processes, for example, the correlation between mean recruit size and the intercept of the growth function is given by

   ```
   > cor.test(m.par.est["grow.int",],m.par.est["rcsz.int",])
   Pearson's product-moment correlation
   data:  m.par.est["grow.int", ] and m.par.est["rcsz.int", ]
   t = 4.4361, df = 18, p-value = 0.000319
   alternative hypothesis: true correlation is not equal to 0
   95 percent confidence interval:
    0.4118899 0.8828641
   sample estimates:
         cor
   0.7226867
   ```

3. Fit bivariate or multivariate statistical models for the parameter distributions where there are significant correlations.

For the *Carlina* model the parameter estimates for the growth intercepts and average recruit size are correlated ($r = 0.72, P < 0.0005$), as expected - the true correlation being 0.77 - and all the other correlations are not significant ($P > 0.2$). In point 1. above it is important to make sure that the different demographic processes are correctly aligned by year. For example, the new recruits that were measured in year t were actually produced in year $t\text{-}1$ so if we simply fit a model for recruit size variation across years we will incorrectly assign the year *t-1*'s parameter to year *t* and so any correlations between recruit size and other demographic processes will be incorrectly estimated.

Before dealing specifically with *Carlina* we present some multivariate models that can be fitted without too much pain. Consider the case where we have two Gaussian responses, say size next year and offspring size, which are linear functions of current size. In addition to this we will assume the intercepts of both relationship vary from year to year and are positively correlated, so in good years individuals are bigger on average next year and produce bigger offspring. Code for simulating and fitting the models is given in Generate and fit year effect models.R. This model can easily be fitted in using nlme package, but before fitting the model we need to arrange the data so the response variables are stacked into a single vector and similarly for the predictor variables, the R-code looks like this

```
both <- data.frame(size1=c(size1,rec),size=c(size,size),
        demo.p=gl(2,n.samp),year=c(years,years))
```

So size1 contains size next year and the recruit sizes, size contains two copies of current size, demo.p are labels for the different demographic processes, and year contains two copies of the year data. We can then fit the model using

```
both.g <- groupedData(size1 ~ size | year, data=both)
fit.trait.1 <- lme(size1~ demo.p/size - 1, random = ~ (demo.p-1)|year,
                weights=varIdent(form = ~ 1|demo.p), data=both.g)
```

This fits a model with separate intercepts and slopes for each demographic process, random = ~ (demo.p-1)|year specifies a correlated, random effect describing yearly variation in the intercepts, and weights = varIdent(form = ~ 1|demo.p) allows the error variance to differ between the two processes. Comparing the estimated variance - covariance matrix for the yearly intercepts with the true values we find

```
> getVarCov(fit.trait.1)
Random effects variance-covariance matrix
        demo.p1 demo.p2
demo.p1 0.2793 0.38120
demo.p2 0.3812 0.58754
  Standard Deviations: 0.52848 0.76651
> Sigma
    [,1] [,2]
[1,] 0.3 0.4
[2,] 0.4 0.6
```

so there is good agreement as expected. We can also fit this model using MCMCglmm; the code is

```
mcmc.data <- data.frame(size1=size1, size=size, rec=rec, year=years)
fit.trait.1.mcmc <- MCMCglmm(cbind(size1,rec)~ trait / size -1,
    family = c("gaussian","gaussian"), random = ~ us(trait):year,
    rcov=~ idh(trait):units, data=mcmc.data)
```

So we fit separate relationships for each trait (note `trait` is a reserved word), both responses are Gaussian, and we have an unstructured variance-covariance matrix for the yearly variation in the intercepts, `random = ~ us(trait) : year` ; this allows the intercepts to be correlated. Finally we specify that the residual variation at the individual level differs between growth and recruitment, and that these are independent using `rcov = ~ idh(trait) : units`. This is the correct model and as before there is good agreement between the estimated parameters and the true values. `MCMCglmm` fits a very wide range of models, so, for example, if we wanted the yearly growth and recruitment intercepts to be independent we would specify `random = ~ idh(trait) : year` or if we thought that individuals might have correlated responses within a year, then we would use `rcov = ~ us(trait) : units`; we can also have responses from different families (e.g., Poisson, binomial).

`MCMCglmm` allows missing values in the response, so continuing our example you might not have measured recruit size for all individuals so there are `NA`'s in `rec` but you can still use all the `size1` data. This is important as it might be difficult to get good data on a particular demographic processes and restricting the analysis to complete cases would then be very wasteful. However, missing data are not allowed in the predictor variables, which is a problem for our *Carlina* growth-recruit size model, so instead we will use JAGS (Just Another Gibbs Sampler, Plummer 2003). JAGS allows a very wide range of models to be specified, but this generality comes at a cost, and setting up and running nonstandard models are not trivial. In order to run these models from R you will need the latest version of R and the appropriate version of JAGS (`http://sourceforge.net/projects/mcmc-jags/files/`). You will also need the `rjags` and `coda` packages. At this point we strongly recommend you befriend a statistician or experienced JAGS user (chocolate, beer, lost puppy look – whatever it takes), as for the inexperienced these models can be tough and we all have, or know people who have, lost weeks/months failing to run JAGS models. This is in no way a criticism of JAGS, which is a remarkable piece of software. But there are many traps for the inexperienced and the models can take a long time to run.

At this point we will assume you have read about JAGS and are happy with the model specification and fitting. The code we provide for fitting a bivariate growth-recruitment size model implements the model of Cam et al. (2002) and the code was kindly provided by Evan Cooch (`Carlina Mixed Effects Demog.R`). This approach seems robust but convergence is slow so a large number of iterations is required.

7.4 Long-run growth rate

One of the most important properties of density-independent stochastic IPMs is the stochastic growth rate, λ_S. There are four main things to know about λ_S: it exists; it can be computed by simulation; it can be approximated in informative ways; and there is a perturbation theory. Because of these, just about any analysis you could do for a density-independent deterministic IPM or matrix model can also be done for a density-independent stochastic IPM. We cover the first three in this section, and the fourth gets its own section below.

It exists. The long-term behavior of the population is still exponential increase or decrease, when there is no density-dependence to put a cap on population increase, or conversely to rescue a population because survival or fecundity improves when numbers are low. λ_S is the long-term growth rate of total population size $N(t)$, exactly as in stochastic matrix models:

$$\lim_{t \to \infty} \frac{1}{t} \log N(t) = \log \lambda_S. \tag{7.4.1}$$

The same is true for any part of the population (e.g., all individuals with z in some specified range), or for a transformed measure of population size. For example, if z is log body mass then the total biomass $B(t) = \int_Z e^z n(z,t) dz$ increases or decreases exponentially at the same long-run rate as the total population.

Some mild assumptions are needed to guarantee this simple behavior. The complete story is in Section 7.8, but the main points are as follows. First, the parameter vectors $\theta(t)$ in equation (7.2.2) must be stationary, ergodic random variables. Stationary means that the distribution of θ is the same each year. Ergodic means that the distant future becomes independent of the past and present; finite Markov chains with a unique stationary distribution, and finite-order autoregressive processes, are examples of ergodic random processes. Second, the kernel $K(z', z, \theta)$ must be a positive, continuous function of its three arguments.[2] Third, the distributions of all model parameters must be bounded. Using an unbounded statistical distribution or state variable to model a finite data set is harmless for short-term forecasting, but it can cause unrealistic long-term behavior of the model (Ellner and Rees 2006). Fitted models for parameter variation therefore have to be modified to have finite limits on the range of possible values, for example, by truncating Gaussian distributions at their the upper and lower 99.99% percentiles.

It can be computed. This is because the long-term growth rate is equal to the average annual growth rate,

$$\log \lambda_S = E \log \left(N(t+1)/N(t) \right)$$

where $N(t)$ is total population size at time t. In most cases, it will be better to look at

$$\log \lambda_S = E \log \left(V(t+1)/V(t) \right) \tag{7.4.2}$$

where $V(t) = \int_Z n(z,t) v(z) dz$ and v is the dominant left eigenvector of the mean kernel $\bar{K}(z', z) = E[K(z', z; \theta(t))]$ (or instead of v use v_0, the dominant left eigenvector for the average-environment kernel $K(z', z; \bar{\theta})$). $V(t)$ is an approximation to the total reproductive value in year t. The advantage of V is that values of $r(t) = \log(V(t+1)/V(t))$ are nearly uncorrelated. So if you have computed the $r(t)$ values from a simulation of the model, you can compute a confidence interval on their mean as if they were an independent random sam-

---

[2] Simple piecewise-continuous models are also OK, for example, a model with a size-threshold for reproduction.

ple (this doesn't say anything about uncertainty in $\log \lambda_S$ due to uncertainty in model parameters, it just tells you if you've done a long enough simulation to estimate the λ_S implied by the parameters that you're using).

It can be approximated using the IPM version of Tuljapurkar's small fluctuations approximation (SFA) to λ_S for matrix models. If year-to-year variability is small, then

$$\log \lambda_S \approx \log \lambda_1 - \frac{Var\,\langle v, K_t w \rangle}{2\lambda_1^2} + \sum_{j=1}^{\infty} c_j. \qquad (7.4.3)$$

The three terms on the right-hand side of (7.4.3) represent the mean environment, the environmental variance, and the effects of between-year correlations in environment. λ_1 is the dominant eigenvalue of the mean kernel $\bar{K}$. In the second term, K_t is shorthand for $K(z', z; \theta(t))$, v and w are the dominant left and right eigenvectors of the mean kernel *scaled so that* $\langle v, w \rangle = 1$, and $K_t w = \int_Z K_t(z', z) w(z) dz$. The second term can be written as

$$-\frac{1}{2\lambda_1^2} \iiiint \mathbf{s}(z_2', z_2)\mathbf{s}(z_1', z_1)Cov\left(K_t(z_2', z_2), K_t(z_1', z_1)\right) dz_1 dz_2 dz_1' dz_2' \quad (7.4.4)$$

analogous to equation (14.71) in Caswell (2001), where $\mathbf{s}$ is the sensitivity function for the mean kernel (equation 4.3.6), and the integrals are over all of $\mathbf{Z}$. See Section 7.8.1 for the derivation of (7.4.4).

Because a variance is always positive, (7.4.3) says that demographic variance *always hurts* population growth. You learned this, really, in your first algebra class: $(1 + x)(1 - x) = 1 - x^2$. One day the value of your retirement portfolio goes up by 10%; the next day it goes down by 10%; and that leaves you with a net loss of 1%.[3] Equation (7.4.4) adds the information that variance hurts most when it's at a high-sensitivity location in the kernel. This is the basis of the *demographic buffering hypothesis* (Pfister 1998) that the highest-sensitivity phases of the life cycle should evolve to have low sensitivity to environmental variation.

It's also illuminating to relate the second term to the idea of measuring population growth by the change in total reproductive value $V(t)$. If we use v to approximate size-dependent reproductive value, and approximate the population structure by w, and assume that environmental fluctuations are small, then

$$E\left[\log\left(V(t+1)/V(t)\right)\right] \approx \log \lambda_1 - \frac{Var\,\langle v, K_t w \rangle}{2\lambda_1^2}. \qquad (7.4.5)$$

We derive this equation in Section 7.8.1. The left-hand side is the average annual growth rate of $V(t)$, so it's the λ_S for $V(t)$ which is the same as the λ_S for population density. The right-hand side is the first two terms of the SFA (7.4.3). So the SFA says when demographic variance is low, and there is no correlation between different years, a population grows *as if* it were always in its stable size distribution.

---

[3] Hopefully offset by income from book sales.

In the third term of the SFA, c_j is the effect of environmental correlations at time-lag j (i.e., between $\theta(t)$ and $\theta(t-j)$). It is given by the unenlightening formula

$$c_j = E \langle v, M_j D^{l-1} M_0 w \rangle \tag{7.4.6}$$

where $M_t = (K_t - \bar{K})/\lambda_1$ and $D^m = (\bar{K}/\lambda_1)^m - P_0$ where $P_0(z', z) = v(z')w(z)$ for v, w scaled so that $\langle v, w \rangle = 1$. Rees and Ellner (2009, Appendix C) give a derivation of (7.4.6), but it's just a translation from Tuljapurkar (1990) and Tuljapurkar and Haridas (2006) into IPM notation.

7.4.1 Implementation

Having fitted models describing how fate depends on state, and how this varies between years we can now implement an IPM. The recipes for kernel- and parameter-selection implementation are very straight-forward:

Kernel selection

1) Construct set of yearly kernels $\rightarrow$ 2) Select a kernel at random $\rightarrow$ 3) Project the population forward one time step, and then repeat from 2).

Parameter selection

1) Simulate a parameter vector from fitted distributions $\rightarrow$ 2) Construct kernel $\rightarrow$ 3) Project the population forward one time step, then repeat from 1).

These approaches are implemented in the `iterate_model` functions in `Carlina Fixed Effects Demog.R` (Kernel selection) and `Carlina Mixed Effects Demog.R` (Parameter selection). For parameter selection, it is important that the fitted models for parameter variation have finite limits. Code to do this for independent and for correlated Gaussian distributions is given in `Carlina Mixed Effects Demog.R`.

The long-term growth rate, λ_S, can be estimated by averaging either the one-step changes in log total population size or log total reproductive value in a simulation of the model. We will illustrate both these approaches using kernel selection, as the calculation of the mean kernel is very quick. Because the kernels are selected at random from the set of fitted year-specific kernels, the mean kernel is simply the average of the year-specific kernels. For parameter selection this would be slower as we would have to generate a large number of kernels from the fitted distributions and average these. With the mean kernel calculated we then extract the dominant left and right eigenvectors, and scale them so $\langle v, w \rangle = 1$,

```
w <- Re(eigen(mean.kernel)$vectors[,1])
v <- Re(eigen(t(mean.kernel))$vectors[,1])
# scale eigenvectors <v,w>=1;
w <- abs(w)/sum(h*abs(w))
v <- abs(v)
v <- v/(h*sum(v*w))
```

The scaling doesn't change the calculation of λ_S, but will be needed for the SFA so we might as well do it now. The one-step growth rates are then

```
rt.V[year.t] <- log(sum(nt1*v)/sum(nt*v))
rt.N[year.t] <- log(sum(nt1)/sum(nt))
```

For the SFA we also need to calculate $Var \langle v, K_t w \rangle$, we already have v and w calculated and scaled, so this is just a matter of evaluating $\langle v, K_t w \rangle$ for each of the kernels, in R we have

```
for(i in 1:n.years) {
                v.Ktw[i] <- sum(v*(K.year.i[i,,] %*% w))*h
        }
```

Putting it all together, let's see how it works, and in particular test if the time series for $r(t) = \log(V(t+1)/V(t))$ really is approximately uncorrelated. Using the fixed-effects models for the parameter estimates, and running `iter <- iterate_model(m.par.est,20,n.est)` with `n.est=20000` to get the time series and averaging we find

```
> cat("log Lambda S using Nt ", Ls.Nt,
      " 95% c.i ", Ls.Nt+2*SE.Ls.Nt, " ", Ls.Nt-2*SE.Ls.Nt, "\n")
log Lambda S using Nt -0.03630683 95% c.i -0.02517611 -0.04743755
> cat("log Lambda S using Vt ", Ls.Vt,
      " 95% c.i ", Ls.Vt+2*SE.Ls.Vt, " ", Ls.Nt-2*SE.Ls.Vt, "\n")
log Lambda S using Vt -0.03634588 95% c.i -0.02621861 -0.0464341
> cat("SFA Stochastic log Lambda = ", approx.Ls, "\n")
SFA Stochastic log Lambda = -0.02255496
```

So both approaches give very similar answers, and the SFA is reasonably close. The 95% confidence intervals are however substantial, suggesting more iterations are required. So let's try `n.est=500000`,

```
> cat("log Lambda S using Nt ", Ls.Nt,
      " 95% c.i ", Ls.Nt+2*SE.Ls.Nt, " ", Ls.Nt-2*SE.Ls.Nt, "\n")
log Lambda S using Nt -0.04565778 95% c.i -0.04343278 -0.04788279
> cat("log Lambda S using Vt ", Ls.Vt,
      " 95% c.i ", Ls.Vt+2*SE.Ls.Vt, " ", Ls.Vt-2*SE.Ls.Vt,"\n")
log Lambda S using Vt -0.04566052 95% c.i -0.04363411 -0.04768692
```

So now the 95% confidence intervals are much tighter and we can be confident the "true" value is ≈ -0.046. Does the $\log(V(t))$ time series approximately follow a random walk? Remarkably it does, the partial autocorrelations for lags 1 and 2 are highly significant for $N(t)$ but much smaller for $V(t)$ despite having a very long time series, see Figure 7.3.

We could alternatively use the posterior modes from the mixed effects models as our parameter estimates, which are stored in `m.par.est.mm`. Using these we find that $\log(\lambda_S) \approx -0.040$ so there is excellent agreement between the approaches. This comes as no real surprise as we have lots of data from all years, and so the two parameter sets are highly correlated ($r = 0.999$). The posterior modes of a mixed model will be preferable to the estimates from a fixed effects model when the sample size varies greatly between years. In that situation all estimates from a fixed effects model are still given equal weight, even though some could be very poorly estimated due to low sample size. In contrast the posterior modes combine information from a particular year with the overall distribution of the

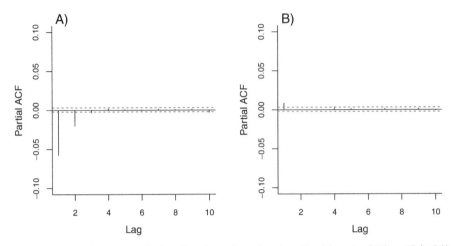

Fig. 7.3 Partial autocorrelation function plots for the A) $r(t) = \log(N(t+1)/N(t))$ and B) $r(t) = \log(V(t+1)/V(t))$ time series. The dashed lines are approximate 95% confidence intervals for an independent time series. Source file: `Carlina Fixed Effects Demog.R`.

parameter across years. The effect of this is to "shrink" poorly estimated yearly parameters back to the population mean. The amount a parameter is shrunk back depends on both how far the parameter is from the mean and how well it is estimated. Imprecisely estimated parameters, a long way from the mean, get shrunk back a lot.

Code for implementing parameter selection is given in `Carlina Mixed Effects Demog.R`, and using this we find $\log(\lambda_S) \approx -0.064$, which is close to kernel selection estimates, but the code takes considerably longer to run. At this point it is worth remembering that the true value of $\log(\lambda_S)$ for the population this dataset is drawn from is ≈ 0.12, so we're predicting extinction for a population that is rapidly increasing.

7.5 Sensitivity and elasticity analysis

Perturbation analysis for stochastic density-dependent IPMs is based on two general formulas. Both of them involve the time-varying population structure w_t and reproductive value v_t. These are computed as follows:

$$\tilde{w}_{t+1} = K_t w_t, \quad w_{t+1} = \tilde{w}_{t+1} / \int_{\mathbf{Z}} \tilde{w}_{t+1}(z) dz \tag{7.5.1}$$

$$\tilde{v}_{t-1} = v_t K_{t-1} = \int_{\mathbf{Z}} v_t(z') K_{t-1}(z', z) dz', \quad v_{t-1} = \tilde{v}_{t-1} / \int_{\mathbf{Z}} \tilde{v}_{t-1}(z) dz$$

Note that these are not scaled to have $\langle \tilde{v}_t, \tilde{w}_t \rangle = 1$.

The first formula is for a general perturbation of K_t to $K_t + \varepsilon C_t$ where $C_t = C(z', z; \theta_t)$ is a sequence of kernels such that the perturbed IPM satisfies our assumptions for small ε. Following Tuljapurkar (1990, p. 89) (or equivalently Caswell (2001, Section 14.4.1)) we have

$$\frac{\partial \log \lambda_S}{\partial \varepsilon} = \frac{1}{\lambda_S} \frac{\partial \lambda_S}{\partial \varepsilon} = E \left[\frac{\langle v_{t+1}, C_t w_t \rangle}{\langle v_{t+1}, K_t w_t \rangle} \right] \tag{7.5.2}$$

Equation (7.5.2) is applied by first computing v_t, w_t from $t = 0$ to a large time T using (7.5.1), with arbitrary positive values of w_0 and v_T. These sequences can then be used to approximate the expectation in (7.5.2) by a time average from $t = k$ to $T - k$ for some moderately large k. There's no general rule about how big T and k have to be; if you double both of them and the answer you get is effectively the same, you're done.

Equation (7.5.2) isn't informative when the right-hand side is 0. That sounds like a very special case, but it happens any time the perturbations C_t have zero mean and are independent of the unperturbed kernel. The small fluctuations approximation (7.4.3) is an example. The fluctuations around the mean kernel $\bar{K}$ are a small perturbation with zero mean, and are independent of $\bar{K}$ because a constant is independent of everything (no variance, no covariance). The effect of the added noise is proportional to its mean square (variance), not its standard deviation. The same is true when noise is added to time-invariant parts of a time-varying kernel. To compare the effects of adding noise to different parts of a time-varying kernel, we need to carry the expansion out to second order in ε. The result is:

$$\log \lambda_S(\varepsilon) = \log \lambda_S(0) - \frac{\varepsilon^2}{2} Var \left[\frac{\langle v_{t+1}, H_t w_t \rangle}{\langle v_{t+1}, K_t w_t \rangle} \right] + O(\varepsilon^3)$$

$$= \log \lambda_S(0) - \frac{\varepsilon^2}{2} E \left[\frac{\langle v_{t+1}, H_t w_t \rangle^2}{\langle v_{t+1}, K_t w_t \rangle^2} \right] + O(\varepsilon^3) \tag{7.5.3}$$

when the perturbation kernels $\{H_t\}$ have zero mean and are independent of the unperturbed kernel sequence $\{K_t\}$. Another way of saying this is that

$$\frac{\partial \log \lambda_S}{\partial \varepsilon} = 0, \quad \frac{\partial \log \lambda_S}{\partial \varepsilon^2} = -\frac{1}{2} E \left[\frac{\langle v_{t+1}, H_t w_t \rangle^2}{\langle v_{t+1}, K_t w_t \rangle^2} \right]. \tag{7.5.4}$$

Equations (7.5.2, 7.5.3) can be used to derive a vast list of perturbation formulas for changes to kernel entries, demographic functions, and parameter values (Table 7.2). One reason for this is that every possible perturbation to a deterministic model is now tripled: we can perturb only the mean, perturb only the standard deviation, or do a fractional perturbation (e.g., a 5% higher value each year) that changes both the mean and the standard deviation (Tuljapurkar et al. 2003, 2004). There are four perturbations we will use, the first three are for time-varying quantities, the fourth for introducing variation into a time-invariant quantity.

Table 7.2 Stochastic sensitivities and elasticities of λ_S. See Rees and Ellner (2009, Appendices E and F)) for derivations.

Sensitivity measure	Notation and Formula
Sensitivity of λ_S to kernel value $K_t(z', z)$ = sensitivity of λ_S to mean of $K_t(z', z)$	$s_S(z', z) = s_S^\mu(z', z) =$ $\lambda_S E \left[\dfrac{v_{t+1}(z')w_t(z)}{\langle v_{t+1}, K_t w_t \rangle} \right]$
Elasticity of λ_S to kernel value $K_t(z', z)$	$e_S(z', z) = E \left[\dfrac{v_{t+1}(z')w_t(z)K_t(z', z)}{\langle v_{t+1}, K_t w_t \rangle} \right]$
Elasticity of λ_S to the mean of kernel value $K_t(z', z)$	$e_S^\mu(z', z) = \bar{K}(z', z)E \left[\dfrac{v_{t+1}(z')w_t(z)}{\langle v_{t+1}, K_t w_t \rangle} \right]$
Elasticity to standard deviation of kernel value $K_t(z', z)$	$e_S^\sigma(z', z) = e_S(z', z) - e_S^\mu(z', z)$ $(= 0 \text{ if } Var(K_t(z', z)) = 0)$
Sensitivity of λ_S to the standard deviation of time-varying kernel value $K_t(z', z)$	$s_S^\sigma(z', z) = \lambda_S e_S^\sigma(z', z)/\sqrt{Var(K_t(z', z)}$
Sensitivity of λ_S to the variance of time-varying kernel value $K_t(z', z)$	$s_S^{\sigma^2}(z', z) = 0.5 s_S^\sigma(z', z)/\sqrt{Var(K_t(z', z)}$
Sensitivity to adding independent variability to a time-invariant kernel value $K(z', z)$	$s_S^{\sigma^2, 0}(z', z) = -\dfrac{\lambda_S}{2} E \left[\dfrac{(v_{t+1}(z')w_t(z))^2}{\langle v_{t+1}, K_t w_t \rangle^2} \right]$
Sensitivity of λ_S to parameter θ_i = sensitivity of λ_S to mean of θ_i	$s_{S,i} = s_{S,i}^\mu =$ $\lambda_S E \left[\left\langle v_{t+1}, \frac{\partial K_t}{\partial \theta_i} w_t \right\rangle / \langle v_{t+1}, K_t w_t \rangle \right]$
Elasticity of λ_S to parameter θ_i	$e_{S,i} =$ $E \left[\theta_i(t) \left\langle v_{t+1}, \frac{\partial K_t}{\partial \theta_i} w_t \right\rangle / \langle v_{t+1}, K_t w_t \rangle \right]$
Elasticity of λ_S to the mean of θ_i	$e_{S,i}^\mu = \bar{\theta}_i E \left[\left\langle v_{t+1}, \frac{\partial K_t}{\partial \theta_i} w_t \right\rangle / \langle v_{t+1}, K_t w_t \rangle \right]$
Elasticity of λ_S to the standard deviation of θ_i	$e_{S,i}^\sigma = e_{S,i} - e_{S,i}^\mu \quad [= 0 \text{ if } Var(\theta_i) = 0]$
Sensitivity of λ_S to standard deviation of $\theta_i(t)$	$s_{S,i}^\sigma = \lambda_S e_{S,i}^\sigma/\sqrt{Var(\theta_i)}$
Sensitivity of λ_S to variance of $\theta_i(t)$	$s_{S,i}^{\sigma^2} = 0.5 s_{S,i}^\sigma/\sqrt{Var(\theta_i)}$
Sensitivity of λ_S to added variance in time-invariant parameter θ_i	$s_{S,i}^{\sigma^2, 0} =$ $\dfrac{\lambda_S}{2} \left(E \left[\left\langle v_{t+1}, \frac{\partial^2 K_t}{\partial \theta_i^2} w_t \right\rangle \middle/ \langle v_{t+1}, K_t w_t \rangle \right] - E \left[\left\langle v_{t+1}, \frac{\partial K_t}{\partial \theta_i} w_t \right\rangle^2 \middle/ \langle v_{t+1}, K_t w_t \rangle^2 \right] \right)$

- $p_t \to p_t + \varepsilon$ changes mean, variance unchanged.
- $p_t \to p_t + \varepsilon p_t$ changes mean and variance with $CV = \sigma/\mu$ constant.
- $p_t \to p_t + \varepsilon(p_t - \bar{p})$ no change to mean, change in variance proportional to ε^2.
- $p \to p + \varepsilon\xi$ where ξ has zero mean and unit variance - no change to mean, change in variance proportional to ε.

7.5.1 Kernel perturbations

There are several formulae in Table 7.2 for perturbations to kernel entries, so let's see where these come from. As in Chapter 4 we will need δ functions to pick out particular transitions, and preserve the smooth properties of the kernel. Because we're dealing with kernels we need the bivariate δ functions $\delta_{z'_0, z_0}(z', z)$, which allow us to select a particular transition $z_0 \to z'_0$; see Box 4.1 for a quick summary of δ functions. Let's first consider a simple additive perturbation, so $C_t = \delta_{z'_0, z_0}(z', z)$ which we can plug into equation 7.5.2, and so

$$\frac{\partial \lambda_S}{\partial \varepsilon} = \lambda_S E \left[\frac{v_{t+1}(z'_0) w_t(z_0)}{\langle v_{t+1}, K_t w_t \rangle} \right] \tag{7.5.5}$$

which is the sensitivity of λ_S to the kernel values. For the elasticity the perturbation is proportional to the kernel, and so $C_t = \delta_{z'_0, z_0}(z', z) K_t(z'_0, z_0)$. Plugging this into equation 7.5.2 and using the definition of an elasticity we find

$$e_S(z'_0, z_0) = \frac{\partial \log \lambda_S}{\partial \log K_t(z'_0, z_0)} = E \left[\frac{v_{t+1}(z'_0) w_t(z_0) K_t(z'_0, z_0)}{\langle v_{t+1}, K_t w_t \rangle} \right]. \tag{7.5.6}$$

We can also look at perturbations to the mean and standard deviation of the kernel. Let $\bar{K}$ be the mean kernel then for a proportional perturbation $C_t = \delta_{z'_0, z_0}(z', z) \bar{K}(z'_0, z_0)$ we find the elasticity to the mean kernel to be

$$e_S^\mu(z'_0, z_0) = \bar{K}(z'_0, z_0) E \left[\frac{v_{t+1}(z'_0) w_t(z_0)}{\langle v_{t+1}, K_t w_t \rangle} \right]. \tag{7.5.7}$$

To calculate the elasticity to the standard deviation of the kernel the perturbation kernel is $C_t = \delta_{z'_0, z_0}(z', z)(K_t(z'_0, z_0) - \bar{K}(z'_0, z_0))$. The reason this is so, is that the variance of εC_t is $\varepsilon^2 Var(K_t(z'_0, z_0))$ and so the standard deviation of the perturbed kernel is proportional to ε. We could plug this into the general perturbation formula but it is simpler to note that this perturbation is the difference between the previous two perturbations and so

$$e_S^\sigma(z'_0, z_0) = e_S(z'_0, z_0) - e_S^\mu(z'_0, z_0) \tag{7.5.8}$$

Rewriting this as $e_S(z'_0, z_0) = e_S^\mu(z'_0, z_0) + e_S^\sigma(z'_0, z_0)$ shows that the stochastic elasticity, e_S is the sum of two terms each with a clear meaning: one describing the effects of varying the mean and the other the standard deviation. Hence, the elasticities e_S being the sum of two terms each describing distinct effects does not have a clear interpretation on its own (Haridas and Tuljapurkar 2005).

To calculate the effect of changing the standard deviation we used the perturbation kernel $C_t = \delta_{z_0', z_0}(z', z)(K_t(z_0', z_0) - \bar{K}(z_0', z_0))$ which is zero when the kernel is time-invariant. In this case we need to use equation 7.5.4, setting the perturbation kernel to $H_t = \delta_{z_0', z_0}(z', z)\xi$ where ξ is a white-noise process with zero mean and unit variance $(=1)$ that is independent of the unperturbed kernel, we find

$$s_S^{\sigma^2, 0}(z_0', z_0) = -\frac{\lambda_S}{2} E\left[\frac{\langle v_{t+1}(z_0')w_t(z_0)\rangle^2}{\langle v_{t+1}, K_t w_t\rangle^2}\right] \tag{7.5.9}$$

Implementation We will illustrate the various calculations using kernel selection as this is simpler and faster to implement. First we will generate the sequence of environments using year.i <- sample(1:n.years,n.est+1, replace=TRUE), where n.years is the number of yearly parameter sets we have, and n.est is the length of the time series we're going to generate. We then calculate the mean kernel and estimate λ_S using the code from Section 7.4.1. For calculating the various expectations we need long time series of w_t and v_t as specified in equation (7.5.1). The R code for this is much as you might expect:

```
### Get wt time series ###
wt <- matrix(1/nBigMatrix, nrow = n.est+1, ncol= nBigMatrix)
for (i in 1:n.est) {
K           <- K.year.i[year.i[i],,]
wt[i+1,]    <- K %*% wt[i,]
wt[i+1,]    <- wt[i+1,] / sum(wt[i+1,])
if(i%%10000 = = 0) cat("wt ", i, "\n")
}
### Get vt time series ###
vt <- matrix(1/nBigMatrix, nrow = n.est+1, ncol = nBigMatrix)
for (i in (n.est+1):2) {
K           <- K.year.i[year.i[i],,]
vt[i-1,]    <- vt[i,] %*% K
vt[i-1,]    <- vt[i-1,] / sum(vt[i-1,])
if (i%%10000 = = 0) cat("vt ", i, "\n")
}
```

We can then evaluate the expectations for the stochastic sensitivities and elasticities. Note, we do this for all combinations of z and z' using vt[i+1,] %*% t(wt[i,]) to generate a matrix of values.

```
for (i in n.runin:(n.est-n.runin)) {
# standard calculations needed for the various formulae
K           <-   K.year.i[year.i[i],,]
vt1.wt      <-   vt[i+1,] %*% t(wt[i,])
vt1.K.wt    <-   sum(vt[i+1,] * (K %*% wt[i,]))
# calculation of the standard sensitivities and elasticities
sens.s <- sens.s + vt1.wt / vt1.K.wt
```

```
elas.s <- elas.s + K * (vt1.wt / vt1.K.wt)
}
elas.s <- elas.s / (n.est - 2 * n.runin + 1)
sens.s <- Ls * sens.s / (n.est - 2 * n.runin + 1)
```

The last two lines add up the values excluding n.runin values at the beginning and end of the series, and then divide by the number of observation to get the average. This code is wrapped up in a function stoc_pert_analysis with arguments specifying the parameter vector (params), length of the time series (n.est) and number of observations to ignore at the beginning and end of the time series (n.runin). Using stoc_pert_analysis we can generated the sensitivity and elasticity surfaces for our IPM, see Figure 7.4. This shows that changes in the mean or standard deviation of the kernel in the size range where flowering occurs (≈ 4 on log scale) have the greatest impact on λ_S.

For parameter selection we use essentially the same procedure except we need to generate and store the sequence of yearly parameter vectors or kernels, rather than year types. Storing parameter vectors save memory but is costly in terms of evaluation time; storing kernels has the opposite properties.

7.5.2 Function perturbations

As you learnt in Chapter 4 the effects of perturbation to kernel functions are kernel-specific. Having said this, the general approach is very similar to Chapter 4, namely: write out the perturbation, identify the perturbation kernel, and substitute into the appropriate sensitivity equation (7.5.2 or 7.5.3). Let's run through an example, say the elasticity of the probability of flowering function at size z_0. So the perturbation is

$$p_b(z, \theta(t)) \to p_b(z, \theta(t)) + \epsilon p_b(z, \theta(t)) \delta_{z_0}(z) \qquad (7.5.10)$$

with $\delta_{z_0}(z)$ being a one-dimensional δ function. The *Carlina* kernel is

$$\begin{aligned}
K(z', z, \theta(t)) =&s(z, \theta(t))(1 - p_b(z, \theta(t)))G(z', z, \theta(t)) \\
&+ s(z, \theta(t))p_b(z, \theta(t))b(z, \theta(t))p_r c_0(z', \theta(t))
\end{aligned} \qquad (7.5.11)$$

and so substituting the perturbed $p_b(z, \theta(t))$ and collecting together the ϵ terms, we find that the perturbation kernel $C_t(z', z)$ equals

$$p_b(z, \theta(t))s(z, \theta(t))\big(p_r b(z, \theta(t))c_0(z', \theta(t)) - G(z', z, \theta(t))\big)\delta_{z_0}(z). \qquad (7.5.12)$$

Plugging this into equation (7.5.2) and using the definition of an elasticity we find

$$e_S^{p_b}(z_0) = \frac{\partial \log \lambda_S}{\partial \log p_b(z_0)} =$$

$$E\left[p_b(z_0, \theta(t))s(z_0, \theta(t))w_t(z_0)\frac{\langle v_{t+1}, p_r b(z_0, \theta(t))c_0(z', \theta(t)) - G(z', z_0, \theta(t))\rangle}{\langle v_{t+1}, K_t w_t\rangle}\right]$$

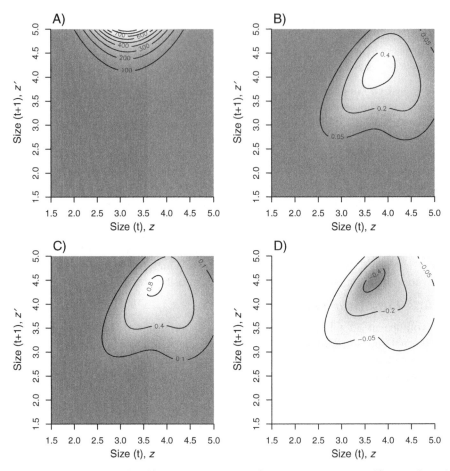

Fig. 7.4 Contour plots for A) kernel sensitivity, B) kernel elasticity, e_S, C) mean kernel elasticity, e_S^μ, and D) elasticity to kernel standard deviation, e_S^σ for the stochastic density-independent *Carlina* IPM. Source file: `Carlina Fixed Effects K pert.R`.

an ugly expression by any measure, but still one that can be calculated using a long time series of w_t and v_t values from the unperturbed model. Substituting $\bar{p}_b(z_0, \theta(t))$ into the preceding equation gives the elasticity to the mean probability of flowering function at size z_0. The difference between these is the elasticity to the standard deviation $(e_S^{p_b,\sigma} = e_S^{p_b} - e_S^{p_b,\mu})$.

Implementation: As the perturbation formulae are kernel-specific we will code the analysis in terms of a general perturbation kernel $C_t(z', z)$. The approach we will use is based on writing the general perturbation formula as

$$\frac{1}{\lambda_S} \frac{\partial \lambda_S}{\partial \varepsilon} = E\left[\frac{\langle v_{t+1}, C_t w_t\rangle}{\langle v_{t+1}, K_t w_t\rangle}\right] = \int \int E\left[\frac{v_{t+1}(z')w_t(z)C_t(z', z)}{\langle v_{t+1}, K_t w_t\rangle}\right] dz' dz \tag{7.5.13}$$

so we evaluate the expectation at each of the mesh points, z', z and then integrate. Continuing with the elasticity of the probability of flowering function at z_0 example, we define a function that represents the perturbation kernel (7.5.12), then construct it for each year using outer,

```
Ct_z1z <- function(z1,z,m.par){
  return( p_bz(z, m.par) * s_z(z, m.par) *
    ( m.par["p.r"]*b_z(z, m.par)*c_0z1(z1, m.par) - G_z1z(z1,z,m.par)) )
}
```

```
C.pert <- array(NA,c(n.years,nBigMatrix,nBigMatrix))
for(i in 1:n.years){
  year.C <- h*(outer(meshpts, meshpts, Ct_z1z, m.par = params.to.use[,i]))
  C.pert[i,,] <- year.C
}
```

The perturbation kernels are stored in a 3-dimensional array, with the first index specifying the year. Note, we ignored the δ function when we constructed $C_t(z', z)$ and so when we use results from calculations using C.pert we will need to sum down columns to obtain elasticities for specific values of z_0. Next we evaluate the perturbation kernels corresponding to the mean function, $\bar{p}_b(z, \theta(t))$, using essentially the same code but with $p_b(z, \theta(t))$ replaced by $\bar{p}_b(z, \theta(t)$ in the kernel function definition. With the two sets of perturbations kernels calculated we then supply these to a function stoc_pert_analysis which evaluates the expectation $E\left[\frac{v_{t+1}(z')C_t(z',z)w_t(z)}{\langle v_{t+1}, K_t w_t\rangle}\right]$ at each of the mesh points, z', z. The code is similar to that used in Chapter 4,

```
vt1.wt    <- outer(vt[year.t+1,], wt[year.t,], FUN="*")
vt1.C.wt  <- vt1.wt * C.t[year.i[year.t],,]
```

Dividing each of the vt1.C.wt by vt1.wt and averaging gives the required expectation at each of the mesh points, z', z. The function stoc_pert_analysis returns a matrix with the expectation at each of the mesh points as entries, and so we sum down the columns to get the elasticity

```
elas.s <- apply(pert.K$elas.s, 2, sum)
```

We use the same approach to calculate the elasticity to the mean function and, as before, the elasticity to the standard deviation is the difference between these $(e_S^{p_b,\sigma} = e_S^{p_b} - e_S^{p_b,\mu})$. The resulting elasticities are plotted in Figure 7.5.

By changing the definition of the perturbation kernel we can evaluate the elasticities for the survival function $s(z)$. The perturbation kernel in this case is the kernel, K_t and so the elasticities integrate to one, providing a useful check on the code. Finally we will look at perturbations to the growth kernel $G(z', z)$, setting the perturbation to $G(z', z) \to G(z', z) + \epsilon G(z', z)\delta_{z'_0, z_0}(z', z)$ we find the perturbation kernel to be $C_t(z', z) = s(z)(1 - p_b(z))G(z', z)\delta_{z'_0, z_0}(z', z)$. Redefining the perturbation kernel allows the same code to be used but now as were interested in elasticities at z'_0, z_0 there is no need to sum over the columns. The resulting elasticities need to be interpreted carefully as $\int G(z', z)dz' = 1$

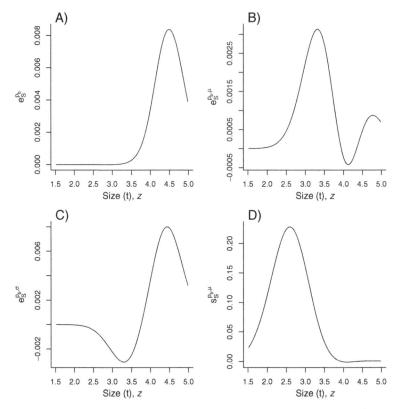

Fig. 7.5 Perturbation analysis for the probability of flowering function p_b. A) Stochastic elasticity, $e_S^{p_b}$, B) stochastic elasticity to mean function, $e_S^{p_b,\mu}$, C) stochastic elasticity to standard deviation, $e_S^{p_b,\sigma}$, and D) stochastic sensitivity to mean, $s_S^{p_b,\mu}$ for the density-independent *Carlina* IPM. Source file: `Carlina Fixed Effects K fun pert.R`.

and so an increase in one growth transition has to be compensated by decreases somewhere else. For locations at which the elasticity is small, increases there that were compensated by decreases elsewhere would generally decrease λ_S. Increases where the elasticity is large, compensated by decreases elsewhere, would generally increase λ_S. With this interpretation, the elasticity analysis of $G(z', z)$ demonstrates the importance of growth transitions into the flowering size classes (Figure 7.6A). Increases in the mean of the growth function increase the stochastic growth rates as expected (Figure 7.6B), while increases in the standard deviation have the opposite effect (Figure 7.6C).

7.5.3 Parameter perturbations

In the preceding sections we have identified the perturbation kernel by substituting the perturbation into K_t and collecting the ε terms. Equally we could have identified the perturbation kernel by calculating $C_t = \partial K_t / \partial \varepsilon$. This is the approach we will use here, as we need the chain rule to identify the perturbation

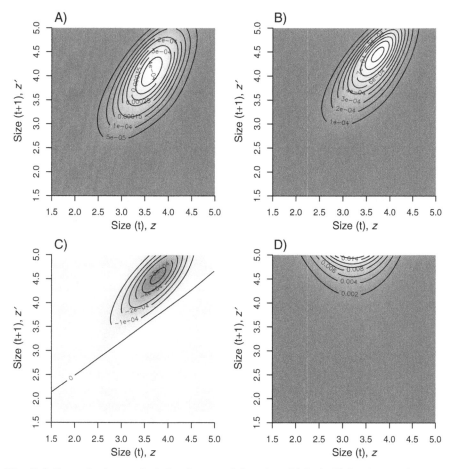

Fig. 7.6 Perturbation analysis for the growth function $G(z', z)$. A) Stochastic elasticity, e_S^G, B) stochastic elasticity to mean function, $e_S^{G,\mu}$, C) stochastic elasticity to standard deviation, $e_S^{G,\sigma}$, and D) stochastic sensitivity to mean, $s_S^{G,\mu}$ for the density-independent *Carlina* IPM. Source file: `Carlina Fixed Effects K fun pert.R`.

kernel when one of the function parameters, $\theta_i(t)$, is perturbed. Specifically we can write

$$C_t = \frac{\partial K_t}{\partial \theta_i} \frac{\partial \theta_i}{\partial \varepsilon}. \tag{7.5.14}$$

Assuming a proportional perturbation so $\theta_i(t) \rightarrow \theta_i(t) + \varepsilon \theta_i(t)$ and using the definition of an elasticity we find

$$e_{S,i} = \frac{\partial \log \lambda_S}{\partial \log \theta_i(t)} = E\left[\theta_i(t) \left\langle v_{t+1}, \frac{\partial K_t}{\partial \theta_i} w_t \right\rangle / \langle v_{t+1}, K_t w_t \rangle\right] \tag{7.5.15}$$

and using $\theta_i(t) \to \theta_i(t) + \varepsilon$ we find the sensitivity to be

$$s_{S,i} = \frac{\partial \lambda_S}{\partial \theta_i(t)} = \lambda_S E\left[\left\langle v_{t+1}, \frac{\partial K_t}{\partial \theta_i} w_t \right\rangle / \langle v_{t+1}, K_t w_t \rangle \right]. \qquad (7.5.16)$$

As with kernel entries this is also the sensitivity of λ_S to the mean of θ_i, $s_{S,i} = s_{S,i}^\mu$. Setting $\theta_i(t) \to \theta_i(t) + \varepsilon \bar{\theta}_i(t)$ we find the elasticity to the mean of θ is

$$e_{S,i}^\mu = \frac{\partial \log \lambda_S}{\partial \log \bar{\theta}_i} = \bar{\theta}_i E\left[\left\langle v_{t+1}, \frac{\partial K_t}{\partial \theta_i} w_t \right\rangle / \langle v_{t+1}, K_t w_t \rangle \right] \qquad (7.5.17)$$

and as before $e_{S,i}^\sigma = e_{S,i} - e_{S,i}^\mu$.

Implementation: Previous calculations for the kernel and kernel functions were based on the perturbation kernels

$$C_t = \frac{\partial K_t}{\partial \varepsilon} \quad \text{and} \quad C_t = \frac{\partial K_t}{\partial f} \frac{\partial f}{\partial \varepsilon} \qquad (7.5.18)$$

respectively. Building on this allows us to reuse our previous code, so we will write the perturbation kernel as

$$C_t = \frac{\partial K_t}{\partial f} \frac{\partial f}{\partial \theta_i} \frac{\partial \theta_i}{\partial \varepsilon}. \qquad (7.5.19)$$

As before we will write the code in terms of the general perturbation kernel, C_t. So let's dive in and look at the elasticity of the intercept of the flowering function, p_b, the perturbation is $\beta_0 \to \beta_0 + \varepsilon \beta_0$, and so various bits we need are

$$\frac{\partial K_t}{\partial p_b} = s(b p_r c_0 - G), \quad \frac{\partial p_b}{\partial \beta_0} = \frac{\exp(\beta_0 + \beta_z z)}{(1 + \exp(\beta_0 + \beta_z z))^2} \quad \text{and} \quad \frac{\beta_0}{\partial \varepsilon} = \beta_0 \qquad (7.5.20)$$

where we have suppressed the function arguments to simplify presentation in the first equation. Defining the perturbation kernel as

```
Ct_z1z <- function(z1,z,m.par) {
return(s_z(z, m.par) *
(m.par["p.r"]*b_z(z, m.par)*c_0z1(z1, m.par) - G_z1z(z1, z, m.par)) *
(1/(1+exp(m.par["flow.int"]+m.par["flow.z"]*z))) * p_bz(z, m.par) *
m.par["flow.int"])
}
```

we can use the function stoc_pert_analysis to evaluate the elasticity at each of the mesh points z', z and then sum these to get the elasticity. Here's the code:

```
> pert.K <- stoc_pert_analysis(params.to.use, n.est, n.runin,
+           C.pert, C.pert.mean)
> elas.s <- sum(pert.K$elas.s)
> elas.s.mean <- sum(pert.K$elas.s.mean)
> elas.s.sd <- elas.s-elas.s.mean
> cat("Stochastic elasticity ", elas.s, "\n")
```

```
Stochastic elasticity -1.417913
> cat("Stochastic elasticity mean ", elas.s.mean, "\n")
Stochastic elasticity mean -1.371312
> cat("Stochastic elasticity sd ", elas.s.sd, "\n")
Stochastic elasticity sd -0.04660165
> cat("Stochastic sens mean ", pert.K$Ls*elas.s.mean/beta.0.mean, "\n")
Stochastic sens mean 0.07966138
> cat("Stochastic sensitivity sd ", pert.K$Ls*elas.s.sd/beta.0.sd, "\n")
Stochastic sens sd -0.04463693
```

Elasticities with respect to underlying parameters are less clearly interpretable than the corresponding sensitivities, because parameters may be negative. A proportional change in a negative parameter makes it go further negative, so the elasticity and sensitivity can have opposite signs. Stochastic sensitivities play an important role in evolutionary demography (Chapter 9) because they can be interpreted as a fitness gradient for evolutionary stability analysis, i.e., for predicting whether a homogeneous population will be open to invasion by a rare alternative type (though not for analysis of short-term selection response in a genetically heterogeneous population (Lande 2007)). So for negative underlying parameters, the sensitivity rather than the elasticity will correctly indicate the predicted direction of long-term adaptive dynamics. This issue doesn't arise in conventional matrix models because the underlying parameters are intrinsically positive, either matrix entries themselves or underlying probabilities of events such as survival, growth, or breeding.

This approach can be used with any of the time-varying parameters simply by redefining the perturbation kernel, C_t. For time-invariant parameters we need to evaluate

$$s_{S,i}^{\sigma^2,0} = \frac{\lambda_S}{2} \left(E\left[\left\langle v_{t+1}, \frac{\partial^2 K_t}{\partial \theta_i^2} w_t \right\rangle \Big/ \langle v_{t+1}, K_t w_t \rangle \right] \right. \tag{7.5.21}$$

$$\left. - E\left[\left\langle v_{t+1}, \frac{\partial K_t}{\partial \theta_i} w_t \right\rangle^2 \Big/ \langle v_{t+1}, K_t w_t \rangle^2 \right] \right) \tag{7.5.22}$$

As an example of this type of calculation let's look at the intercept of the seed production function, $b(z) = \exp(A + Bz)$ which does not vary between years. For the expectations we need $\partial K_t/\partial A$ and $\partial^2 K_t/\partial A^2$ writing out the kernel and differentiating we get

$$\frac{\partial K_t}{\partial A} = sp_r p_b c_0 \frac{\partial b}{\partial A} = sp_r p_b c_0 b, \quad \frac{\partial^2 K_t}{\partial A^2} = sp_r p_b c_0 \frac{\partial^2 b}{\partial A^2} = sp_r p_b c_0 b \tag{7.5.23}$$

where we have suppressed the function arguments. We then implement equation (7.5.22) by calculating the kernels $\partial K_t/\partial A$ and $\partial^2 K_t/\partial A^2$ for each year and passing these to the function stoc_pert_analysis_invariant. Much of this code is the same as stoc_pert_analysis, but the inner products are evaluated directly using

```
vt1.K.t.deriv.wt <- sum(vt[year.t+1,] *
          (K.t.deriv[year.i[year.t],,] %*% wt[year.t,]))
vt1.K.t.deriv2.wt <- sum(vt[year.t+1,] *
          (K.t.deriv2[year.i[year.t],,] %*% wt[year.t,]))
```

and then averaged appropriately. Note, in this case $\partial K_t/\partial A$ and $\partial^2 K_t/\partial A^2$ are the same, but we've written the code as if they were different so it's easier to modify for use with other parameters. In this case $s_{S,A}^{\sigma^2,0} = 0.0905$ so adding variability increases λ_S which is a consequence of Jensen's inequality, which says adding variability increases $b(z)$, as the function is concave up, so in this case we expect $s_{S,A}^{\sigma^2,0} > 0$.

7.6 Life Table Response Experiment (LTRE) Analysis

LTRE is a retrospective analysis that aims to identify which aspects of the variation in vital rates are most important for the population. For stochastic-environment IPMs the "random design" LTRE is most appropriate because it assumes that the observed variation was a random draw from some probability distribution (Caswell 2001, Chapter 10).

The starting point is the joint distribution of all time-varying parameters. In fixed-effects model this is a discrete distribution, consisting of the set of estimated year-specific parameter vectors (assigned equal probability); a mixed-effects model gives you a parametric distribution for the parameter vector. The importance of different vital rates is then measured by how much they contribute to the variation in λ_t, which (in this section) means the dominant eigenvalue of the year-specific kernel K_t. The variance λ_t is decomposed, in ANOVA-like fashion, into contributions from different time-varying parameters. In a matrix model, the "parameters" of interest are usually the matrix entries, but in an IPM it would generally be the time-varying parameters of demographic models.

The simplest LTRE is based on a first-order (linear) Taylor approximation for λ_t as a function of the parameter vector θ:

$$Var(\lambda_t) \approx \sum_{i=1}^{n} \sum_{j=1}^{n} Cov(\theta_i, \theta_j) s_i s_j \qquad (7.6.1)$$

where s_i is the eigenvalue sensitivity $\partial \lambda/\partial \theta_i$ in the mean kernel. Because the kernel is a nonlinear function of the parameters, the mean kernel and $\partial \bar{K}/\partial \theta_i$ should be estimated by drawing θ values from the fitted distribution, and averaging the year-specific K's and $\partial K/\partial \theta_i$'s.[4]

The same approximation can be done at the level of kernel entries using equation (4.3.7): $Var(\lambda_t)$ is approximately

$$\iiiint \mathbf{s}(z_1', z_1)\mathbf{s}(z_2', z_2)Cov(K_t(z_1', z_1), K_t(z_2', z_2))dz_1' \, dz_1 \, dz_2' \, dz_2. \qquad (7.6.2)$$

---

[4] The average value of $\partial K/\partial \theta_i$ equals $\partial \bar{K}/\partial \theta_i$.

A natural way to approximate derivatives is with centered differences, so

$$\frac{\partial f}{\partial x} \approx \frac{f(x + \Delta) - f(x - \Delta)}{2\Delta} \tag{7.6.3}$$

where Δ is a small constant. Selecting Δ is a compromise between accurately approximating the derivative (small Δ is good), and the loss of precision that results from subtracting two large numbers to get a small one (big Δ is good). Luckily this is a standard problem in numerical analysis. The key result is that if the magnitude of f and its second derivative are similar then $\Delta \approx \epsilon^{1/3}$ is about optimal, where ϵ is machine precision; in R `epsilon=.Machine$double.eps`. When the second derivative is needed we use

$$\frac{\partial^2 f}{\partial x^2} \approx \frac{f(x + \Delta) - 2f(x) + f(x - \Delta)}{\Delta^2}. \tag{7.6.4}$$

Box 7.2: Finite differences

We expect that (7.6.1) will be more useful because the components of variability there have immediate biological interpretations.

However, these linear approximations are only accurate for small environmental variation, which is not the norm. The errors can be large. Rees and Ellner (2009) compared linear LTRE for the density-independent stochastic *Carlina* IPM to a nonlinear LTRE based on a generalized additive model for λ as a function of parameters,

$$\lambda_t \approx \beta_0 + \sum_j f_j(\theta_j)$$

where each f_j is a fitted smooth spline function. The nonlinear LTRE accounted for 97% of the variation in λ_t, compared to 77% in the linear LTRE, and the importance of survival variability was under-estimated in the linear LTRE by almost a factor of 2. This issue is not specific to IPMs; we expect that the usual linear LTRE will also be inaccurate for many stochastic matrix models.

Implementation: There are several ways to think about LTREs. We could ask how variation in the estimated model parameters influences the observed sequence of λ_t; this is a classical retrospective analysis, aimed at explaining what has already been observed. Alternatively we could use the fitted mixed effects models to generate a large number of yearly kernels, that have never been seen, and analyze these. This is the approach we will use here but the same code can be used to analyze any observed sequence of kernels. To implement equation 7.6.1 in a general way we will calculate the eigenvalue sensitivities $\partial \lambda_t / \partial \theta_i$, using finite differences, Box 7.2. Finite differences are a very handy way of calculating sensitivities even when exact formulae are available (e.g., Chapter 4), and provide a useful check on code.

The function `LTRE.calcs` takes a parameter matrix (rows = parameters, cols = years) and returns a data frame suitable for analysis (non-varying parameters removed, a column for each parameter, and λ_t as the final column) and the sensitivities. Implementing equation 7.6.1 is then straightforward

```
sens.mat <- outer(d.lambda.d.theta, d.lambda.d.theta, FUN="*")
var.cov <- cov(ps.and.1[,1:6])
var.terms.1.order <- sens.mat*var.cov
```

The variance in λ_t is ≈ 0.98 whereas using the approximation we get ≈ 1.24 as error of about 27%. There are various ways of improving the analysis. For example, if we fit a linear regression model for λ_t we get

$$Var(\lambda_t) \approx \sum_{i=1}^{n} \sum_{j=1}^{n} Cov(\theta_i, \theta_j)\beta_i\beta_j + Var(\epsilon) \qquad (7.6.5)$$

where β_i is the fitted regression slope for parameter i, and $Var(\epsilon)$ the error variance. Implementation is straightforward using lm to fit the regression model and then replacing the sensitivities in the previous code with the estimated slopes. Regression diagnostics can be used to explore the adequacy of the fitted model, and interactions fitted if needed. Where more flexibility is needed Generalized Additive Models (GAMs, in mgcv) can be used. The fitted values of each of the model terms can then be extracted with terms <- predict(fit,type="terms"), and cov(terms) gives the required weighted variance-covariance matrix. The variance-covariance matrix can then be presented either graphically or in a table. Rees et al. (2010) suggest the contribution of variability in θ_i can be assessed by summing along the rows of the weighted variance-covariance matrix and then dividing by the sum, so for the linear regression model we have

$$Cont(\theta_i) = \frac{\sum_{j=1}^{n} Cov(\theta_i, \theta_j)\beta_i\beta_j}{\sum_{i=1}^{n} \sum_{j=1}^{n} Cov(\theta_i, \theta_j)\beta_i\beta_j}. \qquad (7.6.6)$$

This must be interpreted carefully as small values can arise through:

1. little variation in θ_i and so small variance and covariance terms,
2. θ_i having little effect on λ_t, or
3. negative covariance between terms.

As a "one size fits all" approach, we suggest basing LTRE on a totally non-parametric model for the dependence of λ on time-varying parameters. A convenient option is *random forests*, a machine learning algorithm proposed by Leo Breiman based on decision tree regression. The randomForest package (Liaw and Wiener 2002) makes it available in R. We use randomForest to fit a regression model with θ as the predictors and λ_t as the response, and use varImpPlot to plot the relative importance of each parameter.

The first step in fitting the model is to optimize mtry, the number of variables considered for each branch in the decision trees:

```
year.p    <- subset(ps.and.1, select=-lambda.i)
lambda.t <- ps.and.1$lambda.i
```

```
out       <- tuneRF(x=year.p, y=lambda.t,
                    ntreeTry=500, stepfactor=1, improve=0.02)
plot(out) ## tells us to use mtry=4
```

For our example, the plot tells us to use `mtry=4`.

```
lambdaRF <- randomForest(x=year.p, y=lambda.t, ntree=500,
+              importance=TRUE, mtry=4)
set_graph_pars("panel2")
plot(lambdaRF$rsq, type="l")
varImpPlot(lambdaRF, type=1, scale=FALSE, main="")
```

This produces two plots. The first shows that the number of trees is more than enough to reach the maximum possible forecasting accuracy, and the second plots a measure of importance for each variable. Importance is measured by asking: how badly is λ mis-predicted, if the actual value of the focal variable is replaced by a value chosen at random from those in the data frame? It's like dropping one factor in an ANOVA and seeing how much the mean square error goes up (with the difference that the factor isn't actually dropped, instead it's randomized so that it becomes irrelevant).

There is good agreement between the regression-based approaches, with the most important terms being variation in the survival slope, followed by variation in the growth slope and recruit intercept, which are of similar magnitude. In contrast the first order approximation suggests variation in the growth slope is most important followed by variation in the growth intercept and survival slope. For the first order approximation, linear regression and GAM models we ranked the different processes using their contribution to the total variance, (7.6.6). See `Carlina LTRE.R` for the R code for the various analyses.

7.7 Events in the life cycle

Demographic analyses like those in Chapter 3 can also be done when the environment varies over time, using one simple idea: cross-classify individuals by individual state and environment state (Caswell 2009). This ups the dimension of z, but the theory and the formulas stay the same.

That's actually not the entire story, but it handles four of the most important cases: discrete-state Markovian environments (the analog of "megamatrix" models as in Tuljapurkar and Horvitz (2006) and Metcalf et al. (2009a)); periodic environments; continuous-state Markovian environments; and independent, identically distributed random environments.

As an example of the first case, consider our prototype "megamatrix" model for trees classified by size and a two-state Markovian light environment. The survival operator P is given by equation (6.4.9). Picking a typical result from Chapter 3, the variance in lifetime conditional on initial state is given by

$$\sigma_\eta^2(z_0) = \mathbf{e}(2N^2 - N) - (\mathbf{e}N)^2$$

where N is the fundamental operator $N = (I - P)^{-1}$ and $\mathbf{e}(z) \equiv 1$. The dynamic light environment is reflected only in the fact that z_0 is not just size at birth, it's size and light environment at birth. For a midpoint rule implementation using size range $[L, U]$ with m meshpoints $z_i = L + (i - 0.5)(U - L)/m$, the state vector has length $2\,m$ representing sizes $(z_1, z_2, \cdots, z_m, z_1, z_2, \cdots, z_m)$ the first m in low light and the second m in high light. Entries 1 and $m + 1$ in σ_η^2 are the variance in lifespan for individuals born at size z_1 in low and high light environments, respectively. Caswell (2009) presents several calculations of life cycle properties for a four-state megamatrix model of the prairie herbaceous perennial *Lomatium bradshawii* with the four states representing 0, 1, 2, and ≥ 3 years after a fire.

Periodic environments are a megamatrix in which environment state goes from $1 \rightarrow 2 \rightarrow$ some maximum value, and then returns to 0. The "megamatrix" therefore has the layout of a Leslie matrix in which only individuals of the maximum age reproduce, so many calculations can be done without ever constructing the full matrix. As an example, for an environment with period 3, the P operator is schematically

$$\begin{bmatrix} 0 & 0 & P_3 \\ P_1 & 0 & 0 \\ 0 & P_2 & 0 \end{bmatrix}$$

Individuals in environment state 1 have state transitions given by P_1 and are then in environment state 2, and so on. The fundamental operator N can then be expressed explicitly in terms of the P_i using $N = I + P + P^2 + P^3 + \cdots$. Direct calculation shows that

$$P^3 = \begin{bmatrix} P_3 P_2 P_1 & 0 & 0 \\ 0 & P_1 P_3 P_2 & 0 \\ 0 & 0 & P_2 P_1 P_3 \end{bmatrix}.$$

P^{3k} is therefore block diagonal with blocks equal to the kth power of those in P^3. This implies that $I + P^3 + P^6 + P^9 + \cdots$ equals

$$\begin{bmatrix} (I - P_3 P_2 P_1)^{-1} & 0 & 0 \\ 0 & (I - P_1 P_3 P_2)^{-1} & 0 \\ 0 & 0 & (I - P_2 P_1 P - 3)^{-1} \end{bmatrix} \tag{7.7.1}$$

The other terms in N are P and P^2 times (7.7.1); we leave the details to you.

The remaining cases (general Markovian and independent environments) are size$\times$quality models where "quality" is environment state, and the bivariate growth kernel incorporates both the change in individual state and the change in the environment, given current individual and environment states. The theory doesn't care if quality is a property of the individual, or a property of the individual's environment. The continuous size$\times$quality model in Section 6.4.4 can be reinterpreted as a size$\times$environment state model, in which environment state follows a linear autoregressive Markov chain, and it can be implemented using the methods described in Section 6.6.

Cross-classification is not the only possible way to study life cycle properties in stochastic environments (see Tuljapurkar and Horvitz 2006; Metcalf et al. 2009a). But it has the advantage that one general theory and one set of formulas applies to both constant and time-varying environments.

It is important, however, to understand the meaning of variances that are computed for these models (and for the analogous matrix models). They correspond to the experiment of repeatedly simulating the whole life of an individual, including the random change of environment states. So, for example, in the Low light/High light tree IPM, the computed variance in lifetime is the (projected) variability among a set of trees that experience *independent* sequences of light environments generated by the transition probabilities for light' environment. A group of trees that all experience the *same* random sequence of light environments would also have a variance in total lifetime, but it would be a different and probably much smaller variance, and it would depend on what sequence of light environments they all experience. Our methods, and the corresponding ones for matrix models, do not give the among-individual variance for a cohort of individuals who all experience the same sequence of environment states as they grow. There is still much work to be done here, for both matrix models and IPMs.

7.8 Appendix: the details

This Appendix aims to give precise statements of the assumptions and basic results about the density-independent IPMs with environmental stochasticity, based on Ellner and Rees (2007). Density-dependent stochastic IPMs were discussed already in Section 5.4.4.

Assumptions first. The state space $\mathbf{Z}$ is the same as for the general model in the previous chapter. In most applications it is a collection of points, lines, rectangles, cubes, etc., all closed and bounded. Mathematically, we assume that $\mathbf{Z}$ is a compact metric space. The kernel is again defined relative to a Borel measure μ on $\mathbf{Z}$. Stochasticity is introduced by making the kernel depend on a stochastic process $\theta(t) \in \Theta$ representing time-varying environmental conditions, where Θ is a compact metric space. The kernels are then $K(z', z; \theta(t))$ which we sometimes abbreviate as $K_t(z', z)$ or $K(\theta(t))$. We assume that the kernel is continuous as a function on $\mathbf{Z} \times \mathbf{Z} \times \Theta$. The population dynamics are

$$n(z', t+1) = \int_{\mathbf{Z}} K(z', z; \theta(t)) n(z, t) \, dz. \tag{7.8.1}$$

The stochasticity in (7.8.1) is purely environmental; it is still an infinite-population "mean field" model that ignores demographic stochasticity.

We assume that the environment sequence $\theta(t)$ is stationary (the probability distribution of different environment states is the same each year) and ergodic (the distant future becomes completely independent of the present).

Stable population growth requires a kernel positivity assumption. In contrast to the general deterministic model, theory is not available for u-bounded kernels.

We have to assume power-positivity, specifically, that there exists some $m > 0$ such that the $m - step$ kernels

$$K(\theta_m)K(\theta_{m-1}) \cdots K(\theta_1) \gg 0 \tag{7.8.2}$$

is strictly positive whenever $\theta_1, \theta_2, \cdots, \theta_m$ are all in Θ. Here "$\gg 0$" means that the kernel is positive at all points (z', z), and $K_1 K_2$ denotes kernel composition rather than pointwise multiplication, as in Chapter 3. Because of continuity and compactness, it follows that there are positive constants α_i such that

$$\alpha_1 \leq K(\theta_m)K(\theta_{m-1}) \cdots K(\theta_1) \leq \alpha_2 \tag{7.8.3}$$

whenever $\theta_1, \theta_2, \cdots, \theta_m$ are all in Θ. The same assumption is generally made for stochastic matrix models, except that it involves matrix products rather than kernels. Whether or not (7.8.3) is satisfied may depend on the choice of measure μ on the state space, so careful choice of μ may be important in some cases.

As in matrix models, power-positivity is only possible if post-reproductive stages or states (those without any chance of reproducing now or in the future) are removed. The population is modeled as if such individuals are already dead. However, we will explain below that post-reproductives can sometimes be included if the kernels for potentially reproductive states are power-positive. Similarly, the results also hold for "power-smooth" models where the m-step population dynamics are described by smooth kernels, even if the one-step dynamics are not. This situation often occurs if some individual-level state variables remain constant over an individual's lifetime, but differences between parent and offspring are described by a smooth probability density.

Under these assumptions, stochastic density-independent IPMs behave much like stochastic density-independent matrix models (Ellner and Rees 2007). The general results in Ellner and Rees (2007) are essentially a generalization of Lange and Holmes (1981) to IPMs. The original proofs for matrix models, by Joel Cohen in the 1970s, relied heavily on matrix-specific calculations. Lange and Holmes (1981) gave completely different proofs based on the fact that multiplication by a positive matrix is a contraction in Hilbert's projective metric. Ellner and Rees (2007) showed that the general stochastic IPM described above also has the necessary contractive properties, implying the following properties (paraphrasing Chapter 2 of Tuljapurkar (1990))

A. The logarithm of the total population size $N(t) = ||n(t)||_1$ has a long-term growth rate $\log \lambda_S$ which is constant with probability 1,

$$\log \lambda_s = \lim_{t \to \infty} t^{-1} \log N(t) = \lim_{t \to \infty} t^{-1} E \log N(t). \tag{7.8.4}$$

The same is true for any part of the population, or more generally any $N_w(t) = \langle W, n(t) \rangle$ where $W \geq 0$ is a nonzero bounded measurable function on $\mathbf{Z}$.

B. Starting from any nonzero initial population $n(z,0) = n_0(z)$ the population structure $w(t) = n(t)/N(t)$ converges to a time-dependent stationary random sequence of structure vectors $\hat{w}(t)$ which is independent of n_0.

C. The joint sequence of environment states and stationary population structures $(\theta_0, \hat{w}(0), \theta_1, \hat{w}(1), \theta_2, \ldots)$ is a stationary ergodic process.

D. The long-term growth rate can be computed as the average one-step growth rate, $\log \lambda_S = E \log \|K(\theta)\hat{w}\|$ where the expectation is with respect to the joint stationary measure in the last item.

Tuljapurkar (1990) states that properties C and D hold when the the environment process is a countable state Markov chain, but the proofs in Lange and Holmes (1981) show that they hold for any stationary ergodic environment. The intuitive picture behind these properties is that

- A joint stationary distribution (Property C above) must exist due to compactness by a standard argument (see Lemma 1 in Furstenburg and Kesten (1960)).
- Because of contraction in the projective metric, two "copies" of the population structure process with different initial values, running under the same $\theta(t)$ sequence, converge onto each other. So starting from anywhere (copy 1) the population structure converges onto the stationary process (copy 2), which is therefore unique and independent of the initial population, and therefore ergodic.

In general there is no formula for the stationary distribution or stochastic growth rate λ_S. However, the results above ensure that these really do exist, and that they can be estimated as long-term averages in one sufficiently long simulation starting from any initial population state.

The final important general property is that total population size converges to a lognormal distribution. This happens because the log of total population size is approximately a random walk. In the unstructured model $n(t+1) = \lambda(t)n(t)$ with no between-year correlations in the growth rates $\lambda(t)$, $\log n(t)$ *is* a random walk, so its distribution is asymptotically Gaussian, by the Central Limit Theorem. In a structured population this is not quite true because population structure induces correlations between population growth rates at different times. But because of property B, the structure-induced autocorrelations decay rapidly. The random walk approximation is then valid if there are no long-range autocorrelations in $\theta(t)$.

A useful definition of "no long-range autocorrelations" is the *uniform mixing* property. Informally, let A be a statement about $(\theta(k), \theta(k-1), \theta(k-2), \cdots)$ and B be a statement about $(\theta(k+n), \theta(k+n+1), \theta(k+n+2), \cdots)$, and $\phi(n)$ the maximum of $|P(B|A) - P(A)|$ over k, A, and B with $P(A) > 0$. If

$$\sum_{n=1}^{\infty} \phi_n^{1/2} < \infty \tag{7.8.5}$$

then θ is said to be *uniform mixing*. The conclusion (equivalent to Theorem 9 in Lange and Holmes (1981)) is:

E. If θ is uniform mixing, then the asymptotic distribution of the total population size or any part of the population is lognormal, i.e.

$$(\log N(t) - t \log \lambda_s) / \sqrt{t} \;\Rightarrow\; \mathrm{Gaussian}(0, \sigma^2) \qquad (7.8.6)$$

for some $\sigma \geq 0$, where $\Rightarrow$ denotes convergence in distribution.

Formally, let $\mathcal{F}_j^k, j \leq k$ be the σ-algebra generated by $(\theta(j), \theta(j+1), \ldots, \theta(k))$. Then

$$\phi(n) = \sup_k \sup \left\{ |P(B|A) - P(B)| : A \in \mathcal{F}_{-\infty}^k, B \in \mathcal{F}_{k+n}^\infty, P(A) > 0 \right\}. \quad (7.8.7)$$

Uniform mixing is an assumption about how fast the $\theta(t)$ process converges onto its stationary distribution if $\theta(0)$ is not drawn from the stationary distribution. Like many other properties, it is less likely to hold when θ's possible values are unbounded. For example, a first-order autoregressive process $\theta(t+1) = a\theta(t) + Z(t)$, where $|a| < 1$ and $Z(t)$ are independent Gaussian(0,1) random variables, is not uniform mixing. The reason is that for any B and n, $\theta(0)$ can be assigned such an extreme value that B (whatever it is) is very unlikely to occur within the next n time steps. However, results similar to (7.8.6) hold under weaker assumptions about θ, with weaker conclusions. Central Limit Theorems with weaker mixing conditions usually involve scaling by the standard deviation of partial sums. These imply lognormal asymptotic distributions for total population size, but without the conclusion that the variance grows linearly over time.

For sensitivity analysis, we also need the analog of property C for time-dependent reproductive values. This will be true if our assumptions above are also satisfied by the time-reversed transpose kernel, $K_t^* = (K_{-t})^T$ (recall from Chapter 3 that $K^T(z', z) = K(z, z')$). The only assumption that can possibly fail is ergodicity of the time-reversed environment $\theta^*(t) = \theta(-t)$. So for sensitivity analysis and the small fluctuations approximation we also have to assume that $\theta^*(t)$ is ergodic. This will be true, for example, if θ is an ergodic finite-state Markov chain.

Finally, two of our assumptions about the model can be relaxed somewhat. The first is a way to allow senescent (post-reproductive) individuals in the model. Break up $\mathbf{Z}$ as the union of disjoint compact sets $\mathbf{Z}_r$ and $\mathbf{Z}_s$ representing potentially reproductive and senescent individuals, and let kernel components $K_{r,r}, K_{s,s}, K_{s,r}$ describe all the possible state transitions. The model cannot be power-positive because if all individuals at $t = 0$ are senescent, nobody ever gets to $\mathbf{Z}_r$. However, if there is a finite upper limit q to the post-reproductive lifespan, and there is some $m > q$ such that $n_r(z, t)$ and $n_s(z, t)$ are both positive for all z whenever $t > m$ and $n_r(z, 0)$ is not identically zero, then properties A–E above still hold (Ellner and Rees 2007, again based on Lange and Holmes 1981).

The second is directed at models where some traits have deterministic dynamics (see Section 10.4). For example, genotype (for Mendelian traits) and breeding value (for quantitative traits) remain constant during an individual's lifetime. Changes only occur when parents are replaced by offspring. This is not a problem for Mendelian traits with a finite set of genotypes, but a continuous trait that stays constant or changes deterministically during the lifetime is not represented by a smooth kernel. The results in this chapter can still be applied if the model is age-structured with a maximum lifespan M and the parent-to-offspring trait transmission is described by a smooth kernel. In that case the $(M + 1)$-step kernel will often satisfy our general assumptions. Properties A–E then apply to $n(z, t)$ observed at times $t = 0, M+1, 2(M+1), \cdots$, and therefore to $n(z, t)$ for all t.

7.8.1 Derivations

Derivation of (7.4.4). $\langle v, K_t w \rangle$ and $\langle v, (K_t - \bar{K})w \rangle$ differ by a constant so they have the same variance. The latter has zero mean, so

$$Var \langle v, (K_t - \bar{K})w \rangle = E\left[\langle v, K_t w \rangle^2\right] = E\left[\langle v, (K_t - \bar{K})w \rangle \langle v, (K_t - \bar{K})w \rangle\right].$$

Write out the first $\langle v, (K_t - \bar{K})w \rangle$ as a double integral over z_1 and z_1', the second as a double integral over z_2 and z_2', and write their product as a quadruple integral. The only random terms are $K_t - \bar{K}$. Bringing the expectation inside the integral gives (7.4.4).

Derivation of (7.4.5). If we approximate the population structure by w then the population state in year t is $N(t)w$ and the total reproductive value $V(t)$ is approximately $\langle v, N(t)w \rangle$. The population state in the next year is $K_t N(t)w$ so $V(t + 1) \approx \langle v, K_t N(t)w \rangle$, and

$$V(t + 1)/V(t) \approx \frac{\langle v, K_t N(t)w \rangle}{\langle v, N(t)w \rangle} = \langle v, K_t w \rangle.$$

The general small fluctuations approximation for a random variable X with mean μ and (small) variance σ^2 is

$$E[f(X)] \approx f(\mu) + \frac{\sigma^2}{2} f''(\mu).$$

Applying this to $X = \langle v, K_t w \rangle$ and $f = \log$ gives (7.4.5).

Chapter 8
Spatial Models

Abstract Before IPMs were used for size-structured populations, they had a long history as models for spatially structured populations and for spatial spread of infectious diseases and genes. We review some of those models and two recent applications to questions of population persistence. The merger of demographic and spatial structure within one model came much later. The new modeling issue in spatial models is the dispersal kernel that describes individual changes in location. We explain and demonstrate some of the approaches used to model dispersal and estimate the parameters of movement kernels. A major theme in the theory is the difference between bounded and unbounded spatial domains. Models with a bounded spatial domain are "normal" IPMs, and spatial location is just one more trait that differs among individuals. Models with unbounded spatial domain are different. Their long-term behavior is characterized by traveling wave solutions instead of convergence to a stable population structure. We explain how the long-term rate of population spread and its sensitivities can be calculated, and used to determine the factors governing the spread rate of an invasive species.

8.1 Overview of spatial IPMs

Spatial population structure is a natural application of IPMs because spatial location varies continuously, at least at small scales. It's so natural that IPMs for spatial structure came long before IPMs for size structure. A spatial IPM makes a brief appearance in Skellam (1951), but the first substantive ecological applications that we know of were models for spread of infectious disease (Mollison 1972) and spatial variation in gene frequency (Slatkin 1973). The first extended application to spatial population dynamics was by Kot and Schaffer (1986). These papers sparked a long line of subsequent work, more than we can review here (see Kot 2003; Metz et al. 2000; Zadoks 2000). Dieckmann et al. (2000) is a good review on spatial population models in general.

In this chapter we first look at the simpler situation where individuals differ only in spatial location. We then move on to our main topic, models that

© Springer International Publishing Switzerland 2016   229
S.P. Ellner et al., *Data-driven Modelling of Structured Populations*,
Lecture Notes on Mathematical Modelling in the Life Sciences,
DOI 10.1007/978-3-319-28893-2_8

combine spatial and demographic structure, a merger that took surprisingly long (Adler et al. 2010; Jongejans et al. 2011).

With purely spatial structure the population is fully described by its spatial distribution $n(x,t)$. We use x to denote spatial location, which may be bounded or unbounded, but z always denotes a trait with a bounded set of possible values. A population's potential habitat is always bounded in reality, but unbounded space is a useful model for situations where a population can spread over many generations before reaching a boundary.

To introduce spatial IPMs we consider one-dimensional space, so x is a point on the (infinite) line. Sometimes it's reasonable to pretend that space is one-dimensional, such as the spread of zebra mussels along a river or the spread of otters along the west coast of North America. But the main reasons for studying one-dimensional models are analytic tractability and the important fact that they can be used to predict population spread in two-dimensional space, if you're mainly interested in the long-term spread rate, see Section 8.6.

The projection kernel K for a spatial IPM must include two ingredients: a function $k_d(x', x)$ describing the transport of individuals from an initial location x to a subsequent location x' within the spatial domain X, and a model for population growth due to reproduction and mortality. In the absence of density-dependence, the finite rate of population growth is some constant R_0. Then combining population growth and dispersal we have

$$n(x',t+1) = \int_X R_0 k_d(x',x)n(x,t)\,dx. \tag{8.1.1}$$

The *dispersal kernel* k_d expresses our assumptions about how individuals re-locate.[1] The simplest assumption is complete spatial homogeneity, so that the dispersal kernel only depends on the distance between original and subsequent locations, having the form $k_d(|x' - x|)$. Unless otherwise stated, we will assume spatial homogeneity. Commonly used kernels include the Gaussian and lognormal distributions, and the bi-exponential or Laplace kernel

$$k_d(x) = \frac{1}{2L} e^{-x/L}. \tag{8.1.2}$$

Density dependence in local population growth was added by Kot and Schaffer (1986) to study the combined effects of local interactions and dispersal. They assumed that the population's life cycle alternates between phases of local population growth and spatial redistribution within a finite interval $[-m, m]$. This gives models of the form

$$n(x',t+1) = \int_{-m}^{m} k_d(x' - x)f(n(x,t))\,dx, \tag{8.1.3}$$

where $f(n)$ describes local population growth.

---

[1] Calling k_d the dispersal kernel is established in the literature so we use that term even though "kernel" usually refers to $P, F, G,$ or K in this book.

Neubert and Caswell (2000) extended the Kot-Schaffer model by adding demographic structure to the local population dynamics, in the form of a matrix projection model. The obvious next step - using an IPM to describe the local population dynamics - took 10 years (Jongejans et al. 2011) for reasons that are hard to understand in retrospect. The basic density-independent model is a general IPM in which individuals are cross-classified by trait z and spatial location x,

$$n(x', z', t+1) = \iint K(x' - x, z', z) n(x, z, t) \, dx \, dz. \qquad (8.1.4)$$

All of the usual variations (density dependence, environmental stochasticity, demographic stochasticity) can be layered onto this. An important special case is the *Neubert-Caswell model* in which local demography and movement occur sequentially, as in the Kot-Schaffer model. However, the probability distribution for movement distance $x' - x$ is allowed to depend on the initial and final trait values, for example if seeds from taller parent plants disperse farther, or adults that spend less time migrating have more time to forage and therefore grow more. The IPM kernel then has the form

$$K(x' - x, z', z) = k_{d,F}(x' - x|z', z) F(z', z) + k_{d,P}(x' - x|z', z) P(z', z) \quad (8.1.5)$$

where $k_{d,F}$ and $k_{d,P}$ are the probability distributions for movement distance of new recruits and survivors, respectively.

Space is not one-dimensional (we had noticed this fact on our own, even though we are mostly theoreticians). In spatial matrix models and IPMs, two-dimensional dispersal is often described by a probability density $k_R(r)$ for the dispersal distance r, with the direction of movement assumed to be uniformly distributed on $[0, 2\pi]$. This means that $k_R(r)dr$ is the probability of landing somewhere in the annulus between the circles of radius r and $r + dr$, centered on the point of origin. Formally one- and two-dimensional models look exactly the same, except that x is a vector representing two spatial coordinates.

8.2 Building a dispersal kernel

Much of this chapter is about the theory and applications of spatial IPMs. But if your main goal is data-driven modeling of a spatially structured population, the first issue raised by spatial models is, how do I come up with the dispersal kernel? So we begin with that question. The ecological literature on movement modeling is vast and growing, so our presentation is highly selective: entire books could be written about it, and have been. Turchin (1998) is a comprehensive classic with everything from diffusion models to advise about good paints to use for marking insects, and Clobert et al. (2012) is an up-to-date multi-authored overview.

There are two main types of movement model, descriptive and mechanistic. A descriptive model is a probability distribution that represents the outcome of movement, while a mechanistic model represents the process of movement. The essential difference, though, is in the data. A descriptive model may have a mechanistic "story" (e.g., a Gaussian dispersal kernel is the outcome of random walk movements), but it is fitted to data on observed changes in location: where individuals started from, and where they eventually got to. A mechanistic model represents how they got there, and is fitted to data on the steps in the process of moving.

8.2.1 Descriptive movement modeling

Pons and Pausas (2007) studied acorn dispersal by jays. They put radio transmitters in 239 acorns of two oak species in Mediterranean vegetation, put the acorns in bird feeders, and found the final resting point of the 158 acorns that jays stored in caches (acorns that the jays opened and consumed don't contribute to spread of the oak population). These are ideal data for building a descriptive movement model.

By eye, a histogram of the dispersal distances (Figure 8.1A) suggests an exponential distribution. An exponential distribution has only one parameter - its mean - and an exponential distribution with the same mean as the data (dashed curve) looks like a pretty good fit. But the eye is deceived. A quantile-quantile plot (Figure 8.1B) shows that there are too many large distances (relative to the mean) to be consistent with an exponential distribution: the upper quantiles of the data are far outside the 95% confidence envelope for an exponential distribution. A better alternative is suggested by plotting a histogram of log-transformed distances (Figure 8.1C). Now the q-q plot looks good (Figure 8.1D), and a formal test finds no significant departures from normality (Shapiro-Wilk test, $P = 0.55$).

Fitting a lognormal is easy, because its two parameters are the mean and variance of the log-transformed data. In general a descriptive dispersal kernel has to be fitted by maximum likelihood, but that's not much harder. Using mle in the stats4 package, the first step is writing a function that computes the negative log likelihood of the data:

```
J <- read.csv("Pausas2Ellner.csv"); # the data


nllLognormal <- function(m,s) {
    d<-dlnorm(J$dist,meanlog=m,sdlog=s,log=TRUE)
    return( -sum(d) )
}
```

Here m and s are the mean and standard deviation of log distance. In nllLognormal the first line computes the log likelihood of each data point. The second sums those to get the log likelihood of the entire data set, and returns the negative log likelihood.[2] Then mle takes care of finding the maximum likelihood parameter estimates:

```
mulog <- mean(log(J$dist)); sdlog <- sd(log(J$dist));
fit1 <- mle(minuslogl=nllLognormal,start=list(m=mulog,s=sdlog),
        method="Nelder-Mead",control=list(trace=4,maxit=5000));
```

---

[2] In case you've forgotten: if $(x_1, x_2, \ldots, x_n)$ is a set of independent observations, and the likelihood of a single observation x is $L(x)$, then the likelihood of the set of observations is $L(x_1)L(x_2)\cdots L(x_n)$ and the log likelihood is $\log L(x_1) + \log L(x_2) + \cdots + \log L(x_n)$.

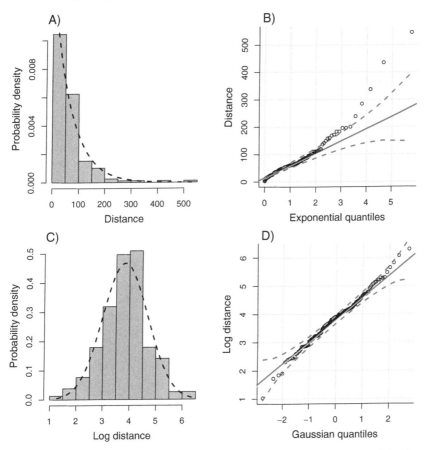

Fig. 8.1 Analysis of Pons and Pausas (2007) data on acorn dispersal by jays. (A) Histogram of the acorn dispersal distances, with (dashed line) an exponential distribution with mean equal to the observed mean distance. We thank J. Pausas for providing the data, which are used here with permission. (B) Exponential quantile-quantile plot of the distances, using qqPlot in the *car* package. Dashed red curves are the pointwise 95% confidence envelope for an exponential distribution. (C) Histogram of log-transformed distances, with (dashed line) a Gaussian distribution having the same mean and variance as the log-transformed data. (D) Gaussian quantile-quantile plot of the log-transformed distances. Source file: PonsPausas.R

Other distributions can be fitted similarly; see PonsPausas.R for code to fit the exponential and gamma distributions (the gamma is a two-parameter family that includes the exponential as a special case). Because the goal of a dispersal kernel is prediction rather than "truth," it is appropriate to compare alternative models with AIC. As Pons and Pausas (2007) found, the lognormal wins hands down ($\Delta AIC = 19.0$ and 3.16 relative to the gamma and exponential, respectively; Pons and Pausas (2007) also found $\Delta AIC = 27$ relative to a Weibull distribution).

Turchin (1998) gives many more examples of mark-recapture studies, meaning that individuals or groups of individuals are given some kind of distinctive mark so that they can be allowed to move and then found again. Jones and Muller-Landau (2008) describe the use of genetic markers for tracking seed dispersal, and Carlo et al. (2009, 2013) describe and apply a method using ^{15}N labeled seeds.

In many studies of wind dispersed seeds, the data are the numbers of seeds that fall into traps at various distances from an isolated source plant. As an example, we consider some data from Bullock and Clarke (2000) on wind dispersal of heather (*Calluna vulgaris*) seeds. Traps were constructed from plant pots, and were designed to capture seeds falling into a circle of radius 4.5 cm. The data we analyze are from the transect of traps extending northeast from a source plant:

```
distance <- c(0.6,0.8,1,1.5,2,3,4,6,8,10,15,20,30,40,60,80);
trapNumber <- c(2,2,2,3,4,6,8,12,16,20,30,40,60,80,120,160);
seedNumber <- c(2004,1323,369,149,86,69,27,17,6,11,3,2,0,0,0,0);
```

Here `distance` is meters from the center of the plant, `trapNumber` is the number of traps at each distance, and `seedNumber` is the total number of seeds collected in all traps. The source plant was 1.2 m in diameter, so `distance`= 0.6 is right at the edge of the source plant.

The data tell us about the number of seeds per unit area - let's call this $S(D)$ - as a function of distance D. As always, Step 1 is to plot the data. A log-log plot is most revealing (Figure 8.2) but note that the last 4 distances have to be omitted because log(0) is undefined:

```
trapSize <- pi*(0.09/2)^2; trapArea <- trapNumber*trapSize;
seedDensity <- seedNumber/trapArea;
plot(log(distance[1:12]),log(seedDensity[1:12]),xlab="Log distance",
     ylab="Log seeds/m2");
```

A straight line fit looks pretty good. It implies $S(D) = aD^{-b}$. We could get a and b from the coefficients of the regression line, but that's not statistically optimal because the data are counts with nonconstant variance. Also, it makes no use of the four distances at which `seedNumber`= 0, and those are also data. Assuming that seeds are dispersed independently, the `seedNumber` at distance x is Poisson distributed with mean equal to $SD(D)$ times the `trapArea` at distance x. The negative log likelihood is then

```
nllPower <- function(a,b) {
    mean=a*(distance^(-b))*trapArea
    d = dpois(seedNumber,mean,log=TRUE);
    return(sum(-d))
}
```

The intercept (10.4) and slope (-2.8) of the regression line are good starting values for maximum likelihood estimation:

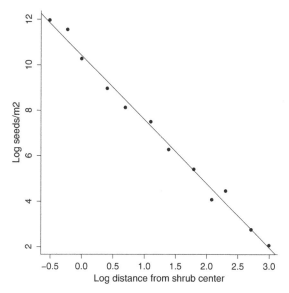

Fig. 8.2 Analysis of data tabulated in Bullock and Clarke (2000) for wind dispersal of *Calluna vulgaris* seeds on the transect extending northeast from the source plant. Plotted points are seed density (number of seeds per m² of trap area) at the distances where the density was nonzero. The straight line is a linear regression fitted to the plotted points by least squares. Source file: `BullockClarke.R`

```
require(stats4);
a0 <- exp(10.4); b0 <-2.8;
fit <- mle(nllPower,start=list(a=10,b=b0),method="Nelder-Mead",
           control=list(maxit=5000,trace=4))
```

With a tiny bit of rounding, the resulting estimates are $a = 37568, b = 3$, giving

$$S(D) = \begin{cases} 0 & D \leq 0.6 \\ 37658 \; D^{-3} & D > 0.6 \end{cases} \tag{8.2.1}$$

As in the previous example, we could go on to compare alternative models. For many of the transects, Bullock and Clarke (2000) found that the data were fitted better by a two-term model, a power-law like equation (8.2.1) plus an exponential decay term.

To finish, we need to figure out how the dispersal kernel is related to $S(D)$. Consider first the radial dispersal kernel $k_R(r)$, the probability density of distance r from the source plant. If a total of N seeds disperse, the number that lands at distances between r and $r+dr$ is $Nk_R(r)dr$ (as usual, this is an asymptotic approximation for small r). In terms of S, the number landing at distance between r and $r + dr$ is $S(r)$ times the area of the annulus bounded by the circles of radius r and $r + dr$, which is $2\pi r dr$. So we have

$$Nk_R(r) = 2\pi r S(r).$$

The total seed number N is the integral of S over the entire plane, i.e., we need to sum up the seeds in annuli of any radius:

$$N = 2\pi \int_0^\infty uS(u)\, du.$$

Combining the last two equations, we have

$$k_R(r) = rS(r)/\int_0^\infty uS(u)\, du. \qquad (8.2.2)$$

For the fitted *Calluna* model, this is

$$k_R(r) = \begin{cases} 0 & 0 \le r \le 0.6 \\ 0.6r^{-2} & r > 0.6 \end{cases}$$

We can similarly derive a fully two-dimensional dispersal kernel $k_d(x, y)$ giving the probability density for moving from coordinates $(0,0)$ to (x,y) in the plane. The number of seeds landing in a rectangle with sides dx and dy centered at (x,y) is $Nk_d(x,y)dx\,dy$, and also is $S(D)dx\,dy$ where $D = \sqrt{x^2 + y^2}$. So we have

$$k_d(x, y) = S(D)/N = S(D)/\left(2\pi \int_0^\infty uS(u)\, du \right) \qquad \text{where } D = \sqrt{x^2 + y^2}.$$

For the fitted *Calluna* model (8.2.1) this works out to

$$k_d(x, y) = \begin{cases} 0 & D \le 0.6 \\ 0.3D^{-3}/\pi & D > 0.6 \end{cases} \qquad \text{where } D = \sqrt{x^2 + y^2}.$$

8.2.2 Mechanistic movement models

A mechanistic model represents the process of movement, and is built from data on each step in the process. A classical example is Patlak's correlated random walk model for animal movements (Turchin 1998). Movement paths in two-dimensional space are assumed to consist of short linear segments. For each segment the speed and temporal duration of motion are chosen from probability distributions, independent of the speed and duration on previous segments. Movement is correlated because the direction of motion is generated by randomly selecting a *turning angle* relative to the previous segment. If turning angles are always small (e.g., uniformly distributed on $[-s, s]$ radians with $s \ll 1$), the direction of motion will change very slowly.

Mechanistic movement modeling has seen enormous developments in the last few decades, so we can only mention a few examples out of many. The models can rarely be solved analytically, but with the growth of computing power this is less and less of an issue.

Beginning with Morales et al. (2004) high-precision data on movements of large animals (e.g., from GPS collars) have been modeled as a combination of several different random walks corresponding to different behavioral states of the individual. Behavioral states are unobserved, and have to be inferred statistically from movement paths. Morales et al. (2004) used Bayesian methods to fit a variety of model with multiple behavioral states to movement data on elk released in Ontario, Canada. Their final model had two states, "encamped" and "exploratory," the latter having longer step lengths and smaller turning angles so that movement is much more directional. For some recent developments in these models, see Cagnacci et al. (2010) and McClintock et al. (2012). Other recent developments include behavior at habitat boundaries, scent marking of territories, and other social interactions affecting individual movement behaviors.

Mechanistic models for animal-dispersed seeds have been developed by layering a probability distribution for seed retention time on top of a model for animal movement (see, for example, Morales and Carlo 2006; Cousens et al. 2010; Bullock et al. 2011). For airborne seed dispersal the WALD model (Katul et al. 2005) has recently become popular because it gives an analytic kernel with only two parameters, both directly tied to the physical processes of seed dispersal, and it has been successfully fitted to a variety of data sets. The kernel can be written as

$$k_R(r) = \sqrt{\frac{\tau}{2\pi r^3}} \exp\left(-\frac{\tau(r-\mu)^2}{2\mu^2 r}\right). \qquad (8.2.3)$$

The location parameter μ equals HU/F and the scale parameter τ equals $(H/\psi)^2$, where H is seed release height, F is the terminal falling velocity of seeds, U is the hourly mean horizontal wind velocity between H and the ground, and ψ is a turbulence parameter that reflects variation in wind speed (Katul et al. 2005). Skarpaas and Shea (2007) fit a WALD kernel by estimating the H, U, F and ψ, and explain how it can be generalized to account for temporal variation in those parameters during the period of seed dispersal. Travis et al. (2011) use the WALD model to derive age-specific dispersal kernels for an invasive shrub *Rhododendron ponticum*, in which seeds from older plants spread further because they have a greater seed release height H.

8.3 Theory: bounded spatial domain

We turn now from the construction of spatial models to their analysis and applications, starting with spatial models for an otherwise unstructured population.

If the spatial domain is bounded, say $X = [-m, m]$, we're back in Chapter 2. For the density-independent model (8.1.1), so long as the kernel $K(x', x) = R_0 k_d(x', x)$ is positive or some iterate of it is positive, the population converges to exponential growth or declines with a stable population structure. The positivity condition has a simple biological meaning: in some finite number of time-steps it is possible for the population to spread from any location in X to any other location. Spatial location within a bounded domain is (mathematically) just another "trait" that can differ between individuals and change over time. All the general theory applies, so a density-independent spatially structured IPM predicts (among other things) an asymptotic spatial structure and space-dependent reproductive value (Chapter 2), the distribution of location at first reproduction (Chapter 3), how location-dependent demographic stochasticity affects population growth (Chapter 10), and so on.

With density-dependent limits to growth, interest focuses on equilibria, their stability, and on oscillations around an unstable equilibrium. An equilibrium of (8.1.3) is a population distribution $n^*(x)$ such that

$$n^*(x') = \int_{-m}^{m} k_d(x' - x) f(n^*(x)) \, dx. \qquad (8.3.1)$$

The stability of the equilibrium can be studied by a linear stability analysis. Let $\eta(x,t) = n(x,t) - n^*(x)$, the deviations from equilibrium. For small deviations, the dynamics of η, to leading order, are

$$\eta(x',t+1) = \int_{-m}^{m} k_d(x' - x) f'(n^*(x,t)) \eta(x,t) \, dx.$$

This is a density-independent linear IPM with kernel $K^*(x',x) = k_d(x' - x) f'(n^*(x,t))$. If all of these IPM's eigenvalues are less than 1 in magnitude, the deviations will shrink eventually to 0, so n^* is locally stable.

If $f(0) = 0$, then one equilibrium is $n^* \equiv 0$, population extinction. The linearized kernel is $f'(0)k_d$, equation (8.1.1) with $R_0 = f'(0) \geq 0$; $f'(0)$ cannot be negative because $f(n) \geq 0$ for all n (i.e., the population can't be negative). Kot and Schaffer (1986) show that the linearized kernel has a unique dominant eigenvalue which is a real number, and because $f'(0) > 0$ the dominant eigenvalue is nonnegative. The zero equilibrium will therefore be unstable if the dominant eigenvalue of $f'(0)$ is > 1. So there are no surprises here: if a near-extinct population tends to grow rather than to decrease further, the zero equilibrium is unstable and the population will persist. As in nonspatial models, Allee effects create the possibility of bistability, so that a very sparse population decreases to extinction but a dense population can persist.

Near a nonzero equilibrium the situation can be more complicated, but it doesn't have to be. Consider the Beverton-Holt model for local population growth, $f(n) = an/(b + n)$ with $a, b > 0$. The difference equation for a totally unstructured population, $n(t + 1) = an(t)/(b + n(t))$ has very simple behavior: if $a \leq b$ the population decreases to extinction, and if $a > b$ it converges to a unique steady state. With spatial structure, if the kernel k_d is power-positive and smooth on $[0, L]$, then the spatial model has the same qualitative behavior. Either the population converges to $n(x) \equiv 0$, if a is small enough that the zero equilibrium is locally stable, or if a is large enough the population converges to a unique, globally stable equilibrium $n^*(x)$.[3] Again, bounded space brings no surprises.

More interesting things can happen if $f'(n)$ can be negative, and in multi-species situations such as host-parasitoid models. These create possibilities for pattern formation driven by dispersal, called *diffusive instability*. In some ways spatial IPMs appear to be intermediate between partial differential equation models (continuous in space and time) and patch models (continuous in time but spatially discrete). Patch models can have many spatially nonconstant stable states (Levin 1974) but these typically don't occur in the analogous spatial IPM (Kot and Schaffer 1986). However, the many possible forms of the dispersal kernel give spatial IPMs more potential for spatial pattern formation than the corresponding partial differential equation model with the same form of local population interactions (Neubert et al. 1995).

8.4 Theory: unbounded spatial domain

Unbounded space changes the picture completely. The population's range can expand without limit, so we no longer expect stationary (or even periodic) spatial patterns as the long-term behavior. The only way to find constancy is to look for it in a moving frame of reference that keeps pace with the spread of the population. Consider a frame of reference moving to the right at constant rate c.[4] An observer in this frame of reference would move from location $x - c$ at time $t - 1$ to location x at time t. The population will appear to be constant (relative to the observer's location) if

$$n(x,t) = n(x - c, t - 1). \tag{8.4.1}$$

That is: one time unit ago everything was the same, only shifted to the left by c units of space. So two time units ago everything was the same, only shifted to the left by $2c$ units of space. Continuing in this way, we eventually get to

$$n(x,t) = n_0(x - ct), \text{ where } n_0(x) = n(x,0). \tag{8.4.2}$$

A population distribution with this property is called a *traveling wave* with speed c.

Can traveling waves really occur? For the density-independent model (8.1.1) with spatial domain $X = (-\infty, \infty)$ the answer turns out to be "yes." In fact it turns out to be "hell yes!" because there are often infinitely many traveling

---

[3] We thank André Grüning, University of Surrey, for suggesting this example and showing us a proof of the result. It follows from the Limit Set Trichotomy (see Hirsch and Smith 2005) for maps which are monotone, strongly sublinear, and order-compact. If the rth iterate of the dispersal kernel is positive, then the rth iterate of the map (8.1.3) is monotone and strictly sublinear because f has those properties, and order compactness follows from the Arzela-Ascoli Theorem when the dispersal kernel is smooth.

[4] The use of c for population spread rate is so consistent in the literature that we use it in this chapter, even though it conflicts with our use of $c(z')$ for offspring size distribution in other chapters.

wave solutions with different speeds. A straightforward calculation (see Box 8.1) shows that

$$n(x,t) = N_0 e^{rt} e^{-sx} \tag{8.4.3}$$

is a traveling wave solution with $r = \log(R_0 M(s))$ and speed

$$c(s) = \frac{1}{s} \log(R_0 M(s)), \tag{8.4.4}$$

where

$$M(s) = \int_{-\infty}^{\infty} e^{su} k_d(u) du \tag{8.4.5}$$

is the two-sided Laplace transform of the kernel, which is called the *moment generating function*. The number $s > 0$ (called the wave *shape* or shape parameter) determines how the traveling wave varies spatially at any one time t: it is a decreasing exponential function of spatial location x, whose rate of decrease is given by s. [5] For any wave shape s such that the moment-generating function $M(s)$ is finite, there is a traveling wave with shape s and speed $c(s)$.

However, these traveling waves all have the property that the population is present everywhere: $n(x,t) > 0$ for all x. As a result, these waves spread faster than any population that was initially limited to a bounded region (see Box 8.1). So if an initially bounded population settles into a traveling wave, its speed c^* must be no larger than the minimum of $c(s)$.

Miraculously, these heuristic arguments lead to the truth. Hans Weinberger and Roger Lui (Weinberger 1982; Lui 1982) showed that under general conditions, a population that initially is limited to a bounded region converges to a traveling wave with speed

$$c^* = \min_s c(s) = \min_s \left(\frac{1}{s} \log(R_0 M(s)) \right) \tag{8.4.8}$$

Even more miraculously, exactly the same is true for a density-dependent spatial IPM (8.1.3) in the absence of Allee effects in the local population growth function f, with $R_0 = f'(0)$. This is the traveling wave speed for the linear IPM obtained by linearizing the kernel at zero population density.

We will not go further into the math, because soon we will move beyond what has been proved rigorously. From here on, we're going to presume that more complicated models act like the simpler ones, in the following sense: if you linearize an IPM at zero population density, and compute the minimum possible wave speed c^* for the linearized model using (8.4.8), then solutions of the original model converge to a traveling wave with speed c^* so long as there are no Allee effects, and the original population was limited to a bounded region of space. This is called the *linearization conjecture*. The empirical evidence (i.e.,

---

[5] Alas, this is one more different meaning for the letter s. Because it is so widely used for wave shape we do the same – but it is only in this chapter that s means wave shape.

To derive $c(s)$ we assume that $n(x,t)$ is given by equation (8.4.3) and look for values of s and c such that $n(x,t)$ is a solution of the model and equation (8.4.1) is satisfied. Because of the assumption that $k_d(x',x) = k_d(x'-x) = k_d(x-x')$, the substitution $u = x' - x$ in (8.4.5) shows that

$$M(s) = \int_{-\infty}^{\infty} e^{s(x'-x)} k_d(x'-x)\, dx$$

for any x'. Then substituting (8.4.3) into (8.1.1) gives

$$n(x',t+1) = \int_{-\infty}^{\infty} R_0 k_d(x'-x) N_0 e^{rt} e^{-sx}\, dx$$

$$= R_0 N_0 e^{rt} \int_{-\infty}^{\infty} e^{-sx} k_d(x'-x)\, dx$$

$$= R_0 N_0 e^{rt} e^{-sx'} \int_{-\infty}^{\infty} e^{s(x'-x)} k_d(x'-x)\, dx$$

$$= R_0 N_0 e^{rt} e^{-sx'} M(s) = R_0 M(s) n(x',t). \qquad (8.4.6)$$

Equation (8.4.3) is therefore a solution of the model if $e^r = R_0 M(s)$, i.e., $r = \log(R_0 M(s))$. It will be a traveling wave (equation 8.4.1) if
$$e^{rt} e^{-sx} = e^{r(t-1)} e^{-s(x-c)}.$$
Solving this last equation for c gives $c = r/s$, which is equation (8.4.4).

Another way of finding the wave speed is to ask: where is the spatial location $x_1(t)$ at which $n(x_1,t) \geq 1$? Equation (8.4.6) implies that $n(x,t) = (RM(s))^t n(x,0) = (RM(s))^t N_0 e^{-sx}$, and a bit of algebra shows that $n(x_1,t) = 1$ at

$$x_1(t) = \frac{1}{s} \log N_0 + \frac{t}{s} \log(R_0 M(s)). \qquad (8.4.7)$$

Thus $x_1(t)$ is moving at exactly the rate c given by (8.4.4).

If a population starts with $n(x,0) = 0$ for all x outside some bounded region, then for any s we can find a value of N such that $n(x,0) < Ne^{-sx}$. Starting from $n_s(x,0) = Ne^{-sx}$, the population $n_s(x,t)$ spreads as a traveling wave with corresponding speed $c(s)$. Because $n(x,0) < n_s(x,0)$ and the kernel in (8.1.1) is nonnegative, $n(x,t) \leq n_s(x,t)$ for all t. So if $n(x,t)$ eventually settles into a traveling wave with speed c^* (in the way that a nonspatial IPM eventually settles into a stable size distribution), then n and n_s will each have a spatial location x_1 that is the largest value of x at which the population density equals 1. If c^* were greater than $c(s)$, then eventually n would have a larger value of x_1 than n_s does, contradicting the fact that $n < n_s$. Therefore it must be the case that $c^* \leq c(s)$ for all s.

Box 8.1: Derivation of the traveling wave speeds $c(s)$ and c^*.

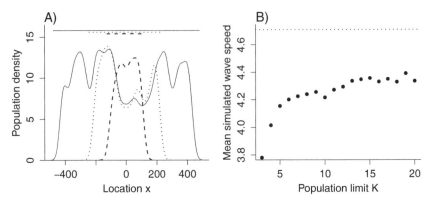

Fig. 8.3 Population spread in an individual-based model for an unstructured population. The population was initiated with 250 individuals near $x = 0$, and run for 500 generations. Individuals in the model produced a Poisson-distributed number of offspring with mean $R_0 = 2$ and then immediately die; offspring are dispersed independently with a Gaussian dispersal kernel (mean=0, sd=4); this was followed by density-dependent mortality that was zero for local population density (individuals/m) below K and then rapidly increased, so that the population density did not get far above K. **A)** Plotted curves are the population density at generations 25 (dashed), 50 (dotted), and 100 (solid) for one model run with $K = 10$, estimated using the density function in R. The horizontal lines at the top show the predicted population ranges based on the theoretical wave speed c^* for the corresponding IPM. **B)** Rate of population spread, averaged over 10 replicate simulations for each value of K. For each replicate, spread rate was estimated by determining at each time step the largest x value at which the local population density was 0.1 or higher, and using linear regression to estimate how quickly that value increased over the last 100 generations. The dotted horizontal line shows the theoretical wave speed c^* for the IPM. Source file: KS 1D Gaussian IBM.R

computer simulations) suggests that the linearization conjecture is true for all the models we discuss in this chapter, and without further comment we will assume that it's true.

Another caveat is the general principle that *demographic stochasticity slows down invasions* (Snyder 2003; Brunet et al. 2006): the population spread rate in an IPM is faster than the spread rate in the corresponding IBM. The higher the population carrying capacity, the smaller the effect of demographic stochasticity. Figure 8.3 shows how the simulated population spread rate in an IBM increases with carrying capacity K (specifically, the population was limited by density-dependent mortality that was zero up to a threshold and then increased rapidly, such that the equilibrium for the local population dynamics is K). Without demographic stochasticity (the limit as $K \to \infty$), the theoretical wave speed c^* for that model is 4.71 m/yr, and that is also the speed that occurred in numerical solutions of the spatial IPM using midpoint rule. Over a wide range of K values, demographic stochasticity decreases the spread rate by about 10% in this case. For some continuous-time diffusion models it is known that the slowdown due to demographic stochasticity is inversely proportional to the squared log of population size (Brunet et al. 2006). If that rate is also valid for IPMs, it implies that the slowdown would not be halved again until about $K = 70$. The intuitive reason for the slow decrease in the slowdown is that population spread is driven by what happens at the invasion front, which is where the population is sparsest.

8.5 Some applications of purely spatial IPMs

In a deservedly famous paper, Kot et al. (1996) explored the ecological implications of traveling waves in purely spatial IPMs. One of the most important was that the speed of population spread could be drastically underestimated by classical reaction-diffusion models. Those models are based on Fickian random diffusion of individuals in continuous time and space, and lead to a Gaussian movement distribution. But movement distributions are often fatter-tailed than a Gaussian distribution – they are *leptokurtic*, meaning that the ratio between their fourth central moment and the square of their second central moment is higher than that of a Gaussian. The lognormal distribution for cached acorn distribution by jays (Figure 8.1) is leptokurtic; seed dispersal kernels are typically leptokurtic, and parent-offspring distances are often well approximated by the bi-exponential distribution (8.1.2) or by a lognormal distribution. Kot et al. (1996) showed that dispersal kernels with fatter tails could lead to substantially faster predicted rates of population spread than a Gaussian dispersal kernel with the same variance (Figure 8.4). The flexibility of an IPM, which can have any dispersal distribution whatsoever, is therefore important for making accurate predictions of population spread when the movement distribution is anything other than Gaussian.

Kot et al. (1996) also suggested that highly fat-tailed kernels might explain cases such as the initial introduction of Japanese beetles and European starlings in the USA, where the rate of population spread initially increased over time, until the population started to encounter its geographical limits. If the tails are so fat that $M(s)$ is not finite for any s, then the purely spatial IPM predicts accelerating population spread rather than a wave with constant speed. Examples include the Cauchy distribution and the "$2Dt$" distribution that was a good fit to seed dispersal by many tree species (Clark et al. 1999). For the kernel

$$k_d(x) = \alpha^2 e^{-\alpha\sqrt{|x|}}/4 \qquad (8.5.1)$$

Kot et al. (1996) showed that the speed of population spread increases linearly over time (in an infinite spatial domain). Rapid spread due to a very fat-tailed kernel has been proposed as an explanation for the rapid northward spread of tree species following the last glaciation ("Reid's paradox"; Clark et al. (1998)). Data on tree seed dispersal are consistent with models in which a small fraction of seeds disperse with a very fat tail (equation 8.5.1), and this greatly increases the predicted speed of population spread (Clark 1998).

Frithjof Lutscher and collaborators have hypothesized that leptokurtic dispersal might help to resolve the "drift paradox" (Lutscher et al. 2010). The "drift paradox" refers to the persistence of populations in streams and rivers despite continual downstream advection (e.g., invertebrate larvae that are car-

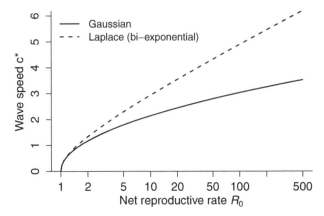

Fig. 8.4 Long-run traveling wave speed c^* for the Gaussian and Laplace (bi-exponential) dispersal kernels with variance=1, as a function of the low-density net reproductive rate $R_0 = f'(0)$. The moment-generating functions for these kernels are $M(s) = e^{\sigma^2 s^2/2}$ for the Gaussian with variance σ^2 and $M(s) = 1/(1 - b^2 s^2)$ for the Laplace with parameter b and variance $2b^2$; these lead to formulas for $c(s)$ that can be minimized to find c^*. The difference between c^* for Gaussian and Laplace kernels is less dramatic here than in Kot et al. (1996, fig.4) because they compared the Laplace kernel with a Gaussian fitted to the same data by least squares, while the comparison here (to a Gaussian with equal variance) corresponds to maximum likelihood fitting of the Gaussian. Source file: GaussLaplaceWaves.R

ried downstream by currents). A general principle from their theoretical studies is that a population can persist in a sufficiently long river if it could invade upstream in an infinitely long river. So if the downstream current is constant, the population will persist so long as the current speed is lower than the wave speed c^* that it would have in the absence of any current. This intuitive conclusion has been proved mathematically for some simple models, and has held up in all numerical tests with more complicated models. So all else being equal, the higher wave speed that results from leptokurtic dispersal promotes persistence in the face of downstream advection.

8.6 Combining space and demography: invasive species

If the spatial domain is bounded, then the spatial IPM (8.1.4) is just a general IPM with spatial location as one of the individual-level traits, and all the theory for those models, deterministic and stochastic, still applies. But beyond that (8.1.4) is too general for much more to be said. So we turn instead to two empirical applications in which local demography and dispersal jointly determine the spread rate of an invading species.

Jongejans et al. (2011) extended the analysis of traveling wave speed by Neubert and Caswell (2000) (who in turn had extended that of Kot et al. (1996)). The kernel is assumed to have exponential or thinner tails so that $M(s)$ is finite in some interval containing 0. In one spatial dimension, the calculations are remarkably similar to those for the basic spatial IPM (8.1.1). A traveling wave is a model solution $n(x, z, t)$ such that $n(x, z, t) = n(x - c, z, t - 1)$, and we find them by considering solutions of the form

$$n(x, z, t) = n(z)e^{rt}e^{-sx}. \tag{8.6.1}$$

Substituting (8.6.1) into (8.1.4), we get a traveling wave where $n(z)$ is the dominant right eigenvector of the transformed kernel

$$H_s(z', z) = \int_{-\infty}^{\infty} e^{sx} K(x, z', z) \, dx \tag{8.6.2}$$

and

$$c = \frac{1}{s} \log(\lambda(s)) \tag{8.6.3}$$

where $\lambda(s)$ is the dominant eigenvalue of H_s. The "wave shape" $s > 0$ again specifies the spatial shape of the wave at any one time t. The population structure (proportional to the eigenvector $n(z)$) is constant across space, but the total population decreases exponentially with increasing distance from the source of the wave. The wave speed (8.6.3) is related to whether the total population grows or shrinks. H_0 is the kernel for the total population $\int n(x, z, t)dx$, so the total population grows if $\lambda(0) > 1$. Then $\lambda(s) > 1$ for s near 0, and (8.6.3) gives traveling wave solutions in which the population is expanding (moving with speed $c > 0$ to the right).

For the Neubert-Caswell model (8.1.5), the transformed kernel is

$$H_s(z', z) = M_F(s; z', z)F(z', z) + M_P(s; z', z)P(z', z) \tag{8.6.4}$$

where M_F and M_P are the moment generating functions of the dispersal kernels for new recruits and survivors (which can depend, in general, on the initial and final sizes of survivors, and on the parent and final sizes of recruits). If only recruits move (e.g., seed dispersal), then $M_P \equiv 1$. Note that the multiplications on the right-hand side of (8.6.4) are element-by-element rather than composition of kernels (i.e., $M_F(s) \circ F + M_P(s) \circ P$ in the notation of Chapter 3).

Jongejans et al. (2011, Appendix A) modeled the spread of *Carduus nutans* using a WALD kernel for seeds dispersal, which gives the probability distribution of dispersal distance r in two dimensions. But theory for population spread rate in one-dimensional models can still be applied, by "marginalizing" a two-dimensional dispersal kernel. "Marginalizing" means using a one-dimensional model in which $n(x, t)$ is the distribution of individuals' x-coordinates (regardless of their y coordinate), and $k_d(x' - x)$ is the probability density for changes in an individual's x-coordinate, regardless of the change in y coordinate. Jongejans et al. (2011, Appendix A) show that if $M_R(s)$ is the moment generating function of k_R, then the moment generating function for the marginalized kernel is

$$M(s) = \frac{1}{\pi} \int_0^\pi M_R(s\cos\theta)\, d\theta. \tag{8.6.5}$$

A traveling wave in two dimensions is an expanding circle whose radius grows at a constant rate c. The one-dimensional spread rate computed using (8.6.5) then gives the asymptotic rate of increase for the radius of the occupied region in the plane.

Analogous to equation (8.4.8), and with exactly the same caveats, the predicted long-run spread rate in an infinite habitat is the minimum possible wave speed,

$$c^* = \min_s \left(\frac{1}{s} \log(\lambda(s)) \right). \tag{8.6.6}$$

These properties are illustrated in Figure 8.5. These are numerical results for a Neubert-Caswell model of a population with "small" and "large" individuals. At each time-step the local demography is described by the transition matrix $A = \begin{bmatrix} 0.5 & 2 \\ 0.5 & 0.8 \end{bmatrix}$. This is followed by dispersal; the dispersal distributions were Gaussian with mean 0 and standard deviations $\sigma_S = 4, \sigma_L = 10$ for small and large individuals, respectively. Dispersal was followed by density-dependent mortality such that the population behind the wave front increased to a stable "carrying capacity."

Box (8.2) shows how the asymptotic wave speed c^* can be computed numerically by applying equation (8.6.6) to the model without density-dependence. Note in particular how the calculations have to correctly reflect the order of events in the life cycle. The model assumes that in each time step local growth and survival are followed by dispersal, so the dispersal kernel depends on the individual's final size z'. This affects how we calculate the transformed kernel H_s whose dominant eigenvalue determines the wave speed.

After 100 simulated generations (Figure 8.5A), the population is expanding as a wave with a well-defined transition between densely and very sparsely occupied areas at about $x = 800$. The predicted spread rate (equation 8.4.8) is based on the leading tail of the population converging to the form $n(z)e^{-sx}$ posited in (8.6.1). A close-up of the transition between low and high density regions (Figure 8.5B) shows the posited form:

- The log-transformed densities of Small and Large individuals are both approximately straight lines with negative slope, indicating exponential decrease in space.
- The separation between the lines is nearly constant, indicating constant proportions of Small and Large individuals.

However, on larger spatial scales (Figure 8.5C), we see that convergence is still not complete. At locations near the wave front, the slope of log-transformed density (corresponding to the exponent s) is near the predicted value that corresponds to c^* (the dotted horizontal line in Figure 8.5C). But far out in the tail the slope ($\partial(\log n)/\partial x$) becomes more negative, indicating that the popu-

The code here is taken from `NC 1D Gaussian DiffEq Sparse Matrix.R`, which draws Figure 8.5. The first step is to specify the model:

```
a11=0.5; a21=0.5; a12=2; a22=0.8;
A=matrix(c(a11,a12,a21,a22),2,2,byrow=TRUE);
sigmaS=4; sigmaL=10;
```

Then we need a function to compute $c(s)$, the wave speed as a function of the shape parameter, using equation (8.6.3). Equation (8.6.4) says that the transformed kernel H_s is computed by multiplying each entry of the survival and reproduction kernels (which in this case are matrices) by the moment generating function (mgf) of the corresponding dispersal kernel. In our model the dispersal kernel depends only on final size, so the top row of the projection matrix is multiplied by the mgf for dispersal of small individuals, and the bottom row is multiplied by the mgf for dispersal of large individuals:

```
## moment generating function for Gaussian(mean=0,sd=sigma)
Gaussmgf <- function(s,sigma) { exp(sigma^2*s^2/2) }

## wave speed c(s)
cs <- function(s) {
    MS=Gaussmgf(s,sigmaS);   # mgf Small
    ML=Gaussmgf(s,sigmaL);   # mgf Large
    Ma11=MS*a11; Ma12=MS*a12; # Top row
    Ma21=ML*a21; Ma22=ML*a22; # Bottom row
    MA=matrix(c(Ma11,Ma12,Ma21,Ma22),2,2,byrow=TRUE);
    L1 = max(abs(eigen(MA)$values));
    return((1/s)*log(L1))# c(s) using the general formula
    }
```

Finally we get c^* by finding the minimum of $c(s)$ using the function `optimize` for one-dimensional function minimization:

```
cs = Vectorize(cs,"s"); # Optimize needs a vectorized function
plot(cs,0.05,2);        # Get a rough idea of where the minimum is
out=optimize(cs,lower=0.01,upper=0.5);
cat(out$objective,"\n"); # Predicted asymptotic spread rate
```

Box 8.2: Computing wave speeds $c(s)$ and c^* for a spatial matrix projection model.

lation tail has not "caught up" with the asymptotic shape (8.6.1), and at the edge of the habitat ($x \approx 1500$) the slope drops precipitously because of individuals dispersing past the habitat boundary. As a result, the population wave speed (average speed 8.21 over the last 25 generations) was slightly below the predicted speed ($c^* = 8.32$). The situation is analogous to an age-structured population in which the mortality rate is constant above some age. The stable age distribution then has an exponential tail of old individuals. A population

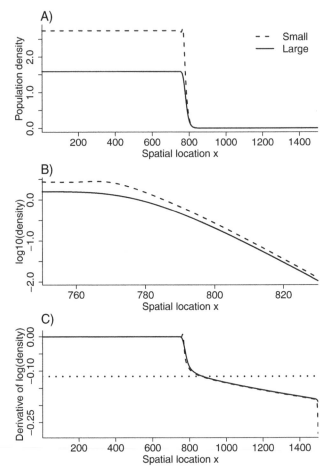

Fig. 8.5 Population spread in the deterministic Neubert-Caswell model described in the text. **A)** Population densities of large and small plants after 100 generations, starting from a small population near $x = 0$. The distribution is symmetric about $x = 0$ so the negative half is not shown. **B)** Log-transformed densities near the wave front. **C)** Slope of the log-transformed densities. The dotted horizontal line is the theoretical slope s^* corresponding to the minimum wave speed c^* that characterizes the long-term spread rate in an infinite habitat. Source file: NC 1D Gaussian DiffEq Sparse Matrix.R

started with only young individuals gets closer and closer to the stable age distribution as time goes on, but it's always missing part of the tail.

Box (8.3) shows how the asymptotic wave speed can be calculated for a spatial IPM, a "hybrid" species in which a WALD dispersal kernel for *Carduus nutans* (Skarpaas and Shea 2007) is grafted onto the size-structured *Oenothera* IPM from Chapter 2; the complete code is in NC 1D WALD IPM.R. The functions tau_z, mu_z in the code specify how the WALD kernel parameters τ, μ depend on parent size z. In the WALD model μ is proportional to seed release height H and τ is proportional to H^2. The (made-up) functions assume that H is proportional to z and give a typical flowering plant ($z \approx 3$) values of τ and μ close to those

```
source("../c2/Monocarp Demog Funs.R") # load the Oenothera IPM
L <- (-2.65); U <- 4; # size range
tau_z <- function(z) {h <- pmax(z/3,0.1); 7*h*h}
mu_z <- function(z) {h <- pmax(z/3,0.1); 3*h};
```

The functions tau_z, mu_z give the WALD kernel parameters τ, μ as a function of parent size z. Next, we need functions to compute the WALD mgf and marginalize it using equation (8.6.5):

```
WALDmgf <- function(s,w.par) {
    tau <- w.par["r.tau"]; mu <- w.par["r.mu"];
    t1 <- (tau/mu); t2 <- 2*(mu^2)*s/tau;
    mgf <- exp(t1*(1-sqrt(1-t2)));
    return(mgf)
}
margWALDmgf <- function(s,w.par) {
  (1/pi)*integrate(function(q) WALDmgf(s*cos(q),w.par),0,pi)$value;
}
```

Using the marginalized mgf we can now compute the transformed kernel, equation (8.6.4). $M_P \equiv 1$ because parents don't move. The dispersal kernel depends only on parent size so the transformed fecundity kernel is $M_F(s; z)F(z', z)$. This means that each column of the iteration matrix F is multiplied by the marginalized mgf for the corresponding parent height.

```
Hs <- function(s,m,m.par,L,U) {
    out<-mk_K(m,m.par,L,U); P <- out$P; Fs <- out$F; zvals <- out$meshpts;
    for(j in 1:m) {
        mu <- mu_z(zvals[j]); tau <- tau_z(zvals[j])
        w.par <- c(mu,tau); names(w.par) <- c("r.mu","r.tau")
        Fs[,j] <- Fs[,j]*margWALDmgf(s,w.par)
    }
    return(P+Fs)
}
```

From here on out, it's the same as for a matrix model:

```
cs <- function(s,m,m.par,L,U) {
    M <- Hs(s,m,m.par,L,U); L1 = abs(eigen(M)$values[1]);
    return((1/s)*log(L1))
    }
cs = Vectorize(cs,"s");
out=optimize(cs,lower=0.05,upper=0.16,m=100,m.par=m.par.true,L=L,U=U);
cat("Wave speed cstar =",out$objective,"\n");
```

Box 8.3: Computing the wave speeds $c(s)$ and c^* for a spatial IPM.

estimated for *Carduus nutans* (Skarpaas and Shea 2007). Again, note how the calculation of the transformed kernel depends on the order of events and the assumption that seed dispersal distance depend only on parent size.

Equation (8.6.6) makes it possible to relate the sensitivity analysis of c^* to the sensitivity analysis of population growth rate $\lambda(s)$. This depends on the following result from calculus. Suppose $F(\theta) = \min_s f(s, \theta)$ where s, θ are real and f is a smooth real-valued function, and $s^*(\theta)$ denotes the value of s at which $f(s, \theta)$ is minimized. Then $\frac{\partial f}{\partial s}(s^*(\theta), \theta) = 0$ and we have

$$\frac{\partial F}{\partial \theta} = \frac{\partial}{\partial \theta} f(s^*(\theta), \theta) = \left[\frac{\partial f}{\partial s} \frac{\partial s^*}{\partial \theta} + \frac{\partial f}{\partial \theta} \right] (s^*(\theta), \theta) = \frac{\partial f}{\partial \theta}(s^*(\theta), \theta). \quad (8.6.7)$$

We can apply this to equation (8.6.6) with θ being any parameter of the model, any entry in the demographic kernels, etc. The result is

$$\frac{\partial c^*}{\partial \theta} = \frac{1}{s^*} \frac{\partial \log(\lambda(s^*))}{\partial \theta} = \frac{1}{s^* \lambda(s^*)} \frac{\partial \lambda(s^*)}{\partial \theta}. \quad (8.6.8)$$

So any perturbation analysis for c^* is directly proportional to a perturbation analysis for the dominant eigenvalue of the transformed kernel $H(s^*)$.

For example, for a Neubert-Caswell spatial model with transformed kernel (8.6.4), the sensitivity of c^* to $F(z_1, z_0)$ is then

$$\begin{aligned}
M_F(s; z_1, z_0) \frac{\partial c^*}{\partial H_s(z_1, z_0)} &= \frac{M_F(s; z_1, z_0)}{s^* \lambda(s^*)} \frac{\partial \lambda(s^*)}{\partial H_s(z_1, z_0)} \\
&= \frac{M_F(s; z_1, z_0)}{s^* \lambda(s^*)} \frac{v_{s^*}(z_1) w_{s^*}(z_0)}{\langle v_{s^*}, w_{s^*} \rangle}
\end{aligned} \quad (8.6.9)$$

where v_{s^*}, w_{s^*} are the left and right eigenvectors of H_{s^*}.

Because $\lambda(s^*)$ is the population growth rate of the transformed kernel H_{s^*}, not that of the demographic kernel $F + P = H_0$ that describes the growth of total population size, the parameters or demographic rates that have the largest effect on population *spread* rate might not be the same as those with the largest effect on population *growth* rate. In practice, however, the key rates for population growth and population spread have generally been similar. For the monocarpic thistle *Carduus nutans* in New Zealand modeled with a spatial IPM, Jongejans et al. (2011) found a very high correlation between the elasticities of population growth rate and population spread rate to demographic parameters and demographic kernel entries. Demography thus influences population growth and population spread in similar ways. However, for dispersal this was not true as the parameters in the dispersal kernel influence population spread rate but have no effect on population growth rate in a density-independent model. For *Carduus* the highest elasticity parameter for population spread rate was the standard deviation of wind speed in the WALD dispersal kernel.

8.7 Space and demography 2: invasive species in a temporally varying environment

Ellner and Schreiber (2012) took the final step (so far) of adding environmental stochasticity to a spatial IPM for invasive species spread, meaning that the demographic and dispersal kernels are allowed to vary randomly over time. The calculations and results are again very similar to the pure spatial model. The transformed kernel is a stochastic IPM $H_{s,t}(z', z)$ which has a stochastic growth rate $\lambda_S(s)$ when the assumptions presented in Chapter 7 are satisfied. The predicted spread rate c^* is

$$c^* = \min_s \left(\frac{1}{s} \log(\lambda_S(s)) \right). \tag{8.7.1}$$

Using (8.6.7) perturbation formulas for (8.7.1), analogous to (8.6.8), can be derived from the perturbation formulas and small-variance approximations for λ_S. Using those, Ellner and Schreiber (2012) showed that temporal variation in the demographic kernel slows down population spread. This is a very general consequence of the small-variance approximation, which shows that adding zero-mean variation to the demographic kernel always decreases the stochastic growth rate. In contrast, variation in mean dispersal distance generally *increases* the rate of population spread. Moreover, when mean dispersal distance varies over time, demographic variation that is positively correlated with the variation in mean dispersal distance can also increase the rate of population spread.

To illustrate the potential importance of temporally variable dispersal, Ellner and Schreiber (2012) developed an IPM for the spread of perennial pepperweed, *Lepidium latifolium*, a Eurasian crucifer that is now widely invasive in wetlands and riparian zones in the western USA (Leininger and Foin 2009). Pepperweed forms dense monospecific stands, so the model describes patches of pepperweed rather than individual plants, with z=log patch radius. Individual growth in this model is expansion of established stands (mainly due to lateral expansion of established genets), and fecundity is establishment of new patches by seeds and water-dispersed root fragments. A density-independent IPM at the level of patches is valid so long as patches are sparse enough that they don't collide with each other, even though individual plants compete with others in their patch. According to the linearization hypothesis, this density-independent model will correctly predict the asymptotic spread rate of the population of patches.

The model was based on studies in the Cosumnes River Preserve (Hutchinson et al. 2007; Viers et al. 2008) and nearby sites in the Sacramento-San Joaquin River delta, California (Andrew and Ustin 2010). Water – soil moisture and flooding – is the most important component of environmental variability. Established patches tended to grow during dry years and shrink during wet years.

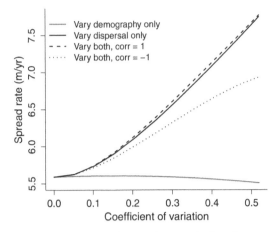

Fig. 8.6 Asymptotic one-dimensional spread rate c^* for the perennial pepperweed model. All time-varying parameters (d, f, g, L) had a uniform distribution centered on their estimated means $(\bar{d} = 0.06, \bar{f} = 0.04, \bar{g} = 0.5, \bar{L} = 15)$. The local demography parameters $(d, f$ and $g)$ were perfectly correlated with each other, while mean displacement L was either perfectly correlated or perfectly anti-correlated with the local demography parameters (correlation coefficient $= +1$ or -1). Source files: `PepperweedIPMFunctions.R`, `PepperweedIPMRun=U.R`

Most new patches appeared during dry years, but these are presumed to have been initiated by seeds or root fragments that had been moved by water flow in wet years (Hutchinson et al. 2007).

The distance between "parent" and "offspring" patches was assumed to have a bi-exponential distribution (equation 8.1.2) based on data in Andrew and Ustin (2010), with mean absolute distance $L(t)$ that varied across years. The mean displacement $L(t)$ for pepperweed patches was about 15 m on average, but it varied roughly from 8 m to 21 m in the 4 years for which "parent-offspring" distances were estimated using remote-sensing data (Andrew and Ustin 2010).

The Laplace-transformed kernel for the model is

$$H_{s,t} = (1 - s^2 L(t)^2)^{-1} F_t + P_t. \tag{8.7.2}$$

where P_t represents survival and growth of established patches, F_t represents production of new patches, and the factor $(1 - s^2 L(t)^2)^{-1}$ is the Laplace transform of the dispersal kernel. Because established patches have dispersal distance $= 0$ with probability 1, the Laplace transform of their "dispersal distribution" is constant with value 1. The demographic kernels F and P are determined by three time-varying parameters $d(t), g(t)$ and $f(t)$ that determine, respectively, the mortality, growth rate, and fecundity (new patch production) of established patches. Details of the demographic models are given in Box 8.4.

Figure 8.6 shows how random variation in demography and dispersal affect the predicted population spread rate. The coefficient of variation in $L(t)$ for pepperweed is about 0.4, based on the four years of data, which is enough to

The individual state variable z is the natural log of patch radius in meters.

Growth is spread of established patches. Because new and established patches were observed to have similar rates of radial growth, patches are assumed to increase in radius by an amount that depends on environmental conditions in year t, but not on patch size. If all patches in year t grow in radius by the same amount $g(t)$, we would have $z' = \log(e^z + g(t))$. To allow for between-patch variation in growth, z' is assumed to have a Gaussian distribution with mean $\log(e^z + g(t))$ and variance σ_g^2.

Mortality is permanent disappearance of a patch. Similar to growth, patches are assumed to disappear with probability $d(t)$ that varies over time but is independent of patch size.

Fecundity is creation of a new patch by seeds or root fragments dispersed from an established patch. This is assumed to be proportional to patch area, or equivalently to patch radius squared, $b(z) = f(t)e^{2z}$. Because patch expansion is so rapid, the size of new patches is immaterial. For convenience, their size distribution $c(z')$ is a $\beta_{3,3}$ distribution shifted to the interval $-3 \leq z' \leq -2$ (radius roughly 5 to 13 cm, representing a founding plant). `plot(function(x) dbeta(3+x,3,3),-3,-2)` will show you the assumed distribution of initial log patch radius.

Combining these assumptions, the demographic kernels for the model are then

$$F_t(z', z) = f(t)e^{2z}\beta_{3,3}(3 + z) \tag{8.7.3}$$
$$P_t(z', z) = (1 - d(t))\phi(z'; \log(e^z + g(t)), \sigma_g^2) \tag{8.7.4}$$

where $\phi(x; \mu, \sigma^2)$ is the density of a Gaussian distribution for x with mean μ and variance σ^2.

Box 8.4: Details of the perennial pepperweed IPM

increase spread rate by about 30% (solid black curve). Variation in just the demographic parameters (solid gray curve) has much less effect, but demographic variation slows spread more substantially if it acts in opposition to dispersal variation (dotted curve), so that a good year for offspring dispersal (large $L(t)$) is a bad year for offspring production (small $f(t)$). Demographic variation that reinforces dispersal variation (large f with large L) has a positive effect, as predicted by the general analysis in Ellner and Schreiber (2012), but it's very small.

These results reinforce previous findings on population spread using integrodifference models. In particular, with multiple dispersal modes such as seeds dispersed by wind and by animals (Neubert and Caswell 2000; Ellner and Schreiber 2012), the models predict that population spread rate is largely determined by the dispersal mode with largest mean parent-offspring distance. This is true even when very few offspring disperse by the long-distance mode. The analogous property illustrated in Figure 8.6 is that population spread rate is governed by the best years for dispersal. More variation in dispersal improves the best years, so the population spreads faster even though the worst years are even worse. Positive correlation between dispersal and demography makes the best years even better, because dispersal and fecundity are both high, while negative correlation decreases the quality of the best years.

The overall intuitive picture is that population spread is "pulled" by the lucky few that go the furthest. This suggests that Allee effects might be extremely important in limiting population spread rates, because the few who go furthest are then doomed. Conversely, any rare combination of events or individual traits that lets some individuals establish far ahead of the main wave of advance will set the long-term pace of population spread.

Chapter 9
Evolutionary Demography

Abstract Patterns of survival and reproduction determine fitness and so it should come as no surprise that there is a rich body of theory linking demography with evolution. Here we provide an overview of these methods showing how evolutionary dynamics and selection can be understood using IPMs. We show how selection can be approximated using sensitivities and how this leads to an approximation for trait dynamics. The endpoints of evolution – what we expect to see in nature – are explored using ideas from *Adaptive Dynamics*, the key methods based on Evolutionarily Stable Strategies and Convergence Stability. Efficient methods for finding ESS are presented. We then extend these ideas to cover stochastic environments and function-valued traits. For function-valued traits we model the entire function rather than the underlying parameters, and so we are not tied to a specific fitted function.

9.1 Introduction

Evolutionary demography uses the demographic theory introduced in Chapters 3, 4, 6, and 7 and applies it to evolutionary questions. These questions fall into two broad categories: those dealing with the traits organisms possess, and those dealing with the dynamics of trait change. For example, we might be interested in the evolution of flowering size in *Oenothera*, and more specifically we might want to predict the observed flowering size and the rate of evolution of this trait. To do this we need a realistic ecological model capturing 1) the size-dependent demography, 2) the action of density dependence, and 3) the genetic control and transmission of the trait. Typically we will assume very simple genetics. We do this for two reasons: 1) for many life history traits we lack any knowledge of their genetic basis, although this is not always true (e.g., Hanski and Saccheri 2006; Gratten et al. 2012; Johnston et al. 2013; Santure et al. 2013) and is rapidly changing, and 2) we are not geneticists. Typical assumptions used are haploid genetics (no sex) and small or very small genetic variance, and when we say small we mean really small as in the case of *Adaptive Dynamics* where we assume that evolution is mutation-limited. This doesn't

© Springer International Publishing Switzerland 2016   255
S.P. Ellner et al., *Data-driven Modelling of Structured Populations*,
Lecture Notes on Mathematical Modelling in the Life Sciences,
DOI 10.1007/978-3-319-28893-2_9

mean a more complex treatment of genetics is not possible, but rather in most cases simple assumptions are good enough and can be extended if suitable genetic data are available. This is especially true when you are less interested in predicting the transient dynamics than you are in predicting the final outcome - the trait value that maximizes fitness, or is an evolutionarily stable strategy.

In contrast to the simplicity of the genetic assumptions, we will allow complex state-dependent demography and density dependence. We assume density-dependent demography as we believe realistic ecological models should include this, and because models ignoring density dependence often have poor ability to predict life history evolution (Hesse et al. 2008; Gremer and Venable 2014). This does not mean that density dependence must always be assumed. For example, in stochastic environments when population size is small, changes in population density may have very little impact on individual performance.

The approach we advocate has been used successfully to predict, for example

1. Flowering size in several plant species (Childs et al. 2004; Rees et al. 2006; Hesse et al. 2008; Kuss et al. 2008; Metcalf et al. 2008; Miller et al. 2012).
2. Seed dormancy (Rees et al. 2006; Gremer and Venable 2014).
3. When to produce twins in Soay sheep (Childs et al. 2011).

A key feature of the approach is that individuals are cross-classified by a dynamic trait, size[1] and by a static trait ("genotype") that is assigned at birth and remains constant over life. This assumption tacitly underlies analyses of optimal or evolutionarily stable life history traits (e.g., Rees and Rose 2002; Childs et al. 2004; Rees et al. 2006; Hesse et al. 2008; Kuss et al. 2008), and is explicit in models for the dynamics of evolving traits across multiple generations developed in this chapter and previously by Rees et al. (1999); Rose et al. (2002); Coulson et al. (2011); Bruno et al. (2011); Vindenes and Langangen (2015); Rees and Ellner (2016).

For the analyses we describe, the demographic costs and benefits of different behaviors must be specified. For example, in monocarpic perennials the costs and benefits of flowering are well defined, as only plants that don't flower have the potential to grow (and so increase their potential fecundity) if they survive to the next year. For other life history decisions it can be less easy to specify the costs and benefits (e.g., Miller et al. 2012). Quantifying trade-offs may be especially difficult in situations where individuals vary greatly in their *acquisition* of resources, so the effects of differences in *allocation* are hard to quantify (van Noordwijk and de Jong 1986).

Throughout this chapter we focus on the methods that we have personally found useful, so we are very selective. This seems reasonable as there are other books covering, for example, evolution in age-structured populations (Charlesworth 1994), adaptive dynamics (Dercole and Rinaldi 2008), quantitative genetics (Falconer and Mackay 1996; Lynch and Walsh 1998), and life history evolution (Stearns 1992; Roff 2002). We also do not cover approaches based on partitioning change in a state variables, such as size, based on the

---

[1] One last time: or any other continuously varying trait or traits.

Price equation (Coulson et al. 2010), or approaches based on applying quantitative genetics methods to dynamic traits (see Chevin 2015, for discussions of how those approaches compare to the ones in this chapter). Continuously varying static traits are often problematic for mathematical analysis of IPMs (see Section 10.4), but in this chapter we use approximations that do not rely on the formal theory.

9.2 Motivation

Consider a model for *Oenothera* like the ones presented in Chapters 2 and 5, but we now assume individuals are characterized by two traits, their size z and the intercept of the regression describing how the probability of flowering depends on size, β_0 (Figure 2.4B). We assume β_0 is under genetic control with no environmental component, so β_0 is effectively "genotype." Changing β_0 moves the probability of flowering function, and so changes the mean size and age at flowering. As in quantitative genetics, genotype determines the probability distribution of phenotype for an individual, but not its precise value. As a benchmark, we first modified the individual-based model (IBM) for *Oenothera* so each seed produced inherits its parent's β_0 plus, as a result of mutation, a random deviation drawn from a Gaussian distribution. We assumed density dependent recruitment, such that some fixed number of new recruits are selected at random from all seeds produced, without regard to β_0.

The evolutionary trajectories for 10 population starting at either $\beta_0 = -20$ or $\beta_0 = -30$ are shown in Figure 9.1. In panel A) all populations converge to a mean genotype of $\bar{\beta}_0 \approx -25$, while in B) the genetic variance increases to ≈ 3.5, where it is maintained by the mutation-selection balance. The strengths of this approach are that we can have complex ecology, for example including size and age-dependent demography, density dependence and environmental stochasticity, and don't have to make any assumptions about evolution maximizing some quantity, say R_0 or λ, as we would if we were using an optimization approach. The weakness is that simulations take a long time, and we have no idea whether the quantities we are interested in are well defined. Before the advent of IPMs this was pretty much all that could be done (Rees et al. 1999; Rose et al. 2002) but now we can do much more. Our aim then in the first part of this chapter is to show how IPMs can be used to predict the properties of systems like our evolving *Oenothera* IBM.

9.3 Evolution: Dynamics

Our approach to studying the evolutionary dynamics is to first translate the IBM into an IPM which we can then solve numerically, and then to approximate the dynamics of the IPM. Translating the IBM into an IPM involves building a 2-dimensional IPM where individuals are characterized by their size, z and genotype, β_0. The IPM also needs to have density dependent recruitment, like the IBM. Fortunately we have already built a density-dependent, size-structured IPM for *Oenothera* in Chapter 5. With $\mathcal{R}$ new recruits, the seedling contribution to the subsequent year's population is $\mathcal{R}c_0(z')$, the number of recruits multiplied by the seedling size distribution. The IPM iteration is then

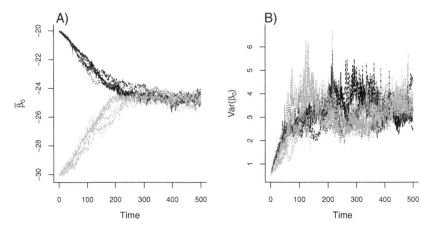

Fig. 9.1 Evolutionary trajectories for the *Oenothera* IBM showing the evolution of A) the mean genotype, $\bar{\beta}_0$, and B) genetic variance of β_0. Five populations were started at -20 and five at -30, in all cases the population evolves to $\bar{\beta}_0 \approx -25$. Source file: Monocarp Evol Demog Dynamics.R

$$n(z', t+1) = \mathcal{R}c_0(z') + \int_L^U P(z', z)n(z, t)\, dz \qquad (9.3.1)$$

with $P(z', z) = (1 - p_b(z))s(z)G(z', z)$ as in Chapter 2. We now need to generalize this so individuals are characterized by both size, z and genotype, β_0. First let's assume all individuals breed true, so all seeds inherit their parent's β_0 exactly. The fraction of recruits captured by individuals with flowering strategy, β_0, is their seed production divided by the total seed production. Defining the total seed production by β_0 plants as

$$S(\beta_0) = \int p_b(z, \beta_0)b(z)n(z, \beta_0, t)\, dz \qquad (9.3.2)$$

gives the IPM iteration

$$n(z', \beta_0, t+1) = \mathcal{R}c_0(z')\frac{S(\beta_0)}{\int S(\beta_0)\, d\beta_0} + \int P(z', z; \beta_0)n(z, \beta_0, t)\, dz \qquad (9.3.3)$$

with $P(z', z; \beta_0) = (1 - p_b(z, \beta_0))s(z)G(z', z)$. To allow for the fact that seeds don't inherit their parent's β_0 exactly we need to define an inheritance kernel, $M(\beta_0, \beta_0^P)$ which describes the probability density function for β_0 in seeds produced by parents with genotype β_0^P. Modifying the previous equation to include this, we obtain

$$n(z', \beta_0, t+1) = \mathcal{R}c_0(z')\frac{\int M(\beta_0, \beta_0^P)S(\beta_0^P)\, d\beta_0^P}{\int S(\beta_0)\, d\beta_0} + \int P(z', z; \beta_0)n(z, \beta_0, t)\, dz.$$
$$(9.3.4)$$

This IPM like the IBM allows us to study evolution by simply iterating the model. Seeds with new genotypes are produced according to the inheritance kernel, $M(\beta_0, \beta_0^P)$. These then compete for establishment, and, if successful, ultimately flower having grown and survived in the intervening period. The IPM projects the density of z and β_0, and so we can characterize the population using the moments of this distribution. However, like the IBM, the IPM is not terribly transparent, and so it is useful to also have approximations for the trait dynamics.

9.3.1 Approximating Evolutionary Dynamics

In Appendix 9.8, following the approach of Iwasa et al. (1991), we derive an approximation for the dynamics of the mean β_0 assuming 1) small variance in β_0, 2) the population is close to the stable size distribution so we can use λ as a measure of fitness, 3) fitness is frequency/density dependent such that the fitness of an individual depends on the population mean, $\bar{\beta}_0$, which we write as $\lambda(\beta_0; \bar{\beta}_0)$, and finally 4) as a result of density dependence and the small variance of β_0 we have $\lambda(\bar{\beta}_0; \bar{\beta}_0) = 1$. With these assumptions and using the approach described in Appendix 9.8 we can approximate the dynamics of the $\bar{\beta}_0$ by the following equation:

$$\bar{\beta}_{0,t+1} \approx \bar{\beta}_{0,t} + G_{\beta_0} \left(\frac{\partial \lambda(\beta_{0,t}; \bar{\beta}_{0,t})}{\partial \beta_{0,t}} \right) \Bigg|_{\beta_{0,t} = \bar{\beta}_{0,t}} \tag{9.3.5}$$

where G_{β_0} is the genetic variance of β_0, and the final term is the sensitivity of λ to β_0 evaluated at $\bar{\beta}_{0,t}$. The importance of this equation cannot be overestimated, as it shows that the eigenvalue sensitivities we calculated in Chapter 4 have a direct interpretation as selection gradients, calculated over the entire life cycle.

As expected the bivariate IPM provides an excellent description of the dynamics of the mean and variance of the flowering intercepts (Figure 9.2). The population mean genotype is determined by the pattern of selection, whereas the variance is determined by the balance between mutation and selection. In order to apply the approximation we needed an estimate of the genetic variance of β_0, and so we used the predicted values from the 2D IPM (lines in Figure 9.2B). Using this we find the approximation (9.3.5) provides a remarkably accurate description of the dynamics of the mean intercept. This is rather surprising given the breadth of the assumptions used to derive the approximation, some of which are clearly not true, for example we do not have small variance (Figure 9.2B). For a discussion of why the approximation works so well, and approximations for variance dynamics, see Rees and Ellner (2016)

The approach used to approximate the dynamics can be extended in several ways (Iwasa et al. 1991):

1. If the phenotype is not completely determined by the genotype, then we have $z = x + e$ where z is the phenotype, x the genetic component, and e is a zero mean environmental component. The dynamics of phenotypic change are then

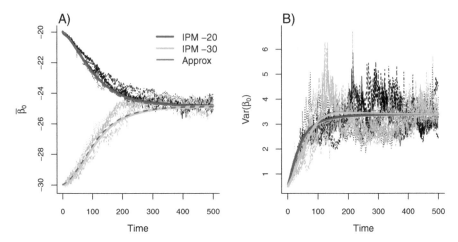

Fig. 9.2 Evolutionary trajectories for the *Oenothera* IBM showing A) the mean genotype, $\bar{\beta}_0$, and B) genetic variance of β_0. Five populations were started at -20 and five at -30, in all cases the population evolves to $\bar{\beta}_0 \approx -25$. In both A) and B) the turquoise and purple lines are from the 2-dimensional IPM. The red lines in A) are from the approximation, equation 9.3.5. Source file: `Monocarp Evol Demog Dynamics.R`

$$\Delta\bar{z} \approx G_x(\partial\lambda(z;\bar{z})/\partial z)/\lambda(\bar{z};\bar{z}), \qquad (9.3.6)$$

where G_x is the genetic variance, not the phenotypic variance. As e has zero mean this is also the change in the genetic component, x.

2. For diploid organism the same equations hold providing genetic effects are purely additive. With fitness interactions between genes there will be nonadditive effects and these must be separated statistically, therefore G_x is the additive genetic variance rather than the total genetic variance.

3. Multiple characters can also be studied using this approach (see the Appendix in Iwasa et al. (1991)), and perhaps not surprisingly we end up with a matrix equation. For two evolving characters x and y we have

$$\begin{pmatrix} \Delta\bar{x} \\ \Delta\bar{y} \end{pmatrix} = \begin{pmatrix} G_x & B_{xy} \\ B_{xy} & G_y \end{pmatrix} \begin{pmatrix} \dfrac{\partial\log\lambda(x,y;\bar{x},\bar{y})}{\partial x} \\ \dfrac{\partial\log\lambda(x,y;\bar{x},\bar{y})}{\partial y} \end{pmatrix} \qquad (9.3.7)$$

where B_{xy} is the additive genetic covariance between x and y, and the partial derivatives are evaluated at $\bar{x}, \bar{y}$.

9.4 Evolution: Statics

The IBM, IPM, and approximation all suggest that a flowering intercept of ≈ -25 is favored by evolution (Figure 9.2). So how can we predict this, short of simulating the model? Our main tools come from *Adaptive Dynamics*, a body of theory that deals with mutation-limited evolution in ecologically realistic models (e.g., Metz et al. 1992; Dieckmann and Law 1996; Geritz et al. 1998; Dercole and Rinaldi 2008). The idea is to imagine evolution of a clonal lineage where mutations occasionally arise. If the new mutation is better, then it spreads through the population and eventually excludes the previous clone. There the population stays until a new clone with higher fitness is produced by mutation, and so the process continues. Fitness in this framework is given by λ, evaluated for a rare mutant with trait value x invading a population where the current (resident) strategy has trait $\bar{x}$. Using the notation from Appendix 9.8, a new mutant with trait x will be successful providing $\lambda(x; \bar{x}) > 1$. Combining this idea with density dependence so that $\lambda(\bar{x}; \bar{x}) = 1$ leads to a large, elegant body of theory (Dercole and Rinaldi 2008). As we noted above, we are primarily interested in predicting evolutionary endpoints, and for this the simplifying genetic assumptions of Adaptive Dynamics are less important than they are for predicting the dynamics of trait change in an evolving population.

9.4.1 Evolutionary Endpoints

In the Adaptive Dynamics framework evolution stops when $\lambda(x; \bar{x}) < 1$ for all strategies other than $\bar{x}$ for which $\lambda(\bar{x}; \bar{x}) = 1$. Thinking about this for a bit leads to the classic conditions for an Evolutionarily Stable Strategy (ESS), namely

$$\left. \frac{\partial \lambda(x; \bar{x})}{\partial x} \right|_{x=\bar{x}} = 0 \quad \text{and} \quad \left. \frac{\partial^2 \lambda(x; \bar{x})}{\partial x^2} \right|_{x=\bar{x}} < 0. \qquad (9.4.1)$$

So at the ESS λ is maximized, in the sense that all strategies other than $\bar{x}$ will have $\lambda < 1$ as invaders into a population at the ESS. However, this does not mean that we can find the ESS by maximizing λ. Under density dependence, any genetically homogeneous population will settle down to a steady state at which the long-term population growth rate is 1, but a rare invader can still have $\lambda > 1$ and displace the dominant type. A very general approach to finding an ESS is to use an iterative approach, as follows:

1. For the resident, $\bar{x}$, find $\bar{\mathbf{N}}$, the equilibrium population size, by iterating the IPM.
2. Substitute $\bar{\mathbf{N}}$ into the kernel and find the value of x that maximizes $\lambda(x; \bar{x})$, call this x_{opt}.
3. If $|\lambda(x_{opt}; \bar{x}) - 1| < \varepsilon$, where ε is the numerical tolerance, then we have an ESS otherwise repeat from 1) using the value of x_{opt} as the new resident, $\bar{x} = x_{opt}$.

In `Monocarp Evol Demog Statics.R` we provide code for finding the ESS β_0 for *Oenothera*. The function `Find_ESS` takes a parameter vector and locates the ESS to some numerical tolerance. Note that we have to use a very wide size range for this calculation, because changes in β_0 affect the size distribution of the population, and the model's size range has to work for all values of β_0. In this case density dependence acts on recruitment and so we initially find p_r such that the resident strategy is at equilibrium, $\lambda = 1$, and then use `optimize` to find the best invading strategy. The output in R looks like this:

```
> find_ESS(m.par.true,0.0000001)
1.079078     -23.96693
1.001688     -24.9013
1.000001     -24.92352
[1] -24.92354
```

So $\beta_0 \approx -24.9$ is the ESS. Starting with $\bar{\beta}_0 = -18$ we find the best invading strategy is $\beta_0 = -23.97$ which has $\lambda = 1.08$. Making this the resident we find the best invader is $\beta_0 = -24.90$, but now the fitness advantage is very small, $\lambda = 1.002$. Finally when the resident is $\beta_0 = -24.90$ the best invader is $\beta_0 = -24.92$ and this has $\lambda = 1.000001$ which is within our tolerance, so we have an ESS.

Evolutionarily stable strategies are the endpoints of the evolutionary process, and their properties are defined in terms of the behavior of new mutants once the ESS has been achieved, equation (9.4.1). However, we would also like to know whether the ESS can be achieved. The key concept here is *convergence stability*, which deals with resident strategies *not* at the ESS and asks whether mutants with strategies closer to the ESS can invade (Eshel and Motro 1981); if they can, then evolution will move towards the ESS. There are algebraic conditions for different kinds of convergence stability (Christiansen 1991; Eshel 1983; Levin and Muller-Landau 2000) but these are not very useful in this case as we don't have an algebraic expression for fitness. We shall therefore use a graphical tool called *Pairwise Invasibility Plots* or *PIPs* for short (Geritz et al. 1997). PIPs are constructed by selecting a range of resident strategies, $\bar{x}$, and for each resident strategy we calculate $\lambda(x; \bar{x})$ and plot those parameter pairs $\bar{x}, x$ where $\lambda(x; \bar{x}) > 1$, Figure 9.3. Clearly on the diagonal of the PIP we have $\lambda(\bar{x}; \bar{x}) = 1$. Where the boundary of $\lambda > 1$ region intersects the diagonal we have an *evolutionarily singular strategy* and these points are of special interest. Near such points the fitness landscape for a rare mutant is locally flat. There are two important cases here, corresponding to a fitness maximum at which there is stabilizing selection, this is therefore an ESS, and a fitness minimum where there is disruptive selection. The singular strategy in the *Oenothera* PIP occurs at -24.9 which corresponds to the ESS we found earlier, and as all invading strategies have $\lambda < 1$ when $\bar{\beta}_0 = -24.9$ this confirms that we do in fact have an ESS.

Let's consider what happens when the resident strategy is not at the ESS. Say the current resident has $\bar{\beta}_0 = -28$ then all invading strategies with $\beta_0 < -28$ have $\lambda < 1$ and so these will fail to invade, Figure 9.3. Now imagine that as a result of mutation a new strategy with $\beta_0 = -26$ emerges. This has $\lambda > 1$ and so will successfully invade and displace the current resident. Likewise this new resident is invadable by new mutants with β_0 closer to the ESS and so the population will evolve through a series of substitutions to the ESS. The singularity in the *Oenothera* PIP is therefore evolutionarily stable and also convergence stable. At this point you would be forgiven for thinking that evolutionary and convergence stability come together – you can't evolve to a fitness minimum, can you? Well, it turns out that you can (Abrams et al.

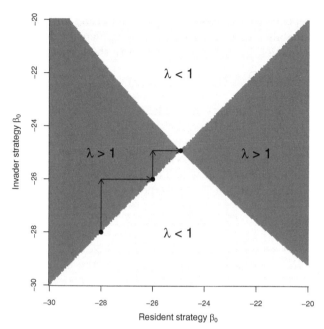

Fig. 9.3 Pairwise Invasibility Plot for the *Oenothera* IPM. Red areas indicate the invading strategy has $\lambda > 1$ while the white areas have $\lambda < 1$. Source file: Monocarp Evol Demog Statics.R

1993), in fact all combinations of evolutionary and convergence stability or instability are possible. Singularities which lack both kinds of stability are called *repellers* as evolution will result in the trait moving away from the singularity. Singularities that are convergence stable but evolutionarily unstable are called *branching points*, evolution pushes the trait towards the singularity but once there the population experiences disruptive selection and so becomes dimorphic (Geritz et al. 1997).

9.4.2 Finding ESSs using an optimization principle

The iterative approach to finding ESSs is very general but can be slow, particularly with a high-dimensional IPM. It is therefore useful to know when more efficient approaches can be used. Several papers have explored the ecological conditions required for an ESS to be characterized by maximizing some demographic quantity, say R_0. This is known as an *optimization principle* (Mylius and Diekmann 1995; Metz et al. 2008), and means that only a single maximization is required. There are several ways of spotting if evolution maximizes some quantity in a model. These depend, perhaps not surprisingly, on the action of density dependence, as this determines how individuals interact. In what follows we assume that density dependence is undercompensatory so the population tends to a stable equilibrium and that density dependence operates independently of the strategy under consideration.

1. Density dependence decreases $F(z', z; \theta)$ uniformly for all z, say as a result of the probability of recruitment, p_r, being density dependent as in the *Oenothera* model. In this case we can write the kernel as

$$K(z', z, \bar{\mathbf{N}}; \theta) = P(z', z; \theta) + g(\bar{\mathbf{N}})F(z', z; \theta), \qquad (9.4.2)$$

where $g(\bar{\mathbf{N}})$ is the density dependence function, and $\bar{\mathbf{N}}$ is the measure of population size that determines the impact of density dependence. A resident strategy will, under our assumptions, reach a stable equilibrium with a corresponding density $\bar{\mathbf{N}}_{\mathbf{R}}$. An invader will be successful if it has $\lambda > 1$, or equivalently $R_0 > 1$. The R_0 for the invader is the dominant eigenvalue of its next-generation kernel (see Section 3.1.1)

$$R_I = g(\bar{\mathbf{N}}_{\mathbf{R}})F(\theta_I)(I - P(\theta_I))^{-1} \qquad (9.4.3)$$

where θ_I is a vector of parameters characterizing the invader. Hence the invader's R_0 is $g(\bar{\mathbf{N}}_{\mathbf{R}})\widetilde{R}_0(\theta_I)$ where $\widetilde{R}_0(\theta)$ is the dominant eigenvalue of $F(\theta)(I - P(\theta))^{-1}$. The resident has $R_0 = 1$, because it is at a stable equilibrium, neither increasing nor decreasing. Consequently, the invader has $R_0 > 1$ and hence $\lambda > 1$ if and only if $\widetilde{R}_0(\theta_I) > \widetilde{R}_0(\theta_R)$, where θ_R is the vector of parameters characterizing the resident strategy. Therefore the ESS is characterized by maximizing $\widetilde{R}_0(\theta)$.

2. Density dependence decreases $P(z', z; \theta) + F(z', z; \theta)$ uniformly for all z. In this case we can write the kernel as

$$K(z', z, \bar{\mathbf{N}}; \theta) = g(\bar{\mathbf{N}})(P(z', z; \theta) + F(z', z; \theta)), \qquad (9.4.4)$$

the invader's λ is $g(\bar{\mathbf{N}}_{\mathbf{R}})\widetilde{\lambda}(\theta_I)$ where $\widetilde{\lambda}(\theta)$ is the dominant eigenvalue of $P(z', z; \theta) + F(z', z; \theta)$. The invader has $\lambda > 1$ if and only if $\widetilde{\lambda}(\theta_I) > \widetilde{\lambda}(\theta_R)$, and so the ESS is characterized by maximizing $\widetilde{\lambda}$. So, for example, if mortality acted before reproduction and the probability of survival was density dependent, $g(\bar{\mathbf{N}})s(z)$, then the kernel could be written as equation 9.4.4 and we could maximize $\widetilde{\lambda}$ to find the ESS.

3. If density dependence is modeled as a function of a single quantity $\mathbf{M}$ (say the total number of seeds produced, total number of adults in the population, or total size of all adults) with $\partial K / \partial \mathbf{M} < 0$ then evolution maximizes this quantity. This is known as a *pessimism principle*, because the resident which maximizes this quantity has $\lambda = 1$ while all other strategies have $\lambda < 1$ because the environment is worse than any other strategy can sustain at equilibrium (Metz et al. 2008). Conversely, if the kernel is an increasing function of some quantity whose steady-state value is a function of the resident's strategy (such as the steady-state availability of a limiting resource), then evolution minimizes this quantity, as in the R^* rule of competition for a single limiting resource.

Finally we can also use PIPs to identify if some quantity is maximized by evolution. If the PIP is skew symmetric, so flipping the PIP over the main

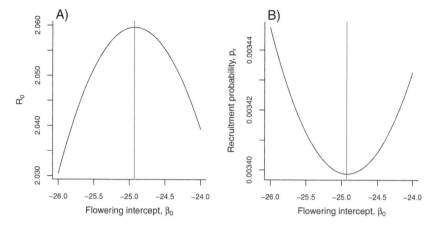

Fig. 9.4 Illustrating the use of A) maximizing R_0 or B) minimizing the recruitment probability, p_r for finding the ESS using the *Oenothera* IPM. Red lines indicate the ESS found previously using the iterative approach. Source file: `Monocarp Evol Demog Statics.R`

diagonal only changes the $+$ and $-$ signs, then we know that there is an *optimization principle*, although we don't know what is maximized (Metz et al. 2008). So, for example, in Figure 9.3 the PIP is skew-symmetric so we know there's an *optimization principle*. In this case density dependence acts on initial offspring survival so we know evolution maximizes R_0, and also because density dependence is modeled as a function of the probability of establishment this is minimized at the ESS. These approaches are illustrated in Figure 9.4.

A crucial part of these cases is the one-dimensionality of competition (Metz et al. 2008): resident trait value determines the steady-state value of *one quantity*, which in turn determines the success of an invader. Without that, optimization principles are harder to find and may not exist. In a stochastic environment, even if competition is determined by one quantity (such as total number of individuals), resident trait value determines the *distribution* of this quantity, which is described by at least two attributes (mean and variance). Several different strategies may then coexist through the storage effect (Chesson 1982, 2000). And instead of an ESS there may be an *evolutionarily stable coalition* (ESC), a collection of strategies that jointly resist invasion by others (e.g., Ellner and Hairston 1994). Another example is nontransitive competition ("rock, paper scissors" dynamics: P beats R, S beats P, R beats S), so that any monomorphic population can be invaded. Such situations would be revealed by making a PIP, seeing that it is not skew-symmetric and that all evolutionarily singular strategies can be invaded.

9.5 Evolution: Stochastic Environments

The models we have considered in the previous sections have assumed the world does not vary from year to year, which is clearly not a realistic assumption for many populations. Indeed in many of the best studied field populations there is enormous variation in growth, survival, and reproduction, and in most cases this variation is linked to measures of an individual's state (Coulson et al. 2001; Rees et al. 1999, 2006). The inclusion of environmental stochasticity is discussed extensively in Chapter 7, and so here we assume the model has been fitted and ask how to study evolution in this context. Luckily many of the ideas we developed for constant environment models carry over to variable environment ones with some minor changes.

To illustrate the various methods we will develop an IBM for *Carlina vulgaris* which has been the focus of numerous studies (Rose et al. 2002; Childs et al. 2003, 2004; Rees and Ellner 2009; Rees et al. 2006; Metcalf et al. 2008). As in the previous *Oenothera* constant-environment models we will assume each individual is characterized by its size, z, and genotype, β_0. The model is very similar to the *Oenothera* IBM but with yearly variation in growth, recruit size, survival, the probability of flowering, the number of successful recruits, and recruitment is density dependent. Years that are good for growth of established plants are also good for growth of recruits, and so the intercept of the growth function and mean recruit size are positively correlated, see Chapter 7 for a more detailed description of the demography. As offspring inherit their parent's flowering strategy plus some small deviation, we can simply iterate the model to see how the trait evolves, Figure 9.5. The individual-based simulations all rapidly converge on $\bar{\beta}_0 \approx -14$ with the variance in $\beta_0 \approx 3$. As with the *Oenothera* example we can develop an IPM for *Carlina* where individuals are characterized by their size and flowering intercept, so we have a 2D stochastic, density-dependent model. As expected the IPM provides an excellent description of the IBM capturing both the evolution of the trait and the maintenance of genetic variation, through the mutation-selection balance (Figure 9.5).

In models such as those discussed in this chapter, where individuals are cross classified by their state, z and genotype x, the change in the mean genotype from one year to the next is given by

$$\Delta \bar{x} = \frac{1}{\bar{\lambda}(t)} Cov(\lambda(x,t), x) = Cov\left(\frac{\lambda(x,t)}{\bar{\lambda}(t)}, x\right) \tag{9.5.1}$$

where $\lambda(x,t)$ is the one time step population growth of genotype x, that is

$$\lambda(x,t) = \int\int K(z', z; x) n(z, x, t)/n(x, t) \, dz dz' = n(x, t+1)/n(x, t), \tag{9.5.2}$$

and $\bar{\lambda}(t)$ is the average of $\lambda(x,t)$ with respect to the population's distribution of x. This is the IPM version of the Secondary Theorem of Natural Selection (Robertson 1966; Price 1970), which is a general result and can be applied to any IPM (i.e., density dependent or independent, constant or stochastic environment; Rees and Ellner 2016).

In a stochastic environment we might want to know the expected change in mean genotype, conditional on $n_t = n(z, x, t)$. $\lambda(x,t)$ and $\bar{\lambda}(t)$ are functions of the environment state (the year-specific kernel parameters) but because the environment is independent of population state, equation (9.5.1) implies

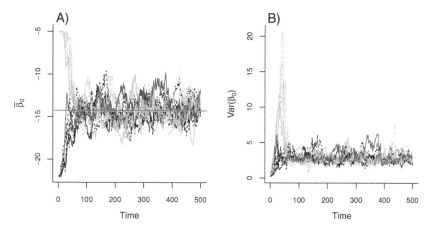

Fig. 9.5 Evolutionary trajectories for the stochastic environment *Carlina* IBM showing A) the mean genotype, $\bar{\beta}_0$, and B) genetic variance of β_0. Five populations were started at -5 and five at -22, in all cases the population evolves to $\bar{\beta}_0 \approx -14$. The red line in A) is the ESS prediction, the turquoise line is the simulation mean over the last 400 years, the purple and green lines the solution of the 2D stochastic IPM. Source file: Carlina Evol Demog Dynamics.R

$$E_e[\Delta \bar{x}] = Cov\left(E_e\left[\frac{\lambda(x,t)}{\bar{\lambda}(t)}\right], x\right) \qquad (9.5.3)$$

where E_e is expectation with respect to the random environmental variation.

The arguments leading from the Secondary Theorem to equation (9.8.5) in Appendix 9.8, namely

$$\Delta \bar{x} \approx G_x(\partial W(x; \bar{x}_t)/\partial x)/W(\bar{x}_t; \bar{x}_t),$$

where W is fitness, also apply in stochastic environments. However the right-hand side is then a function of the environment parameters in year t, so the left-hand side (the trait change) is a random, environment-dependent quantity. One quantity of interest is the average of $\Delta \bar{x}$ conditional on the current trait mean. The right-hand side is the gradient of log W, so $E[\Delta \bar{x}]$ is the gradient of $E[\log W]$. For an unstructured, density-independent population, Lande (2007) observed that $E[\log W]$ is the long-term population growth rate, and that this quantity therefore defines an adaptive topography ("fitness landscape") for the expected trait change under fluctuating selection. For IPMs this is complicated by the dependence of population growth on population structure. The relevant fitness measure is $\lambda(x,t)$, as in equation (9.5.3), and this depends on the genotype-specific population structures. So for IPMs (and structured populations in general) in stochastic environments, it appears that there is not a constant adaptive topography describing the expected response to selection (Rees and Ellner 2016). Of course for low levels of stochasticity, there is an approximately constant adaptive topography because populations grow approximately as if they are at the stable structure for the mean kernel.

Another quantity of potential interest in stochastic environments is the expected change in trait mean given the environment state. Once a population has settled into steady-state fluctuations (e.g., $t > 300$ in Figure 9.5) trait change is driven mainly by the year to year changes in environment and therefore in selection. As a rough approximation for this situation, W as a function of environment state and trait value x can be approximated by the population growth rate that results when the kernel for the environment state acts on the trait-specific steady state population structure. The environment-specific expected trait change is then approximated by the right-hand side of equation (9.8.5), either averaged over the steady-state fluctuations in trait mean or simply evaluated the long-term trait mean.

Moving onto ESSs we simply substitute λ_S for λ in the ESS conditions and we're done, so the ESS conditions become

$$\left.\frac{\partial \lambda_S(x; \bar{x})}{\partial x}\right|_{x=\bar{x}} = 0 \text{ and } \left.\frac{\partial^2 \lambda_S(x; \bar{x})}{\partial x^2}\right|_{x=\bar{x}} < 0 \qquad (9.5.4)$$

which in words says that if a new mutation arises it will be doomed to extinction as the resident strategy has $\lambda_S = 1$ and all new mutations have $\lambda_S < 1$. A very general method for finding an ESS is to use an iterative invasion approach rather like the one we used in a constant environment. The steps are as follows:

1. For the resident, $\bar{x}$, iterate the IPM, and store the model parameters, and population size or other properties that influence density dependence for each year. This defines the resident environment.
2. Using the resident environment find the value of x that maximizes $\lambda_S(x; \bar{x})$, call this x_{opt}.
3. If $|\lambda_S(x_{opt}; \bar{x}) - 1| < \varepsilon$, where ε is the numerical tolerance, then we have an ESS otherwise repeat from 1) using the value of x_{opt} as the new resident, $\bar{x} = x_{opt}$.

We implemented this approach using the stochastic *Carlina* IPM described in detail in Chapter 7. We iterate the stochastic IPM using the function (iterate_model) which generates yearly parameters from the fitted demographic models, then constructs the kernel and iterates the population forward. This function also stores the sequence of environments defined by the parameters of the model in each year. For a given sequence of environments defined for some resident strategy we calculate the stochastic growth rate of a rare invader using (invader_gr), and use optimize to find the best invading strategy. Finally we wrap all this up in a function find_ESS which has the resident parameters and numerical tolerance as arguments. Here's what the output looks like

```
> ESS.iterative.invasion <- find_ESS(m.par.true,0.001)
iterate: 1000
iterate: 2000
iterate: 3000
iterate: 4000
iterate: 5000
0.1083766    2.761073
iterate: 1000
iterate: 2000
iterate: 3000
iterate: 4000
iterate: 5000
0.01119516   -0.9109975
iterate: 1000
iterate: 2000
iterate: 3000
iterate: 4000
iterate: 5000
> cat("ESS flowering intercept ",ESS.iterative.invasion,"\n")
ESS flowering intercept -14.33883
```

So we take the estimated parameters and iterate the model for 5000 years. This defines the resident environment and an invader with flowering strategy, $\bar{\beta}_0+2.76$ is the best invader into this environment, having $\log(\lambda_S) = 0.108$. This becomes the new resident and then we repeat this process until $|\log(\lambda_S)| < \varepsilon$. This gives an ESS flowering strategy of -14.34 which is not significantly different from the estimated value of -16.1 ± 1.65, and in excellent agreement with the IBM simulation (Figure 9.5). The predicted mean arithmetic flowering size is also in reasonable agreement with that observed in the field

```
> ESS.size.fl
[1] 51.92773
> mean(exp(size.fl),na.rm=TRUE)
[1] 51.99065
```

We can check our ESS flowering strategy is indeed an ESS by calculating a fitness landscape assuming the resident flowering strategy is at the ESS, and then calculating $\log(\lambda_S)$ for a range of alternative strategies. Figure 9.6 shows the resulting fitness landscape; clearly our predicted ESS is not invadable by alternative strategies and the fitness losses when not following the ESS can be very substantial.

This calculation provides a nice example of how the use of multiple cores can speed things up considerably. On a UNIX-based operating system (Linux or OS X) we can calculate the fitness landscape using

```
Ls <- simplify2array(mclapply(flow.inter.diffs,
      stoch_lambda_beta,params=iter$params.yr,mc.cores=24))
```

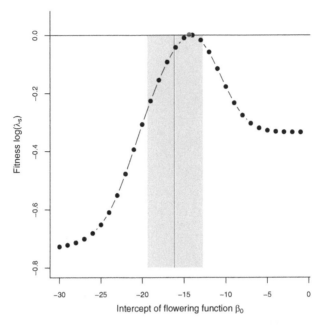

Fig. 9.6 Fitness landscape for the stochastic *Carlina* model assuming the resident strategy is at the ESS (red filled circle). The vertical turquoise line indicates the strategy observed in the field and the grey box is 2 standard errors. Source file: `Carlina Evol Demog Statics.R`

where `mclapply` is the multicore version of `lapply`.[2] On a current Mac Pro this results in ≈ 9 fold increase in speed. We can also calculate a PIP for the *Carlina* model, and here the use of `mclapply` is critical as this is a very time-consuming calculation. The approach is exactly as in the constant environment case but using λ_S to determine if a new mutation can invade, the resulting PIP is shown in Figure 9.7. As expected the ESS we found by the iterative invasion process is indeed an ESS, and it is also *convergent stable* as resident strategies not at the ESS will evolve towards the ESS. This PIP is not skew symmetric and so we know there is no *optimization principle*; we really do have to find the ESS by iterative invasion. We can also calculate a new type of plot called a *Mutual Invasibility Plot* or MIP. To construct an MIP we look for pairs of strategies, say A and B, where A can invade B and B can invade A, so they are mutually invasible. In R this is easily done by element-wise multiplication of the matrix containing the $\log(\lambda_S)$ values for the PIP with its transpose. Any elements that are positive after doing this satisfy our condition for mutual invasibility. The MIP for *Carlina* is plotted in Figure 9.7B, this demonstrates that there are pairs of flowering strategies which form protected polymorphisms, and these consist of one strategy on either side of the singularity (Metcalf et al. 2008). The

---

[2] The corresponding function for Windows is `parLapply` in the `parallel` package, which works a bit differently.

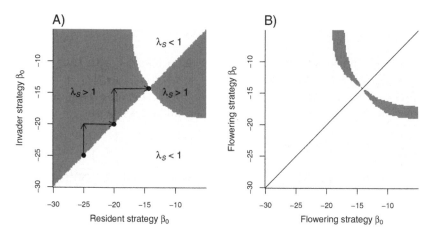

Fig. 9.7 A) Pairwise invisibility plot for the stochastic *Carlina* model, B) Mutual invisibility plot showing the area of parameter space (red) where the strategies are mutually invasible, so there is a protected polymorphism. Source file: `Carlina Evol Demog Statics.R`

stability of the protected polymorphism can be deduced from the geometry of the MIP (Geritz et al. 1999). If the angle between the two coexistence boundaries is narrower than 90°, the protected polymorphisms lacks evolutionary stability, as in this case.

In a stochastic environment, two distinct processes, namely nonlinear averaging and non-equilibrium dynamics, influence fitness, and Rees et al. (2004) develop methods for decomposing these effects. For *Carlina* the absolute effect of temporal variation on the evolutionarily stable flowering strategy is substantial (ca. 50% of the evolutionarily stable flowering size) but the net effect is much smaller (ca. 10%) because the effects of temporal variation do not influence the evolutionarily stable strategy in the same direction (Rees et al. 2004). For example, fluctuations in recruitment select for larger sizes at flowering, whereas fluctuations in survival and growth have the opposite effect.

The calculation of ESSs in stochastic environments is time consuming even for simple IPMs and so, as before, it's useful to know when there's an optimization principle. Lande et al. (2009) and Engen et al. (2013) derived optimization principles for unstructured populations with some forms of weak density-dependent stochastic selection, but at present there is no similar analysis for structured populations. For the case where density dependence decreases $P(z', z; \theta) + F(z', z; \theta)$ uniformly for all z, we can write the kernel as

$$K(z', z, \mathbf{N_t}; \theta(t)) = g(\mathbf{N_t})(P(z', z; \theta_t) + F(z', z; \theta_t))$$
$$= g(\mathbf{N_t})K_0(z', z; \theta_t)$$

where $g(\mathbf{N_t})$ is the density dependence in year t, $K_0(z', z; \theta_t)$ the kernel for a population in the absence of density dependence, and θ_t the model parameters in year t. Total population size then changes as

$$\mathbf{N}_{t+1} = g(\mathbf{N_t}) \int \int K_0(z', z; \theta_t) n(z, t) \, dz \, dz' \qquad (9.5.5)$$

Consider now an invader with trait x_I facing a resident with trait x_R. Equation (9.5.5) holds for the resident with $K_0(z', z; \theta_t, x_R)$ as the kernel. We can then write the invader's λ_S as follows, where $\mathbf{N}_t^{(I)}$ denotes the invader's total population size and $g(\mathbf{N}_t)$ is the density dependence set by the resident:

$$
\begin{aligned}
\log(\lambda_S^{(I)}) &= E\left[\log\left(\mathbf{N}_{t+1}^{(I)}/\mathbf{N}_t^{(I)}\right)\right] \\
&= E\left[\log\left(g(\mathbf{N_t}) \int \int K_0(z', z; \theta_t, x_I) n_I(z, t) \, dz \, dz' / \mathbf{N}_t^{(I)}\right)\right] \\
&= E\left[\log(g(\mathbf{N_t})) + \log \int \int K_0(z', z; \theta_t, x_I) n_I(z, t) \, dz \, dz' - \log \mathbf{N}_t^{(I)}\right] \\
&= E\left[\log(g(\mathbf{N_t}))\right] + \log(\lambda_{S,0}(x_I))
\end{aligned}
$$

where $\lambda_{S,0}(x_I)$ the stochastic growth rate of an invader in the absence of density dependence. This applies also when $x_I = x_R$, in which case it gives the growth rate for the resident invading itself, which is 0. A resident with trait x_R that maximizes $\lambda_{S,0}(x)$ will therefore be uninvadable, and so the ESS is characterized by maximizing $\lambda_{S,0}(x)$.

9.6 Function-valued traits

We noted above that the approaches in previous sections extend straightforwardly to model joint evolution of two or more quantitative traits. However many important traits are best described as *functions*: the shape of a wing or leaf; clutch size or flowering probability as a function of body size; or any continuous norm of reaction such as the response of growth rate to temperature (e.g., Kingsolver et al. 2001; Dieckmann et al. 2006; Stinchcombe et al. 2012). Here we describe an IPM-based approach for finding evolutionary endpoints (ESSs) of function-valued traits. We emphasize endpoints rather than dynamics because fewer assumptions are needed, but we also discuss the prospects for modeling the dynamics.

To explain the approach, we will again use the example of flowering strategy evolution for a monocarpic perennial. The flowering "strategy" is the relationship between size z and flowering probability: a function, $p_b(z)$. Up to now we've described it by an analytic formula, logistic regression. But logistic regression, or any other simple formula is really too limiting. It limits variation to a few "degrees of freedom," instead of the complete freedom of saying "probability of reproducing is some continuous function of body size," and this creates the risk that the optimum within those limits may be highly suboptimal (Dieckmann et al. 2006). If the optimal life history is a threshold strategy - grow without breeding up to some critical size, and then breed but not grow – that can't possibly be discovered if your model assumes an allometric relationship between body size and reproductive effort.

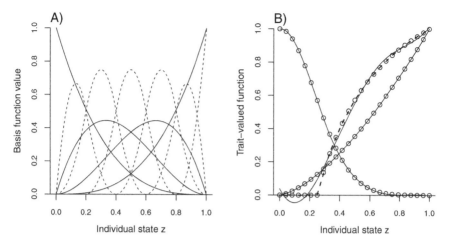

Fig. 9.8 Function approximation by spline basis expansion. A) Cubic spline basis with $d = 4$ basis functions on [0,1] (solid curves) and quadratic spline basis with $d = 6$ basis functions and no intercept (dashed curves). B) Examples of approximating nonlinear functions: Gaussian $y = e^{-2x^2}$, allometric $y = x^{1.5}$ and saturating with threshold $y = max(0, (3.6x/(.2 + x)) - 2)$. Solid curves are the least-squares spline approximation using cubic splines with $d = 6$ basis functions; dashed curve is a fit to the threshold function using quadratic splines with 12 basis functions and no intercept. Source file: `PlotSplines.R`

Ideally we would have no restrictions at all on the possible relationship between size and strategy. For some deterministic models restrictions can be avoided using optimal control theory or the calculus of variations (Parvinen et al. 2006, 2013). For stochastic IPMs with density or frequency dependence, and many other models, we don't know of any mathematical theory that can do the job, so we take a computational approach. To make this feasible, we borrow methods from *functional data analysis* in statistics (Ramsay and Silverman 2005; Ramsay et al. 2009). The simplification that leads to a practical solution is to let the degrees of freedom be finite but arbitrarily large. B-splines are a convenient and popular way to do this. Figure 9.8A shows the cubic B-spline basis with $d = 4$ basis functions and evenly spaced breakpoints (spline functions are piecewise polynomials, and breakpoints are where the pieces meet up), and a quadratic spline basis with $d = 6$ basis functions and no intercept, which could be used to satisfy the constraint that $f(0) = 0$.

Then for a trait that is a function $f(z)$ of size, we assume that

$$f(z) = \sum_{j=1}^{d} c_j \varphi_j(z) \qquad (9.6.1)$$

for some coefficients c_j, where the functions $\varphi_j(z)$ are a B-spline basis on the size range $[L, U]$. Using the fda library, it takes a few steps to create the plotted basis function values:

```
B3 <- create.bspline.basis(rangeval=c(0,1), nbasis=4, norder=4)
px <- seq(0,1,length=200);
py3 <- eval.basis(px,B3);
```

The first line creates a "basis object" with information about the kind of basis: the range of values for the independent variable, the number of basis functions, and the spline *order*. norder=4 creates cubic splines. The spline basis functions consist of segments, each of which is a cubic polynomial, joined up in such a way that each function has a continuous first and second derivative. The second and third lines of code cause all of the basis functions to be evaluated at 200 points between 0 and 1, yielding py3 as a 200×4 matrix that matplot can plot.

To approximate a nonlinear function with B-splines, we find the coefficients c_j in (9.6.1) that minimize the sum of squared deviations from the target at a grid of points. This is easily done by using the matrix of basis values py3 as the independent variables in a linear regression, and the fitted regression coefficients are the least-squares estimate of c_j. Figure 9.8B shows three examples using cubic splines with $d = 6$ basis functions. Function values are plotted as open circles, and the solid curves are the fitted spline. The Gaussian and allometric functions are fitted well, but not the threshold function. One way to improve the fit is by increasing the value of d (nbasis in the code above), so the spline approximation has more degrees of freedom to work with. Another is to use lower-order splines. Cubic splines have a continuous second derivative; the threshold function isn't even differentiable at the threshold. To get rid of that mismatch we can use linear splines, which are continuous but not differentiable (dashed curve), and with $d = 12$ basis functions there is a good fit.

Very similar 'mechanics' let us put function-valued traits into an IPM and then find optimal or ESS coefficient values. Specifically, we want to let flowering probability in *Oenothera* be any smooth function of size with values between 0 and 1. We can do that by setting

$$\text{logit } p_b(z,t) = \sum_{j=1}^{d} c_j \varphi_j(z).$$

We can code that as follows (from Monocarp Demog SplineFuns.R):

```
p_bz_spline <- function(z, B, cj) {
linear.p <- eval.basis(B,z)%*%cj; # linear predictor
p <- 1/(1+exp(-linear.p))          # logistic transformation
return(p)
}
```

Here B is a pre-constructed basis object, and we evaluate the linear predictor by evaluating each basis function, and then summing them with the coefficients.

For the *Oenothera* model with density-limited recruitment, we saw in Section 9.4.2 that the ESS flowering strategy maximizes $\widetilde{R}_0$, the dominant eigenvalue of $F(\theta)(I - P(\theta))^{-1}$ where θ is the parameter or set of parameters characterizing the flowering strategy. Maximizing $\widetilde{R}_0$ as a function of β_0 we found that

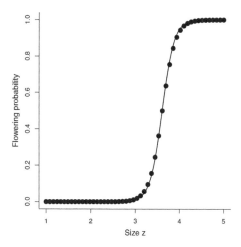

Fig. 9.9 Comparison of optimal flowering strategies in the *Oenothera* model with constant total recruitment. The solid curve is the slope-constrained optimal function-valued strategy using a 6-function spline basis. The overlaid circles are values of the logistic regression model with the ESS value of the intercept parameter β_0. Source file: Monocarp FunctionValued Flowering.R

$\beta_0 = -24.9$ was the ESS. Now, using p_bz_spline the IPM kernels are functions of the coefficients cj,

```
F_z1z_spline <- function (z1, z, m.par, B, cj) {
    p_bz_spline(z, B, cj)*b_z(z, m.par)*m.par["p.r"]*c_0z1(z1, m.par)
}
P_z1z_spline <- function(z1, z, m.par, B, cj) {
    (1 - p_bz_spline(z, B, cj))*s_z(z, m.par)*G_z1z(z1, z, m.par)
}
```

Much as before, we can construct $F(I - P)^{-1}$ as a function of cj and compute its dominant eigenvalue, then optimize the eigenvalue numerically to find the ESS flowering strategy. R code for this is displayed in Box 9.1. One subtle point is that we have to constrain the slope of the spline-defined flowering strategy, to match the previous ESS analysis in which the slope parameter β_1 was held constant. Because we know a step function with $\beta_1 = \infty$ is the best, we do this by rejecting any spline coefficients in which the slope of the linear predictor is much above β_1.

The results are comforting (though it would be more fun if they weren't). In Figure 9.9, the solid curve is the spline ESS flowering strategy, and the circles show values of the logistic regression strategy with the ESS value of β_0. The differences are minor, and not surprisingly the $\widetilde{R}_0$ values are nearly identical (2.0598 versus 2.0597). Increasing from 6 to 12 basis functions had no visible effect. So we can be confident that logistic regression is not just a good model for the actual flowering strategy, it is also a good model for the optimal strategy in this model.

The code below finds the ESS criterion $\widetilde{R}_0$ as a function of the spline coefficients, for a given spline basis B that covers the model's size range.

```
source("Monocarp Demog SplineFuns.R"); # functions to make the kernel
L <- (-6); U <- 7; # WIDE size range
B <- create.bspline.basis(rangeval=c(L,U),nbasis=6, norder=4)


# Function to compute R0-tilde with spline flowering
Rtilde0_spline <- function(cj,parms,B) {
    IPM <- mk_K_spline(100,m.par=parms,L=L,U=U,B,cj)
    N <- solve(diag(100)-IPM$P);
    R0 <- abs(eigen(IPM$F%*%N)$values[1])
    return(R0)
}
```

We want the optimal coefficients in a situation analogous to optimizing β_0 with β_1 held constant. We do that by writing an objective function that "rejects" any coefficients that produce a slope much above β_1 at any point, and returns the value $-\widetilde{R}_0$ for acceptable coefficients. Minimizing this function using optim gives the ESS spline coefficients subject to the slope constraints. There are more efficient ways to do constrained optimization, but this one is easy and it works.

```
X <- eval.basis(seq(L,U,length=250),B); # for computing slope
dz=(U-L)/250;


objfun=function(cj,parms,B) {
    slope <- diff(X%*%cj)/dz; # slope of linear predictor
    if(max(slope) > 0.25+parms["flow.z"]) {
        return(1e12)
    }else{
        R0 <- Rtilde0_spline(cj,parms,B)
        return(-R0)
    }
}
pbfit <- optim(par=p0,fn=objfun,control=list(maxit=2500),
               parms=m.par.true,B=B);
cj.opt <- pbfit$par; R0.opt <- pbfit$value;
```

To make sure that the optimum has been found, the entire process should be repeated with larger values of nbasis to make sure the results are unchanged.

Box 9.1: Finding the ESS function-valued flowering strategy using a B-spline basis representation.

9.6.1 Solving the ESS conditions for function-valued strategies

The example just above was simplified greatly by knowing the optimization principle that characterized the ESS. What if there is no optimization principle, or we don't know what it is?

Actually, there always is a kind of optimization principle: when the resident uses the ESS, no nearby invader can succeed. This leads to the classical ESS conditions for one trait, equations (9.4.1). When multiple traits evolve independently, the first-order condition is that the partial derivative of invader fitness with respect to each invader trait value must equal zero, when the resident and invader have the ESS trait value:

$$\left. \frac{\partial \lambda(\mathbf{x}; \bar{\mathbf{x}})}{\partial x_i} \right|_{\mathbf{x}=\bar{\mathbf{x}}} = 0, \tag{9.6.2}$$

where $\mathbf{x} = (x_1, x_2, \cdots, x_m)$ is the vector of evolving traits. We can find ESSs by finding solutions of (9.6.2), and verifying that they are uninvadable. Usually we don't have an explicit formula for λ, so the left-hand side is approximated by simulating first a resident with strategy $\bar{\mathbf{x}}$ and then invaders with slightly different strategies.

To illustrate this approach, we will consider the stochastic *Carlina* model with a function-valued flowering strategy. The demographic models are the same as the *Oenothera* model above; it's only the parameter values that differ. The script `Carlina Spline ESS.R` illustrates how to find ESS values of the spline coefficients c_j by solving (9.6.2). Ideally this would be done using a root-finding algorithm, but the only one currently available in R is Newton-Raphson which is prone to fail if it starts too far from a solution. So instead we use the workhorse optimization function `optim` to minimize the sum of squared partial derivatives

$$\Phi(\mathbf{x}) = \sum_{i=1}^{M} \left(\left. \frac{\partial \lambda(\mathbf{x}; \bar{\mathbf{x}})}{\partial x_i} \right|_{\mathbf{x}=\bar{\mathbf{x}}} \right)^2. \tag{9.6.3}$$

$\Phi = 0$ is the smallest possible value, and any solution of $\Phi(\mathbf{x}) = 0$ satisfies the first-order conditions for an ESS. Your friends in math departments will frown on this approach, but it's the path of the least resistance within R's current capabilities.

For this model the λ in (9.6.3) is the invader's stochastic growth rate λ_S in the competitive environment determined by the resident with trait $\bar{\mathbf{x}}$. λ_S is a well-defined number, but when we estimate it from a finite-duration simulation the value depends on the sequence of random year-specific parameter values. This random variation in function values will be fatally confusing to `optim`. We prevent this by first simulating one series of year-specific parameter vectors, and using the same series in *all* simulations. We create a matrix `params.yr` with named columns to hold all the values, and then (using variables defined in `Carlina Demog Funs.R`),

```
X <- eval.basis(seq(L,U,length=250),B); dz=(U-L)/250; #for computing slope

### Penalized lambda_S of invader
lambdaS.inv=function(cj,params.yr,B,m,L,U,max.slope,X) {
    slope <- diff(X%*%cj)/dz;
    mslope <- max(slope);
    pen <- ifelse(mslope > max.slope, 1000*((mslope-max.slope)^2), 0)
    lam<-spline_invader_gr(params.yr=params.yr,B=B,cj=cj,
            m=m,L=L,U=U)$lambda.hat
    return(lam-pen)
}


### Objective function: 1 + rms(gradient of lambda_S)
### Gradient is computed by forward finite difference
gradnorm.inv <- function(cj,params.yr,B,m,L,U,max.slope,X) {
    res<- iterate_spline_resident(params.yr=params.yr,B=B,cj=cj,
            m=m,L=L,U=U)
    params.yr <- as.matrix(res$params.yr)
    npar=length(cj); mat=diag(1+npar); out=numeric(1+npar);
    for(i in 1:(1+npar)){
        out[i] <- lambdaS.inv(cj+0.01*mat[1:npar,i],params.yr,
                B=B,m=m,L=L,U=U,max.slope=max.slope,X=X)
    }
    grad <- 100*(out[1:npar]-out[1+npar])
    return(1+sqrt(sum(grad^2)));
}
```

Box 9.2: Finding the ESS function-valued flowering strategy in *Carlina* by using optim to find a solution of the first-order ESS conditions; excerpted from Carlina Spline ESS.R.

```
for(gen in 1:n.iter){
    m.par.year <- m.par.true + qnorm(runif(12,0.001,0.999))*m.par.sd.true
    m.par.year[c("grow.int","rcsz.int")] <-
        m.par.true[c("grow.int","rcsz.int")] +
        matrix(qnorm(runif(2,0.001,0.999)),1,2) %*% chol.VarCovar.grow.rec
    params.yr[gen,] <- m.par.year;
}
params.yr[,"p.r"] <- 1;
```

The last line is so that Fn will be the total number of seeds, which then gets scaled down by density-dependent establishment. Similarly, we compute once and for all the recruit numbers for each year (i.e., the R for each year in (9.3.1)). As in Box 9.1, to specify the flowering strategy we create a spline basis object B on the size range. The function

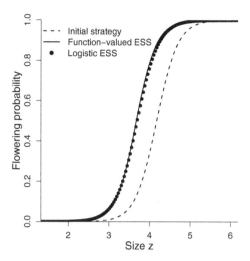

Fig. 9.10 Finding the ESS function-valued flowering strategy in the stochastic *Carlina* model with density-dependent recruitment by minimizing Φ, equation (9.6.3), as a function of the spline coefficients defining the flowering strategy. The dashed black curve shows the initial flowering strategy, a spline approximation to the flowering strategy estimated from the data. The solid black curve is the minimum-Φ flowering strategy with 6 spline basis functions. Overlaid circles points are the ESS logistic regression flowering strategy with $\beta_0 = -14.34$. Source file: `Carlina Spline ESS.R`

```
iterate_spline_resident <- function(params.yr,B,cj,m,L,U)
```

uses the IPM to simulate a resident population with flowering strategy `cj`, and it returns `params.yr` with `params.yr[,"p.r"]` set to the recruitment probabilities resulting from the resident's seed production each year. Using those, the function

```
    spline_invader_gr <- function(params.yr,B,cj,m,L,U)
```

simulates the growth of an invader and returns its population growth rate that we use as an estimate of λ_S for the invader. The most time-consuming step is computing all the year-specific kernels, so in `Carlina Spline ESS.R` we use a few tricks to speed that up by avoiding loops and minimizing spline evaluations.

Those are the building blocks; Box 9.2 shows how they are put together. The function `lambdaS.inv` imposes the slope constraint by severely penalizing a strategy that violates the constraint, but does this in such a way that the penalized invader λ_S is a differentiable function of the strategy parameters `cj`. `gradnorm.inv` computes the partial derivatives of penalized λ_S and combines them into $1 + \sqrt{\Phi}$. We find a minimum of Φ by turning this function over to `optim` with initial parameter guess `cj`:

```
out=optim(par=cj,f=gradnorm.inv,params.yr=params.yr,
    B=B,m=60,L=L,U=U, max.slope=1.05*m.par.true["flow.z"],X=X)
```

Starting from parameters `cj` that correspond to the estimated true flowering strategy (the fitted logistic regression), the result is fairly quick convergence onto an uninvadable strategy (Figure 9.10). As in the *Oenothera* model, the

ESS is very accurately approximated by logistic regression, validating our initial analysis with β_0 as the only evolving parameter.

With one evolving trait β_0, we found the ESS by simulating successive invasions until a strategy was reached that resists all invaders. That approach is less successful when several traits interactively affect fitness. If traits are initially near a convergence-stable ES, then a sequence of invasions should reach the ESS if each new resident is similar to the previous one. But large steps can't be permitted, or the sequence is likely to oscillate or diverge. This is a common feature of frequency-dependent selection. If the frequency of Hawks is above the mixed-strategy ESS, the highest-fitness invader is the "all Dove" strategy. Then, the fittest invader of "all Dove" is "all Hawk," and so on *ad infinitum*. But steps also can't be too small, unless your patience is very large. So in our experience it is much more efficient to numerically find a solution of the first-order ESS condition.

9.7 Prospective evolutionary models

Our case studies of evolutionary IPMs, and most others in the literature, are retrospective analyses. They start with observed traits, and try to understand them as the result from selection subject to tradeoffs and constraints: when should a monocarp flower? Which sheep should bear twins? Are these what we actually observe, and if not, why not?

But ecologists are increasingly called upon to predict the future. In a rapidly changing world, forecasting evolutionary responses to environmental change is part of that challenge. Evolutionary models like the ones in this chapter make it possible to integrate a population's ecology, life history, and the dynamics of quantitative or Mendelian traits in a varying environment. Coulson et al. (2011) used an eco-evolutionary IPM to predict how the wolf population in Yellowstone Park would respond to changes in environmental mean or variance. Their model classifies wolves by sex, age, weight, and Mendelian genotype at the diallelic K locus that affects coat color, survival, and fecundity (Coulson et al. 2011, Table S2). By altering the distributions of parameters, Coulson et al. (2011) found that changes in variability had much smaller effects on population than changes in mean, suggesting that environmental trends would be more important than projected changes in environmental variability.

Linking evolutionary forecasts to specific scenarios about future environments is more challenging. If we know how growth, fecundity, and other demographic rates depend on temperature, we can predict the ecological effect of rising temperature. To predict the evolutionary effect, we also need to know the range of possible norms of reaction to temperature. That by itself is enough to predict evolutionary endpoints, but to predict the actual path of evolutionary change across generations, we also need to know about the inheritance of temperature response curves. The norm of reaction to an environmental variable is a function-valued trait, so simple Mendelian inheritance seems unlikely, but quantitative trait models may be possible (Kirkpatrick and Heckman 1989; Beder and Gomulkiewicz 1998; Kingsolver et al. 2001; Stinchcombe et al. 2012), if we know enough to estimate the genetic variance-covariance function.

There's a lot of "if we know..." in that. The evidence so far is that many details affect the rate and limits of evolutionary response to changing ecological conditions (Ellner 2013). The tools to predict are available; the question is whether they can succeed given the practical limits on data collection and our incomplete knowledge of the constraints and tradeoffs that are likely to be important limits to adaptation. At present we can't answer that question, but we are collectively acting as if the answer is "yes" and it's worth our while to make forecasts about the ecology of a future Earth. It would be very valuable to get an empirical answer. As with many ecological questions, laboratory experiments on small, short-lived organisms may be the best way to get a first-cut answer (Benton et al. 2007; Bell and Collins 2008; Datta et al. 2013; Collins 2013). If we can't predict evolutionary futures in the lab, success in the field is unlikely. And learning how to succeed in the lab will increase our odds of getting it right when it might actually matter (if anybody is listening).

9.8 Appendix: Approximating evolutionary change

We wish to approximate the change in the mean of a trait, say x, from one time step to the next. We assume the trait, x, is continuous, under genetic control with no environmental component, and that selection is frequency dependent so the fitness of an individual, $W(x; \bar{x}_t)$ depends on the mean population value of the trait, $\bar{x}_t$. To simplify what follows we assume $n(x, t)$ is a probability density function, so it integrates to 1. After selection occurs the new mean, $\bar{x}_{t+1}$, is given by

$$\bar{x}_{t+1} = \frac{\int x n(x, t) W(x; \bar{x}_t)\, dx}{\bar{W}} \tag{9.8.1}$$

where the denominator $\bar{W} = \int n(x, t) W(x; \bar{x}_t)\, dx$ normalizes the post-selection trait distribution $n(x, t) W(x; \bar{x}_t)$, so that it integrates to 1. So the change in mean trait value is

$$\begin{aligned}
\Delta \bar{x} &= \bar{x}_{t+1} - \bar{x}_t \\
&= \frac{\int x n(x, t) W(x; \bar{x}_t)\, dx - \bar{x}_t \bar{W}}{\bar{W}} \\
&= \frac{\overline{xW} - \bar{x}\bar{W}}{\bar{W}} = Cov(x, W(x; \bar{x}_t))/\bar{W}.
\end{aligned} \tag{9.8.2}$$

Next we Taylor expand $W(x; \bar{x}_t)$ around $x = \bar{x}_t$ and assuming the variance of x is small, we ignore all higher order terms, giving

$$W(x; \bar{x}_t) \approx W(\bar{x}; \bar{x}_t) + (x - \bar{x}_t)\frac{\partial W(x; \bar{x}_t)}{\partial x}. \tag{9.8.3}$$

Substituting this into the formula for $\bar{W}$ gives $\bar{W}(x; \bar{x}_t) \approx W(\bar{x}_t; \bar{x}_t)$, and substituting into the Cov formula gives

$$Cov(x, W(x; \bar{x}_t)) \approx G_x \frac{\partial W(x; \bar{x}_t)}{\partial x} \qquad (9.8.4)$$

where G_x is the variance of x. Combining these we obtain

$$\Delta \bar{x} \approx G_x(\partial W(x; \bar{x}_t)/\partial x)/W(\bar{x}_t; \bar{x}_t) = G_x(\partial \ln W(x; \bar{x}_t)/\partial x)|_{x=\bar{x}_t} \qquad (9.8.5)$$

For age-structured (Lande 1982) and stage-structured matrix models (Barfield et al. 2011) it can be shown that, under certain assumptions, the fitness $W(x)$ associated with trait value x is given by $\lambda(x)$, the dominant eigenvalue of the Leslie or Lefkovitch matrix for individuals with trait value x. This is because, under weak selection, all genotypes will be at their stable structure (weak selection is equivalent to our assumption that x has small variance, because these both mean that fitness is nearly constant within the range of trait variation). The subpopulation with trait value x therefore increases by a factor of $\lambda(x)$. This gives

$$\Delta \bar{x} \approx G_x(\partial \ln \lambda(x; \bar{x}_t)/\partial x)|_{x=\bar{x}_t}. \qquad (9.8.6)$$

When the environment varies over time, the calculations leading to (9.8.5) hold for any given year if W is replaced by year-specific fitness W_t. However, in a structured population the kernel eigenvalue is not a good predictor of population growth because the population structure varies over time. A more accurate approximation is to calculate fitness using the stationary year-specific population structure for the mean trait, $w_t(\bar{x}_t)$ (because under weak selection all traits will have approximately the same population structure) and the year- and trait-specific kernel $K_{t,x} = K_t(z', z; x)$. The changes in mean trait in year t are then approximated by equation (9.8.5) with $W_t(x, \bar{x}_t) = ||K_{t,x}w_t(\bar{x}_t)||/||w_t(\bar{x}_t)||$.

Chapter 10
Future Directions and Advanced Topics

Abstract This chapter is a collection of some advanced and cutting-edge topics that we think are important directions for future research on IPMs. (1) Fitting more flexible IPM kernels using nonlinear and/or nonparametric regression. (2) Putting demographic stochasticity into IPMs. (3) IPMs where some of the individual state variables have deterministic dynamics or are subject to deterministic constraints. Theory and numerical methods are totally undeveloped for these models. (4) More kinds of data. We revisit our assumption that the modeler has accurate observations on state and fate for marked individuals who can be re-found at each census. Better methods for fitting IPMs using mark-recapture data where recapture is not certain, or data on unmarked individuals, will greatly expand the scope for IPM applications, especially applications to animal populations.

It's time to wrap up and look ahead. IPMs are now solidly established in population biology. We hope that we have shown you why that has happened, and that we have given you the tools to build and use IPMs yourself. This chapter is targeted at users who want to push the edge of what can be done, and mathematical and statistical ecologists who can move the edge forward. In previous chapters we have identified some areas where more work is needed. But we see lots of mathematical and statistical questions whose answers will broaden the potential scope of IPM applications.

10.1 More flexible kernels

The majority of IPMs, in this book and in the literature, use relatively simple parametric models to describe the demographic processes, linear or generalized linear models in deterministic IPMs, and relatively simple mixed effects models in stochastic IPMs. But some data sets need more flexibility. In this section we identify some ways to give IPMs more flexibility, focusing on cases where the coding is relatively simple. To be concrete we will think about parameterizing a growth kernel $G(z', z)$, depending only on individual size, by specifying the conditional mean $m(z) = E[z'|z]$ and the pattern of individual variation around

© Springer International Publishing Switzerland 2016

S.P. Ellner et al., *Data-driven Modelling of Structured Populations*,

Lecture Notes on Mathematical Modelling in the Life Sciences,

DOI 10.1007/978-3-319-28893-2_10

the mean. As we noted in Chapter 2, standard regression analysis focuses on the mean function, but for an IPM it is just as important to do a good job of modeling the between-individual variation around the mean.

10.1.1 Transforming variables

A venerable approach in applied statistics is to seek a transformation such that a simple linear model provides a good fit to the transformed data. The ubiquitous log-transformation is one example, and in many cases that does the trick. When that fails, a power transformation is sometimes effective, $z = x^\lambda$ where x is the raw size measurement (e.g., Bruno et al. 2011). Maximum likelihood can be used to guide the choice of λ. For a linear model $y^\lambda \sim a + bx^\lambda + \text{Normal}(0, \sigma^2)$ with $x, y, \lambda > 0$, the profile negative log-likelihood of λ is (Box and Cox 1964, p. 215)

```
bcNLL=function(lambda,x,y) {
    xl <- x^lambda; yl <- y^lambda;
    fit <- lm(yl ~ xl); s2hat <- mean(fit$residuals^2);
    return( 0.5*length(x)*log(s2hat/lambda^2) - (lambda-1)*sum(log(y)) );
}
```

In the following test case, artificial data are created for which $z = \sqrt{x}$ follows a standard linear regression model:

```
z <- runif(N,1,10); z1 <- 1 + 0.5*z + 0.2*rnorm(N);
x <- z^2; x1 <- z1^2;
```

To estimate λ we minimize bcNLL with x=size in year t, y=size in year $t + 1$,

```
lambdaHat <- optimize(bcNLL,c(0.01,3),x=x,y=x1)$minimum
```

This gives λ close to 0.5, as it should; in 5000 replicate simulated data sets with N= 100 data points, the mean and median estimates were 0.50 and 90% of values were between 0.38 and 0.61 (the code for this is in BoxCoxExample.R). Note, that bcNLL is not equivalent to boxcox in R's MASS library. In boxcox only the dependent variable is transformed. In an IPM growth kernel, the independent and dependent variables are the same trait, so both have to be transformed in the same way.

   If a transformed variable is used in the growth kernel, it is simplest to use the same transformation for the entire IPM, but other demographic models can still be fitted on other scales. For example, suppose x is measured size, a linear growth kernel fits well using $z = \log x$, but survival is fitted best by logistic regression on untransformed size, $\text{logit}(s(x)) = a + bx$. The survival function on log scale is just $\text{logit}(s(z)) = a + be^z$, so it's easy to use z for growth and survival. Transforming growth and offspring size models to different scales is also possible, see Appendix 2.9.

10.1.2 Nonconstant variance

In many empirical applications the conditional mean (expected size next year) is fitted well by a linear model, but the variance in growth depends on initial size. This relationship needs to be modeled, so that the IPM produces realistic size distributions and growth trajectories.

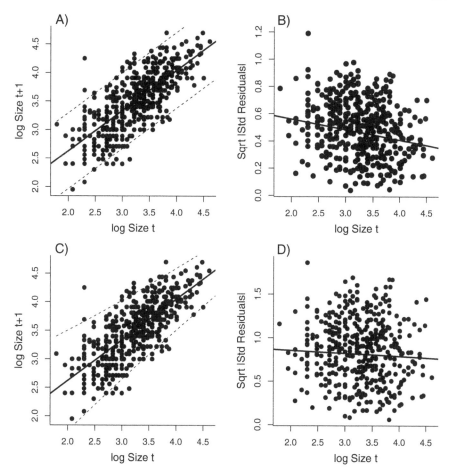

Fig. 10.1 Example of fitting a linear model with nonconstant variance using `mle`. A) The data points are log size $t+1$ versus log size t for *Carlina* including all years. The solid line is the fitted ordinary least square linear regression, the dashed lines are ± 2 estimated standard deviations from the regression line. B) Scale-location plot of the residuals with fitted regression line. C) as in A) with nonconstant variance, $\sigma = \sigma_0 e^{-dz}$ where z is log size t. D) Scale-location plot of residuals from the nonconstant variance regression, with fitted regression line. Source file: `FitGrowthData.R`

Any parametric variance function can be fitted by maximum likelihood. For example, if the mean and standard deviation of z' are both linear functions of z, the negative log likelihood is

```
-sum(dnorm(z1,mean=a+b*z,sd=c+d*z,log=TRUE))
```

and the parameter values can be estimated by `mle` in `stats4`. Pre-built fitting functions are available for some nonconstant variance model (e.g., in `gls`), but there are compelling advantages to using maximum likelihood. A non-Gaussian distribution for growth variation can be fitted just by substituting a different probability density for `dnorm`, such as a t distribution to accommodate fatter tails. And it is easy to compare models of varying complexity (e.g., constant variance versus linear, quadratic, or exponential dependence on size) using AIC, BIC, or likelihood ratio tests.

Fitting the *Carlina* growth data by simple linear regression (Figure 10.1A), the scale-location plot (Figure 10.1B) shows nonconstant variance (for the sake of using real data, we pool all years in this analysis even though we know there is year-to-year variation). The trend isn't large – the regression plotted in the scale-location plot only explains about 4% of the variation in residual magnitude – but it's easy to do better. In FitGrowthData.R we use mle to fit a model with linear mean and exponential size-dependence in the growth variability:

```
LinearLogLik2=function(a,b,sigma0,d) {
  loglik=sum(dnorm(size2,mean=a+b*size1,sd=sigma0*exp(-d*size1),log=TRUE))
  return(-loglik)
}
fit2=mle(LinearLogLik2,start=list(a=1,b=1,sigma0=sqrt(mean(err^2)),d=0),
      method="Nelder-Mead")
```

The estimated value of d is 0.24, and using confint(fit2) the 95% confidence interval is $(0.13, 0.36)$ confirming that there is evidence for nonconstant variance. Figure 10.1C, D shows the fitted model and scale-location plot. The trend in residual magnitude is reduced, and the fitted regression is not statistically significant ($p = 0.23$) and explains under 0.5% of the variance in residual magnitude.

10.1.3 Nonlinear growth: modeling the mean

If your data cannot be transformed so that mean growth is a linear function of present size, a nonlinear mean function (such as a polynomial) can also be fitted by maximum likelihood. But unless you have some biological basis for specifying a particular nonlinear growth function, it is probably preferable to fit a flexible model whose shape is dictated by the data. This is a strength of R, and many options are available. If growth variance appears to be independent of size, the gam function in mgcv can fit a spline whose degree of nonlinearity is automatically chosen based on the data. For the Soay data, this is just

```
require(mgcv); gamGrow <- gam(z1 ~ s(z),data=sim.data)
```

Note that s(z) in the line above is *not* the survival function, it's how you tell gam to fit the mean of z1 as a spline function of z. Figure 10.2 illustrates that this works surprisingly well with a moderate amount of data, despite high variance about the mean.

The residuals from the fitted mean curve provide an estimate of the growth distribution's standard deviation,

```
sse <- sum(resid(gamGrow)^2);
sdhat <- sqrt(sse/df.resid(gamGrow));
```

In 1000 replicate simulated data sets with the same structure as in Figure 10.2 and $\sigma = 5$, the mean estimate of σ was 4.98 and 90% of estimates were between 4.5 and 5.5.

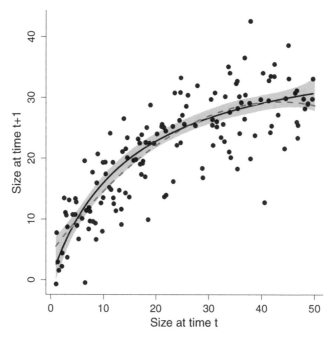

Fig. 10.2 Example of fitting a nonlinear mean growth function with gam. The solid curve (black) is the true mean function $40z/(15+z)$, and the circles are one typical simulated data set (Gaussian with standard deviation $\sigma = 5$, 150 data points). The dashed curve (red) is the estimate of the mean function from the one simulated data set plotted here. The shaded region (blue) shows the pointwise 5th to 95th percentiles of the fitted mean function over 1000 replicate simulated data sets. Source file: gamExample.R

Nonparametric regression functions like gam don't give you a formula for the fitted mean function. But they can still be used in an IPM by using the predict method for the model:

```
Gz1_z <- function(z1,z) {
    Gdata <- data.frame(z=z);
    z1bar <- predict(gamGrow,newdata=Gdata,type="response")
    return(dnorm(z1,mean=z1bar,sd=sdhat))
}
```

10.1.4 Nonlinear growth: parametric variance models

Unfortunately, one data set can have several different complications. In the growth model just above, the mean is nonlinear but variation around the mean is assumed to be Gaussian with constant variance. It is essential to check if growth variability really fits that assumption.

Size-dependent variance and many non-Gaussian distributions can be fitted using the gamlss package. The mean, variance, and up to two additional shape parameters can be fitted as either parametric or nonparametric functions of the independent variables. There are too many options for us to review here, but we give one example to illustrate the possibilities. Suppose the growth variability in the last example is instead a t distribution with $df = 5$ and nonconstant variance. R's rt function generates a standard t distribution with variance $\sigma_{df}^2 = df/(df - 2)$, so below we use a z-dependent factor scale.z to create artificial size data with size-dependent variance in growth:

```
z <- runif(250,1,50); # uniform distribution on [1,50]
z1bar <- 40*z/(15+z); scale.z <- 4*exp(-0.5+z/50);
z1 <- z1bar + scale.z*rt(150,df=5)
```

Figure 10.3 shows results from fitting these "data," estimating df (assumed to be size-independent), a nonparametric mean function (cubic splines) and either a nonparametric $\sigma(z)$ (cubic splines) or the true $\sigma(z)$ function (log-linear). The key point is that nonparametric fitting of $\sigma(z)$ was almost as good as (somehow) knowing the true functional form. The only cost is a slightly higher risk of missing the fat tails in the growth distribution (i.e., estimating $df \gg 5$, Figure 10.3C).

10.1.5 Nonparametric models for growth variation

As a final level of generality, it is not even necessary to specify a distribution for the scatter around the mean growth function. This may be useful if no standard distribution can capture all the features in the data, or it may let you avoid a difficult choice between several candidates for the "right" distribution.

Kernel density estimates are convenient for this because they are easy to use in an IPM growth kernel. A fitted kernel density function is the average of a set of probability densities (typically Gaussian) centered at the data values. That is, each observation is replaced by a probability density representing other observations that might have been obtained instead. Adding these up gives a smooth curve that approximates the underlying distribution from which the data were drawn.

To model growth variability with a kernel density estimate, we just use the residuals from a fitted mean growth function as the "data points." The one subtle point is that replacing each residual by a probability density increases the growth variance, so the residuals should be shrunk to offset this. This sounds complicated, but the code is simple.

For simplicity we assume first that growth variance is independent of initial size, and use gam to estimate the mean function. The residuals give us an estimate of the growth variance:

```
Resids <- resid(gamGrow); # extract residuals
sse <- sum(Resids^2);   # sum of squared errors
sdhat <- sqrt(sse/gamGrow$df.residual) # estimated error Std Dev
```

The only thing we need from the kernel density estimate is h, the standard deviation of the Gaussian probability densities centered at each residual (called the "bandwidth" of the kernel):

```
        h <- bw.SJ(Resids);
```

The estimated probability density for growth is then the average of Gaussian densities centered at the shrunken residuals, which is computed by dfun in the code below:

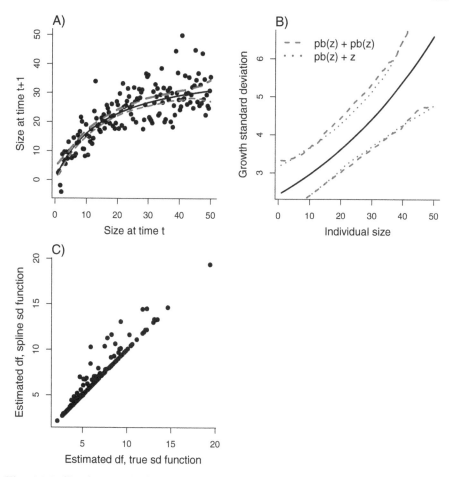

Fig. 10.3 Simultaneously fitting the mean, variance, and shape parameter of a t distribution. (A) The solid black curve is the true mean function $40z/(15 + z)$, and the circles are a typical data set. The dashed curves (red) are the pointwise 5th and 95th percentile estimates of the mean function across 250 simulated "data" sets of size 150, using nonparametric splines (function pb in gamlss) for the mean and standard deviation functions. The nearly identical dotted curves (blue) are the same, but result from using the correct parametric form of the standard deviation. (B) The solid black curve is the true standard deviation; the dashed and dotted curves are pointwise 5th and 95th percentiles, as in panel (A). (C) Estimates of the shape parameter df in the t distribution, with values > 20 not shown. Source file: gamlssExample.R

```
alpha <- sdhat/sqrt(sdhat^2+h^2); # shrinkage factor
hResids <- alpha*Resids              # shrinking the residuals
dfun <- function(z) mean(dnorm(z,mean=hResids,sd=h))
dfun <- Vectorize(dfun)
```

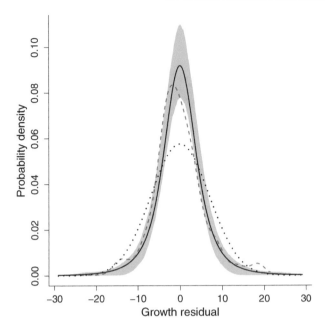

Fig. 10.4 Kernel density estimation, for the same growth function as Figure 10.2 but
t-distributed variation of growth around the mean with $df = 3$. The solid black curve is
the true probability density and for comparison the dotted curve is a Gaussian with the
same variance. The dashed curve (red) is a typical estimate of the density from one set
of simulated data on growth of 150 individuals, obtained as described in the text from
the residuals of fitting the mean growth function with gam. The shaded area (blue) are
the pointwise 5th and 95th percentiles of the density estimates in 500 replicate simulated
data sets on growth of 150 individuals. Source file: kernelExample.R

Figure 10.4 shows an example. The "data" follow the model of Figure 10.2
except that the growth variance was a t distribution with $df = 3$. Because
the error distribution is symmetric, the fat tails won't bias the estimate of the
mean, so we fitted the mean using gam and fitted a kernel density estimate to the
residuals. The resulting dfun was more peaked and fat-tailed than a Gaussian
with the same standard deviation, as it should be. But even with a decent
amount of data (here, 150 data points) the kernel density estimate can be erratic,
especially in the tails. So unless the data set is quite large, a nonparametric
estimate of growth variation is probably best used as an exploratory step to
identify the qualitative shape of the distribution. The final model can then use
a parametric distribution (t, beta, lognormal, etc.) that has the right shape.

A kernel density estimate can also be used if the variance is size-dependent,
but the shape of the distribution is independent of size. To illustrate this, we
use the *Carlina* growth data in CarlinaNonconstantVarkernel.R. The first step
is to estimate the size-dependent mean and variance of growth, which we've
already done (Figure 10.1) to get parameters a,b,sigma0,d specifying the mean

and standard deviation. The second step is to scale the residuals by the standard deviation function

```
sdG <- function(z) sigma0*exp(-d*z)
scaledErr <- (size2-(a+b*size1))/sdG(size1);
```

This removes the size-dependence in growth variability, and gives values that can be used to estimate the distribution shape, exactly as above (in small data sets, the scaled residuals would also be standardized to remove any effects of leverage, but kernel density estimates would not be recommended in that case). Exactly as above, we estimate the bandwidth h and compute shrunken residuals hResids. The probability density for growth is then obtained by restoring the size-dependent variance while preserving the shape of the distribution:

```
dfun <- function(z1,z) (1/sdG(z))*mean(dnorm(z1/sdG(z),mean=hResids,sd=h))
dfun <- Vectorize(dfun,"z1");
```

The growth kernel is then

```
G_z1z <- function(z1,z) dfun(z1-(a+b*z),z);
```

In this case this works reasonably well (Figure 10.5). The scaled residuals are very close to Gaussian, so the nonparametric growth kernel is nearly the same as a parametric Gaussian kernel.

These kernel density estimates all use the residuals from a "pilot" fit, and to do that initial fit we have assumed some parametric distribution of growth variability. This is conceptually unsatisfying, because if the kernel density estimate is right, then the pilot fit was based on an incorrect likelihood function. The nonparametric growth distribution bites its own tail: if it's right, then it's wrong.

A brute-force solution is to iterate the procedure until the pilot and final distributions are identical (re-estimate the mean and variance functions using the nonparametric growth distribution in the likelihood function, get a new set of residuals and a new dfun, etc.). But in principle it should be possible to do a one-step fit that minimizes cross-validated error or an analytic approximation. This seems like an area where considerable improvement is possible. Eventually, fitting a nonparametric growth kernel to data should be as easy and reliable as fitting a kernel density estimate using density.

10.2 High-dimensional kernels

The numerical integration methods presented in Chapter 6 are adequate for IPMs with one- or two-way classification. For three-way they are unpleasantly slow, and beyond that they fail. The problem is that when m mesh points are used for each classifying variable, there are m^d mesh points in d dimensions and 100^5 is a large number. This problem is not unique to IPMs (search on "curse of dimensionality" AND "numerical integration") so there are already some potential solutions. When a function is high-dimensional but otherwise simple (no bumps or ridges), *sparse grid* integration methods (Heiss and Winschel

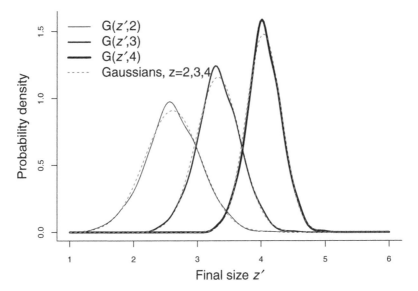

Fig. 10.5 Nonparametric estimate of growth variability in *Carlina* using a kernel density estimate with size-dependent growth variance. The solid curves are the fitted growth kernel $G(z', z)$ for $z = 2, 3, 4$. The dashed red curves are the fitted Gaussian kernel from Figure 10.1 for $z = 2, 3, 4$. Source file: `CarlinaNonconstantVarKernel.R`

2008) can give accurate results with many fewer than m^d mesh points. The `SparseGrid` package by J. Ypma makes these available in R. In practice it's the same as using Gauss-Legendre quadrature, except that you call a different function to get the quadrature nodes and weights; an example is in the script `SizeQualityB2B.R` from Chapter 6.

Bayesian statistics uses Markov Chain Monte Carlo to deal with high-dimensional integrals. Monte Carlo integration is used in many areas of science where high-dimensional integrals need to be evaluated, and a lot is known about how to do it efficiently.[1] The basic idea is simple: $\int_a^b f(x)\, dx$ can be interpreted as the expectation of $f(X)$ when X is a random variable with a uniform distribution on $[a, b]$. So `mean(f(runif(m,a,b)))` converges to the value of the integral as m goes to infinity, with error variance proportional to $1/m$. This remains true if x is a vector and the random numbers are chosen from a multidimensional uniform distribution – there is no intrinsic "curse of dimensionality." However, it takes some tricks to make it work well in practice.

So far as we know, nobody has tried sparse grids for an IPM, and only de Valpine (2009) has used Monte Carlo integration for otherwise intractable integrals in a high-dimensional IPM. So we cannot offer any advice on how or when to use these methods or any others for high-dimensional numerical integration. More work here is needed.

---

[1] We recommend the sections on Monte Carlo integration in the *Numerical Recipes* for your programming language of choice.

10.3 Demographic stochasticity

"A large shortcoming with the current theory of integral projection modeling is that the other main source of stochasticity in population dynamics, demographic stochasticity, has to our knowledge so far been ignored" (Vindenes et al. 2011). In this section we present a new development - a simple approximation for the dynamics of an IPM with demographic stochasticity - and then review existing theory on how demographic stochasticity affects population growth and extinction risk in a density-independent IPM. Much more remains to be done, but these are important first steps.

Demographic stochasticity can always be included in simulations, because an IPM's demographic models can be implemented as a stochastic individual-based model (IBM) that uses random numbers to determine survival, growth, and fecundity for discrete individuals. The price is losing all the analytical machinery for IPMs (deterministic or stochastic growth rate, equilibrium stability, sensitivity analysis, ESS analysis, etc.).

For matrix models, there is a middle ground: stochastic matrix models with a large-sample approximation to demographic stochasticity. These are valid for populations large enough to justify a Poisson or Gaussian approximation to the net effect of random events at the individual level, but small enough that this variance matters. A good example is the stochastic LPA model for laboratory populations of *Tribolium* flour beetles (Dennis et al. 2001).

We can similarly approximate the effects of demographic stochasticity in an IPM viewed as a mean field approximation to an IBM, and implemented using midpoint rule.[2] We consider a medium-density population: small enough that demographic stochasticity matters, but not so small that the fate of one individual has a large effect. Let nt denote the population vector at time t and nt1 the population vector at time $t+1$, with both scaled to represent the absolute number of individuals in each size interval (rather than the number per area, or some other scaled measure of abundance). The survival contribution to the population at time $t + 1$ is approximately

$$\texttt{rpois(length(nt),lambda=P\%*\%nt)} \tag{10.3.1}$$

The intuition behind (10.3.1) is that the number of individuals who survive and have subsequent state in a small interval $(z_k - h/2, z_k + h/2)$ is the sum of independent "coin tosses" by the individuals at time t, where each toss is 1 with probability $p = hP(z_k, z_i)$ for individual i with size z_i individual, and 0 otherwise. Each toss has mean p_i and variance $p_i(1 - p_i) \approx p_i$ because h is small and $p_i = O(h)$. Because tosses are independent, their sum also has variance≈mean= $\sum p_i$. This suggests that the sum should be approximately Poisson with mean $\lambda = \sum p_i$, and it is so long as all of the p_i are small (Hodges and Le Cam 1960). So equation 10.3.1 is a good approximation to the marginal distribution for survivors in each size interval, when h is small.

---

[2] This material is new, and untested apart from what we do here, so by the time you read this it may be superseded or discredited.

The other thing assumed in (10.3.1) is that different intervals are independent. In fact they aren't, but the covariances are $O(h)$ smaller than the mean and variance (see Box 10.1). This makes intuitive sense. If you know how many individuals survive into one interval at time $t + 1$, what does that tell you about a second interval? Almost nothing. When h is small, any individual has only a minute (order h) chance of surviving into the first interval. So the ones who do are only a minute fraction of the individuals who might survive into the second interval. Knowing the exact number in the first interval therefore leads to a minute correction to the expected number in the second interval. So the covariance between the two intervals is small relative to their means, which equal their variances.

However, as discussed at the end of Box 10.1, the approximation can be improved by computing the correlations between different size intervals (equation 10.3.5). Combining these with the marginal Poisson distributions, equation 10.3.1, the R package corcounts (Erhardt 2009) can be used to generate Poisson random vectors with the correct mean and the correct variance-covariance matrix for the survival contribution to the next year's population.

The fecundity contribution can be approximated in much the same way, sometimes. The essential conditions are that the numbers of offspring in disjoint size intervals have mean≈variance and are approximately independent. This is exactly true if a size-z parent has a Poisson-distributed number of offspring with mean $b(z)$, whose sizes are chosen independently; the numbers of offspring in disjoint size intervals are then independent Poisson random variables (this is a general property of Poisson point processes). It will be approximately true when there is a smooth density for offspring size, for very small h. The meaning of "very" small h is that $F(z', z)h \ll 1$ for all z and z'. This will not be valid if all offspring are about the same size, unless h is much smaller than is actually needed to accurately iterate the deterministic model. And it will not be valid for models where all offspring go into a discrete category for newborns. However, when $F(z', z)h \ll 1$ is valid, it will typically mean that each parent puts either 1 or 0 offspring into a size interval of width h, so that fecundity is just like survival with F instead of P. In short, for small h it is often reasonable to approximate the survivors and new recruits in each size interval as two vectors of Poisson random variables, with means given by applying the deterministic IPM to the previous population state. If survival and fecundity are independent, their sum is again a vector of independent Poisson random variables. But even in models where survival and fecundity are not independent, calculations like those in Box 10.1 show that their covariance is $O(h)$ smaller than either variance. Consequently, in a midpoint rule implementation the population state at time $t + 1$ is approximately

$$\texttt{rpois(length(nt),K\%*\%nt)} \qquad\qquad (10.3.6)$$

Let A and B be two disjoint subsets of $\mathbf{Z}$ with length $O(h)$. Our goal is to compute the covariance between $n_A(t+1)$, the number of individuals who survive into A at time $t+1$, and the analogously defined $n_B(t+1)$, conditional on the population state at time t, in a "coin tossing" IBM with the IPM's demographic models. For individual i alive at time t with size z_i let $a_i = 1$ if the individual survives to time $t+1$ and its state is in A, 0 otherwise. So $a_i = 1$ with probability $\gamma(z_i(t), A)$ where for any set S,

$$\gamma(z, S) \equiv \int_S P(z', z)\, dz'.$$

Define b_i the same way for survival into B. Then

$$n_A(t+1) = \sum_i a_i, \quad n_B(t+1) = \sum_j b_j.$$

a_i and b_j are independent for $i \neq j$, so their covariance is 0. Therefore

$$Cov[n_A(t+1), n_B(t+1)] = Cov(\sum_i a_i, \sum_j b_j) = \sum_{i,j} Cov(a_i, b_j) = \sum_i Cov(a_i, b_i).$$

$$(10.3.2)$$

Because an individual can't survive into both A and B, $a_i b_i = 0$. Therefore

$$Cov[a_i, b_i] = E[a_i b_i] - E[a_i]E[b_i] = -E[a_i]E[b_i] = -\gamma(z_i, A)\gamma(z_i, B). \quad (10.3.3)$$

Combining (10.3.2) and (10.3.3) we have

$$Cov[n_A(t+1)n_B(t+1)] = -\sum_i \gamma(z_i, A)\gamma(z_i, B). \quad (10.3.4)$$

Each term in the sum is $O(h^2)$ whereas the mean and variance of n_A and n_B are both of order h. The covariance terms in the variance-covariance matrix for the numbers surviving into each grid cell are therefore vanishingly small compared to the variances, in the limit $h \to 0$.

This asymptotic independence doesn't hold when A and B are not small sets; for example, there is no asymptotic independence between the numbers of individuals above and below the long-term average size. Similarly, although the sum of independent Poisson variables is Poisson, the "true" (i.e., IBM-generated) total population size need not be Poisson because the number of between-cell covariances goes up as h goes down, and all of them contribute to the total population size. To approximate larger-scale properties like these, the between-cell covariances cannot be neglected. For the case of width-h size intervals in a midpoint rule implementation, equation (10.3.4) can be used to estimate the covariances as

$$Cov[n_i(t+1)n_j(t+1)] = -h^2 \sum_k n_k(t) P(z_i, z_k) P(z_j, z_k). \quad (10.3.5)$$

where $n_k(t)$ is the number in interval k, and $n_i(t+1), n_j(t+1)$ are the numbers that survive into intervals i and j in year $t+1$.

Box 10.1: Covariance calculation for demographic stochasticity in survival.

As discussed at the end of Box 10.1, it is also possible, and sometimes be necessary, to take into account the correlations between different size intervals, instead of using (10.3.6). As an example, in Appendix 10.7 we show how these correlations (accounting for survival and fecundity) can be calculated for a post-reproductive census kernel $s(z)G(z', z) + s(z)b(z)c(z')$, if parents who survive to reproduce have independent, Poisson-distributed numbers of offspring. The covariance formulas are not especially illuminating, but they show that the variance-covariance matrix can in fact be calculated in realistic kernels. Because offspring numbers are Poisson and offspring states are independent, the numbers of offspring that an individual produces in two disjoint size intervals are independent. Without this property, there would be additional covariance terms that are much more difficult to calculate.

Now let's see if it works. To make it challenging we use the basic size-structured monocarp model of Chapter 2, where survival and fecundity are tightly coupled. The IBM was simulated for 100 years, and 1000 individuals were selected from the population in year 100 to serve as the "population in year t." We then repeated 5000 times the process of iterating the IBM for one additional year, and calculating the number of individuals in each of 100 equal-length size intervals spanning the range of sizes.

Figure 10.6 shows that the simple Poisson approximation without covariances, equation (10.3.6), is fairly accurate. Visually, the distribution (across replicates) of the number in a size interval is approximately Poisson (Figure 10.6A), and the mean and variance for each interval are very similar (Figure 10.6B). To test more formally that the distribution is approximately Poisson, we did a Monte Carlo test of the 5000 replicates for each size interval against a Poisson distribution with the same mean, using the Kolmogorov-Smirnov test statistic (we needed to evaluate p-values by Monte Carlo because the standard K-S test is not valid in this case because of ties). Under the null hypothesis of Poisson distributions, the p-values should be uniformly distributed with about 5 out of 100 smaller than 0.05, and that's what we see (Figure 10.6C). Finally, the correlations between different size intervals (Figure 10.6D) are all small (94% are 0.03 or smaller in magnitude) with perhaps a weak tendency for nearby intervals to have slightly stronger correlations.

The Poisson approximation also gives fairly accurate projections of the longer-term impacts of demographic stochasticity. Figure 10.7 compares the distributions of projected total population size after 20 years, from IBM simulations and the Poisson approximation, from the same initial population of 1000 individuals. Differences exist, but they are relatively small compared to the range of stochastic variation among replicate projections with demographic stochasticity. The Poisson approximation correctly predicts the skewed distribution of population growth, so that the most likely outcome with demographic stochasticity is smaller than the deterministic projection.

The figures show results for just one randomly constructed starting population, but the results are nearly identical every time the scripts are run. A more important limitation is that they are for one set of parameters, for one model. So before you trust a Poisson approximation, do some similar tests on your model. It is most likely to fail when offspring production is far from Poisson,

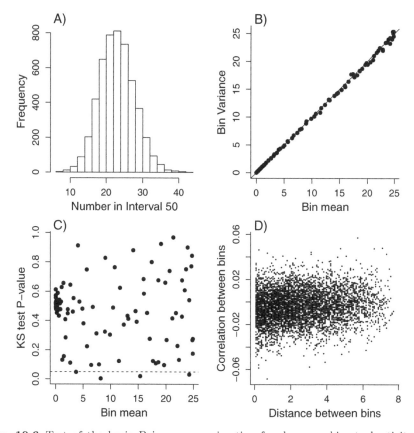

Fig. 10.6 Test of the basic Poisson approximation for demographic stochasticity, in the basic size-structured *Oenothera* model. The simulation experiment is described in the text. (A) Distribution across 5000 replicates of the number of individuals in size interval 50 out of 100. (B) Variance (across 5000 replicates) versus mean for the 100 size intervals. (C) Histogram of *p*-values from Monte Carlo tests against the hypothesis of a Poisson distribution with the observed mean, for the 100 size intervals. (D) Pairwise correlations between different size intervals as a function of the distance between their centers. Source file MonocarpDemogStoch.R

but it might not. For example, in our monocarp model the results were nearly identical to those in Figures 10.6 and 10.7 when we made recruitment deterministic (i.e., each reproducing plant had the expected number of seeds given its size, and the number of surviving seeds was the number produced times the seed survival probability, rounded to an integer). You can see for yourself by setting constantSeeds==TRUE in the script files.

10.3.1 Population growth rate

Vindenes et al. (2011, 2012) studied the effect of demographic stochasticity on population growth in an IPM with a one-dimensional individual state such as size. The key insight, by Engen et al. (2007), is the extension to stochastic

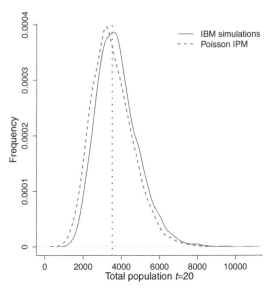

Fig. 10.7 Comparison of 20-year projections by the *Oenothera* monocarp IBM and IPM with Poisson approximation. Starting from an initial population of 1000 drawn randomly from year 100 of an IBM simulation, 5000 20-year projections were done using the IBM and IPM with Poisson approximation. The plotted curves are density estimates for the total number of individuals in year 20 (using density with a common bandwidth selected using bw.SJ). The dotted vertical line is the deterministic projection without demographic stochasticity. Source file IterateMonocarpDemogStoch.R

population growth of R.A. Fisher's observation that the total reproductive value of a density-independent population always grows at rate λ, even when the population is not at its stable distribution. That is: define $V_t = \langle v, n_t \rangle$ where n_t is the state distribution in time t and v is the dominant left eigenvector of K, scaled so that $\langle v, w \rangle = 1$ and $\int w(z)dz = 1$ where w is the dominant right eigenvector. Then for constant K

$$V_{t+1} = \langle v, Kn_t \rangle = \langle vK, n_t \rangle = \langle \lambda v, n_t \rangle = \lambda V_t.$$

We usually don't think of it this way, but demographic stochasticity is equivalent to having small random fluctuations in K that cause the actual number of survivors to deviate from the expected number as a result of individuals tossing coins to decide their survival or death, and that similarly cause the actual number of offspring to deviate from the expected number. So if the kernel in year t is $K_t = \bar{K} + \epsilon D_t$ with ϵD_t the small fluctuations representing demographic stochasticity, the population state n_t will fluctuate around the stable distribution w, say $n_t = N_t w + \epsilon u_t$ where N_t is the total population in year t. Here ϵ is just a "marker" to indicate that terms including ϵ are small compared to other terms. We then have

$$V_{t+1} = \langle v, K_t n_t \rangle = \langle v K_t, n_t \rangle$$
$$= \langle v(\bar{K} + \epsilon D_t), n_t \rangle = \langle v\bar{K}, n_t \rangle + \epsilon \langle v D_t, N_t w + O(\epsilon) \rangle \quad (10.3.7)$$
$$= \lambda V_t + \epsilon N_t \langle v, D_t w \rangle + O(\epsilon^2)$$

Dropping the final $O(\epsilon^2)$ term, equation (10.3.8) in Box 10.2 shows that $V_{t+1} = \lambda V_t$ plus a random noise term that is small (compared to V_t) and independent over time. This contrasts with the dynamics of total population size N_t, whose deviations from exponential growth are correlated over time due to the variation in population structure.

In short: counting total reproductive value, instead of counting individuals, simplifies the dynamics by "filtering out" the correlated component of year-to-year changes (Engen et al. 2007). However, it has no effect on the long-term growth rate. Total population and total reproductive value have to grow or shrink roughly in parallel, because V_t/N_t can't be smaller than the minimum value of v (which is positive under our Chapter 7 assumptions about stochastic IPMs) or larger than the maximum value of v, which is finite.

Vindenes et al. (2011, 2012) analyze V_t using the fact that V_{t+1} is the sum of contributions from individuals at time t. So if individuals are independent, then the variance of their total contribution is the sum of the variances of the individual contributions. An individual's contribution is the sum of its own reproductive value next year (if it lives) and the reproductive values of its offspring (if there are any). Letting $\sigma_d^2(z)$ denote the variance of this random sum,

$$Var(V_{t+1}|V_t) = \int_Z \sigma_d^2(z) n_t(z)\, dz. \quad (10.3.9)$$

To leading order $n_t = N_t w = V_t w$ giving (to leading order)

$$Var(V_{t+1}|V_t) = V_t \int_Z \sigma_d^2(z) w(z)\, dz \equiv V_t \sigma_D^2. \quad (10.3.10)$$

So $V_{t+1} = \lambda V_t + \Delta_t$ where $E[\Delta_t] = 0, Var(\Delta_t) = V_t \sigma_D^2$. Taking logs and using a Taylor series expansion, $\log V_t$ is approximately a random walk with mean step $\log \lambda - \frac{\sigma_D^2}{2\lambda^2 V_t}$ and variance $\frac{\sigma_D^2}{\lambda^2 V_t}$, and $\log V_t$ is approximated well by a diffusion process with these as the infinitesimal mean and variance, so long as the yearly changes in $\log V_t$ are small (Vindenes et al. 2011); if you aren't familiar with diffusion processes, see Box 10.2 for a quick introduction. This is a remarkable reduction in complexity because it shows that population growth is largely determined by just two numbers, λ and σ_D^2.

Vindenes et al. (2011) also show that when there is also (small) environmental stochasticity, exactly the same holds, except that the infinitesimal mean and variance of the diffusion approximation are

$$\mu = \log \lambda - \frac{1}{2\lambda^2} \left(\sigma_D^2/V + \sigma_E^2 \right), \quad \sigma^2 = \frac{1}{\lambda^2} \left(\sigma_D^2/V + \sigma_E^2 \right) \quad (10.3.11)$$

Diffusions are a kind of continuous-time stochastic model that has been widely used in ecology and evolutionary biology because they often provide tractable approximations to complicated models. Here we consider only the simplest case, a one-dimensional diffusion $X(t)$. $X(t)$ is specified by two functions, its *infinitesimal mean* $\mu(X)$ and its *infinitesimal variance* $\sigma^2(X)$. These specify the mean and variance of changes in $X(t)$ over brief time periods. That is, for a small time step τ, given $X(t)$

$$E[X(t+\tau)] = X(t) + \tau\mu(X(t)) + o(\tau)$$
$$Var(X(t+\tau)) = \tau\sigma^2(X(t)) + o(\tau) \tag{10.3.8}$$

where $o(\tau)$ denotes a quantity such that $o(\tau)/\tau \to 0$ as $\tau \to 0$.

The solution of a diffusion model is described by the time-dependent probability distribution of $X(t)$, $\phi(X,t)$. ϕ is the solution of a second-order partial differential equation (the Fokker-Planck equation) which involves μ and σ^2, but except for very simple models (such as linear models with constant mean and variance functions) it cannot be solved. However, it is often possible to calculate many useful properties of X, such as the probability of extinction, mean time to extinction, etc. (Karlin and Taylor 1981; Lande et al. 2003; Ewens 1979, 2004). Many well-known results in population genetics are based on diffusion approximations for finite-population genetic models.

When a diffusion model is used to approximate a discrete-time process such as V_t, the formal definitions of μ and σ^2 above can't be used, because the smallest possible τ is $\tau = 1$ (i.e., one discrete time step). However, a diffusion approximation can be defined by using (10.3.8) with $\tau = 1$ and ignoring the $o(\tau)$ terms. That is, $\mu(X)$ and $\sigma^2(X)$ are defined to be the mean and variance of $X(t+1) - X(t)$ for the discrete-time model, conditional on $X(t) = X$.

This rough-and-ready approach is appealing, but it can be ambiguous because it is not invariant under variable transformations. For example, here are three ways to get a diffusion approximation for $\log V_t$:

1. Compute the mean and variance of $\log V_{t+1} - \log V_t$, and use those to define a diffusion model.
2. Compute the mean and variance of $V_{t+1} - V_t$, use those to define a diffusion model, and then apply a log-transformation to the diffusion model (as explained in section 2.3.2 of Lande et al. (2003)).
3. Compute the mean and variance of $V_{t+1} - V_t$, approximate (by Taylor series) the corresponding mean and variance of $\log V_{t+1} - \log V_t$, and use those to define a diffusion model (this is what Vindenes et al. (2011) did).

Generally these give three different results, and other considerations must be used to choose the "right" one (and none of them is exactly right). For example, in some simple models it is possible to compute the mean and variance of the change during one time interval for the diffusion approximation, and compared those to the discrete-time model.

We recommend Lande et al. (2003) for a quick practical introduction to diffusion models, and for more depth Ewens (1979, 2004). For the full story, see Karlin and Taylor (1981). Allen et al. (2008) give a general recipe for constructing a diffusion approximation to a continuous-time individual-based population model with discrete population structure.

Box 10.2: Diffusion models

where σ_E^2 is a parameter summarizing the effect of environmental stochasticity (we explain below how its value can be estimated). The fact that σ_E^2 is not divided by V reflects the essential difference that demographic stochasticity becomes unimportant in large populations, while the effect of environmental stochasticity is undiluted.

10.3.2 Extinction

Demographic stochasticity is a major determinant of extinction risk for small populations, for density-dependent populations with small carrying capacity, and for populations with λ or λ_S not much above 1. Properties such as the probability of extinction, mean and variance of time to extinction, and duration of the final decline to extinction can be studied using the diffusion approximation for the dynamics of total reproductive value. Lande et al. (2003, chapters 1,2,5) explain in detail how these properties can be calculated for unstructured populations, with several examples. Here we consider just one example to illustrate the wrinkles that continuous population structure introduces, using the *Carlina* model with environmental stochasticity from Chapter 7.

The diffusion model with infinitesimal moments (10.3.11) is accurate when the demographic and environmental variance are small, and λ is near 1 (Lande et al. 2003; Vindenes et al. 2011). Neither of those is true for the *Carlina* model. To give the approximations a fighting chance, *in this section* we reduce the standard deviations of all time-varying parameters by a factor of 4, and reduce the establishment probability to $p_r = 0.0004$ and so that $\lambda_S \approx 1.05$. We also consider a difference equation approximation for V_t, which uses the results of Vindenes et al. (2011) but does not take the final step of going to a continuous-time process. This approximation is

$$V_{t+1} = \lambda_t V_t + \sigma_D Z_t \sqrt{V_t} \qquad (10.3.12)$$

where Z_t is a sequence of independent random variables with mean 0 and variance 1. The variability of λ_t represents environmental stochasticity, and $\sigma_D Z_t \sqrt{V_t}$ adds the variance due to demographic stochasticity, equation (10.3.10). Equation (10.3.12) defines a Markov Chain whose properties can be studied analytically using the methods of Chapter 3. But simulation is very fast with vectorization (1000 replicates of 100 time steps takes seconds), so it's easy to estimate quantities of interest by simulation.

To implement the diffusion and difference equation approximations we need to estimate σ_D^2, σ_E^2, and the mean and variance of λ_t. For σ_D^2, it is possible to use (10.3.10) following Vindenes et al. (2011), but the $\sigma_d(z)$ function is highly model-specific. Instead, we suggest Monte Carlo simulation to estimate $\sigma_d^2(z)$ function. The recipe is to start the IBM with one size-z individual, simulate one time-step using the mean parameters, and compute the total reproductive value of the population (the starting individual, if it survives, and all its offspring). Doing this repeatedly for the same z, and computing the variance of the total reproductive value across replicates, gives an estimate of $\sigma_d^2(z)$.

Because they involve the random environment, σ_E^2 and λ_t usually have to be studied by simulation. This can be done with the IPM, which has $\sigma_D = 0$.

Equation (10.3.8) shows that to leading order λ_t equals the value of V_{t+1}/V_t for a population in the stable state distribution for the mean kernel, which is $\lambda_t = \langle v, K_t w \rangle / \langle v, w \rangle$. The variance of λ_t is an estimate of σ_E^2 (Vindenes et al. 2011, Appendix A). For the difference-equation approximation we compute the mean and variance of $\log \lambda_t$ and simulate $\log(\lambda_t)$ from a Gaussian distribution with those parameters. We use a lognormal distribution for λ_t because the asymptotic distribution of total population size (and therefore of V_t) is lognormal in an IPM with only environmental stochasticity (Chapter 7).

The script LowNoiseCarlinaDiffusionPars.R shows how these estimates are carried out. After modifying the parameters as described above, mk_K with the mean parameters gives K, v and w. We scale these so h*sum(w) and h*sum(v*w) both equal 1. Then we use v to define a function giving reproductive value as a function of size,

```
vfun <- approxfun(meshpts,v)
```

We then use vfun and the IBM to estimate $\sigma_d^2(z)$ at the IPM mesh points, as described above:

```
nreps=2500; vVar <- numeric(m);
for(j in 1:m) { #loop over meshpoints
  V1<-numeric(nreps);
  for(k in 1:nreps) { #loop over replicates
    z <- oneIBMstep(meshpts[j],m.par.true,constantSeeds);
    V1[k] <- sum(vfun(z)); # Total reproductive val after one time step
  }
  vVar[j]=var(V1); # Variance across replicates
}
sigma2D <- sum(h*w*vVar)
```

For environmental stochasticity, we implement the formula $\lambda_t = \langle v, K_t w \rangle / \langle v, w \rangle$ as follows:

```
V.init <- h*sum(v*w); lam.t <- matrix(NA,2500);
for(j in 1:2500){
    m.par.year=m.par.true + qnorm(runif(12,0.001,0.999))*m.par.sd.true
    m.par.year[c("grow.int","rcsz.int")] <-
        m.par.year[c("grow.int","rcsz.int")] +
        matrix(qnorm(runif(2,0.001,0.999)),1,2) %*% chol.VarCovar.grow.rec
    K.year <- mk_K(m, m.par.year, L, U)$K;
    lam.t[j] <- h*sum(v*(K.year%*%w))/V.init;
}
sigma2E <- var(lam.t);
mean.loglambda <- mean(lam.t); sd.loglambda <- sd(log(lam.t));
```

Figure 10.8 shows numerical results. Simulations of the difference equation approximation (Figure 10.8A) starting from a small population behave qualitatively like the IBM. Most of the time there is either quick extinction, or else V_t grows exponentially like the IPM and the risk of extinction decreases over time (Figure 10.8B). The quantitative accuracy is decent (Figure 10.8C), but diffusion approximations under-estimate the extinction risk due to demographic

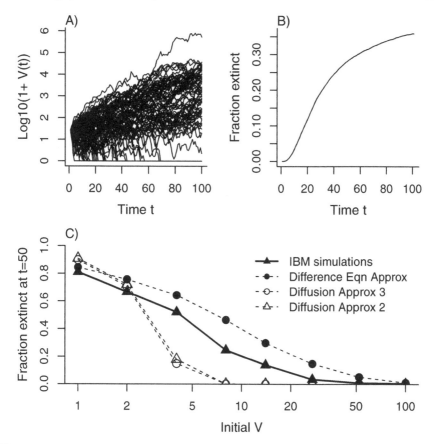

Fig. 10.8 Extinction risk in the stochastic *Carlina* IBM with modified parameters. A) Simulations of the difference equation approximation starting from $V_0 = 25$. B) Cumulative probability of extinction, based on 10,000 replicate simulations of the difference equation approximation. C) Probability of extinction at $t = 25$ as a function of initial total reproductive value V_0, for the IBM (based on 500 simulations for each V_0), the difference equation approximation, and two of the diffusion approximations defined in Box 10.2. Source files LowNoiseCarlinaDiffusionPars.R and CarlinaExtinctionProb-DiffEq.R

and environmental stochasticity in the IBM, sometimes quite severely, while the difference equation over-estimates it.

These results underscore the fact that the diffusion and difference-equation approximations for V_t are only quantitatively accurate when $V_{t+1} - V_t$ is always small. This means that stochasticity must be low, and λ must be very near 1. This was true for all of the examples in Vindenes et al. (2011), and in those cases the diffusion approximation was very accurate. For simple unstructured population models, more accurate approximations have been derived using a different approach which remains accurate when λ is not near 1, the WKB approximation (Ovaskainen and Meerson 2010).

For very small density-independent populations, an exactly analysis of the probability of establishment vs. extinction has been performed very recently by Schreiber and Ross (2016), using the theory of continuous-state branching processes (Harris 1963). The analysis includes time-dependent and asymptotic extinction risk, and sensitivity formulas for the latter, for models with only demographic stochasticity and for varying-environment models conditional on the sequence of environment states. These have to be evaluated numerically, like the life cycle properties derived in Chapter 3, but they allow quick, exact evaluation of properties that would be time-consuming, or impossible, to estimate accurately by simulation.

We remind readers that demographic stochasticity in IPMs is a "postcard from the edge." Everything is new, testing is very limited, and it all needs further development.

10.4 IPM meets DEB: deterministic trait dynamics or constraints

There's another approach to studying structured populations, which is also deservedly popular among ecologists. It generally goes by the name of "physiologically structured population models" because of the book (Metz and Diekmann 1986) that sparked their widespread use. Much like IPMs they are extremely flexible and can describe many different kinds of differences among individuals. The individual state z can be a physiological variable such as mass, a vector of such variables (protein, carbohydrate, lipid, etc.) as in Dynamic Energy Budget models or a more abstract variable like "developmental state" tracking the amount of progress from one stage transition to another in an insect or crustacean (e.g., Nisbet et al. 1989; McCauley et al. 1996; Nisbet et al. 2000; Kooijman 2010; de Roos and Persson 2013). And like IPMs, the models are derived by specifying how the state z of an individual changes over time, and working out what these imply for the population's state distribution $n(z, t)$.

Biologically, an important feature of these models is that changes in individual state are grounded in processes that occur within the individual. In the IPMs in this book, the equations describing growth, mortality, and reproduction are (as the title says) data-driven: they summarize what we saw. In physiologically structured models, the underlying equations are mechanistic models for within-individual processes. Growth, mortality, and reproduction are not modeled directly, but instead are outcomes of those processes. There are challenges in this approach. What happens inside individuals is complicated, and hard to observe and quantify, so the models can (at best) be simplified descriptions for a few key processes. But there are also potential advantages in being more mechanistic, especially for extrapolating models beyond the range of conditions for which historical data are available.

Another challenge is that the existing mathematical theory for IPMs cannot handle models where the individual-level rules are deterministic, which is typically true in physiologically structured population models. For example, individual growth may be specified by an ordinary differential equation

$$\frac{dz}{dt} = g(z),$$

in which case the population-level model is the McKendrick-von Foerster equation

$$\frac{\partial n}{\partial t} + \frac{\partial gn}{\partial z} = -\mu(z)n.$$

More complicated models involve systems of differential equations for multiple state variables, and the population dynamics can be coupled to environmental variables (e.g., the abundance of some limiting resource) that feed back to affect vital rates, or coupled to the dynamics of other populations.

All of the IPMs in this book have smooth kernels, at least piecewise continuous. This implies that the dynamics of all individual-level state variables are all stochastic. Present state determines the distribution of future states, but what actually happens to an individual involves an element of chance. This is not an optional ingredient. Virtually all of the theory underpinning the use of IPMs depends on the smoothness properties of the kernels. Kernel smoothness also goes away if there are deterministic constraints on the changes in state variables; for example, if $z_1(t+1)$ and $z_2(t+1)$ are both Gaussian conditional on $z_1(t)$ and $z_2(t)$, but $z_1(t+1) + z_2(t+1) = \theta_i(t+1)$ where θ is the time-varying parameter vector in a random-environment IPM.

So we are missing a framework for situations in which some individual-level state variables are stochastic while others are deterministic or subject to deterministic constraints. These are realistic and biologically important cases. Body size may change stochastically, while breeding value for body size remains constant over the lifetime. Growth in different body components (protein, carbohydrate, lipid) may be random, but subject to a fixed constraint on their sum (or a constraint that depends on environment conditions).

The limitation is not in writing down the models, but in understanding their behavior. For example, consider the following assumptions:

1. $z = (x, y) \in \mathbf{Z} = [L_x, U_x] \times [L_y, U_y]$
2. x has deterministic dynamics $x' = w(x)$ where w is an invertible function with inverse function α.
3. y has stochastic dynamics specified by a continuous kernel $H(y'|z)$ analogous to the P kernel in a standard IPM (i.e., its integral with respect to y' equals the survival probability of state-z individuals).
4. Fecundity is described by a smooth kernel $F(z', z)$ specifying the reproductive output of individuals alive at time t (i.e., it includes any parental mortality that affects offspring number).

It can then be shown (S.P. Ellner, *unpublished*) that the population dynamics are

$$n(z', t+1) = \int_{\mathbf{Z}} F(z', z) n(z, t)\, dz$$

$$+ \left| \frac{d\alpha}{dx}(x') \right| \int_{L_y}^{U_y} H(y'|x'_*, y) n(x'_*, y, t)\, dy. \qquad (10.4.1)$$

where x'_* is shorthand for $\alpha(x')$ and the second term is set to 0 when $x' \notin$ $[\omega(L_x), \omega(U_x)]$ and x'_* is therefore undefined. Similar equations can be derived for multivariate x and y, and for cases where $x' = \omega(x, y)$ (S.P. Ellner, *unpublished*).

In several test cases, equation (10.4.1) and its extensions do what they ought to: numerical solution of the IPM matches up very closely to the corresponding IBM with a large population size. But every possible question about these IPMs is unanswered. What space of functions does it naturally operate on (i.e., if $n(z, t)$ is in the space, then $n(z, t+1)$ is also)? Does it preserve smoothness or can n develop "shock waves" like the solutions to some partial differential equations? Under what conditions is there a unique long-term population growth rate λ and a corresponding stable state distribution? What kind of perturbation theory is possible - e.g., can we find the sensitivity of λ to ω or to $\omega(x_0)$? What are the fundamental operator, mean lifetime, and variance of lifetime reproductive success? In other words: if you want an open question to work on, pick any section of this book and try to figure out whether it's true for (10.4.1), or what else is true instead.

10.5 Different kinds of data

Plants stand still and wait to be measured. So if plant 334 isn't standing next to its tag next year, it's dead. It's nearly that good with sheep on a small island, or radio-collared wolves. But if it's fish 334 or bird 334, you may never know if it's dead, gone, or hiding.

So it's no accident that the early IPM literature was almost exclusively about plants, followed by especially well-studied animal populations. We believe that one of the most important future directions for IPMs is further development of fitting methods that can cope with less than perfect data. Some important steps have been taken, but an enormous amount remains to be done.

10.5.1 Mark-recapture-recovery data

Mark-recapture methods use data from sequential observations of marked individuals, but they assume that a marked individual might not be either *recaptured* (if alive) or *recovered* (if dead) at a subsequent census. Methods exist for open populations, time-varying demographic parameters, and (most important for IPMs) individual or group covariates. In particular, there is now a large literature (e.g., Nichols et al. 1992; Williams et al. 2002)) and software programs (MARK[3] and M-SURGE[4] (Choquet et al. 2004)) for estimating survival in multi-state capture-recapture populations. "Multi-state" means that individual are classified by a discrete state variable with Markovian dynamics – in other words, a stage-classified matrix model. The theory has seen 20 years of development, and the software is mature, free, and admirably well documented.[5]

---

[3] www.phidot.org/software/mark.

[4] www.cefe.cnrs.fr/en/biostatistics-and-biology-of-populations/software.

[5] www.phidot.org/software/mark/docs/book/.

There is much less theory for time-varying continuous individual covariates; which we will call "size" in what follows, with our standard disclaimers. The challenge in mark-recapture data is the abundance of missing data, because at each census many individuals are neither remeasured nor recovered dead. One solution is to only use *complete cases*, data on individual fate at a census immediately after one when the individual was measured. The assumed possible fates are recapture and remeasure, recover dead (in some studies), or fail to observe.[6] Because the outcome probabilities only depend on the initial size, the likelihood for the complete cases can be computed, and maximized to estimate parameters.

However, if failures to recapture or recover are common, using just complete cases throws away a lot of information and reduces the precision of parameter estimates. An individual measured at times 1 and 3 (but not at time 2) still provides information about growth. And though you don't know its exact size at time 2 you might have a pretty good idea, and you know that it survived to time 3, which is informative about size-dependent survival.

Bonner et al. (2010) compared three approaches for fitting a logistic regression model of size-dependent survival with mark-recapture data, using the St. Kilda Soay data and simulated data based on that system:

- Using only complete cases.
- Simple imputation of unobserved sizes, such as linear interpolation between successive measurements.
- Bayesian imputation, i.e., a standard Bayesian approach to missing data. If a parametric model for growth dynamics is assumed, the joint posterior likelihood for survival and growth model parameters and for unobserved sizes can be sampled by MCMC using BUGS or JAGS.

The three methods produced similar parameter estimates on the real data, probably because missing data are relatively infrequent. Applied to simulated data, the results were what you would expect. If failures to recapture or recover are rare, imputation isn't worth the trouble because most cases are complete. If missing data are more common, Bayesian imputation was more precise than the complete cases approach, so long as the parametric growth model is correctly specified.

Langrock and King (2013) used methods for Hidden Markov Models to optimize an approximate joint likelihood for survival and growth based on discretizing size into many small intervals. Worthington et al. (2014) proposed a simpler two-stage multiple imputation method. A growth model is fitted to all available observations, and then missing data are imputed by sampling from their distribution in the fitted growth model, conditional on the observed sizes. Each imputation creates a complete data set on size-dependent survival that can be fitted by standard methods, and parameter estimates are then averaged over multiple imputations.

---

[6] In some studies, such as the Soays, individuals may be seen and known to be alive, but not recaptured and remeasured.

The methods that use all of the data are computationally intensive, but entirely feasible now using standard tools. The principal limit is that they depend on having a properly specified growth model. Recapture probability therefore has to be high enough that many individuals are measured in two successive years, or it will be hard to tell if the growth model is good. At present the methods have to be coded "by hand," and we don't know which ones are most effective. But it is only a matter of time until the winners are identified and incorporated into mark-recapture software.

Multiple imputation is the simplest and fastest of these methods, so we use it to give a practical example. Box 10.3 describes a computational method for multiple imputation that can be used with any time-invariant growth model, in fact with any time-invariant Markov chain. The script UngulateImputation.R implements the method for our ungulate case study, the deterministic IPM with purely size-dependent demography.

The first step in this illustration is generating the artificial data that we will analyze, by iterating the individual-based model to obtain size trajectories for a cohort of newborn individuals (function sim_growth). Then we randomly replace some of the size measurements by NA (representing an individual not observed at a census), and extract all blocks of data where size needs to be imputed (function prune_data). A typical block of the artificial data then looks like this:

```
> imputeData[[15]]
id a         z     zorig zimpt
21 1 2.911837 2.911837    NA
21 2       NA 2.976656    NA
21 3       NA 3.076020    NA
21 4       NA 3.263010    NA
21 5 3.201483 3.201483    NA
```

Here id is animal ID number, and a is age. The zorig values are true (simulated) sizes of sheep 15; the z values represent here being measured at ages 1 and 5 but not in between. Using the method in Box 10.3 (function imput_missing) the script imputes sizes zimpt for ages 2, 3, and 4. This produces

```
id  a        z     zorig zimpt
21  1 2.911837 2.911837    NA
21  2       NA 2.976656 3.1125
21  3       NA 3.076020 3.2075
21  4       NA 3.263010 3.1375
21  5 3.201483 3.201483    NA
```

The zimpt values are a plausible fill-in of the unobserved growth trajectory. Replacing all NA's in z with the corresponding zimpt values gives a complete data set on each individual, which can be used to estimate the parameters of a survival model using our usual regression approaches. The theory of Multiple Imputation tells us that by doing this process over and over, and then averaging the resulting parameter estimates, we obtain survival model parameters that make full use of the data.

Consider a mark-recapture study conducted over years 1 to S. The size data on an individual will be a sequence of length S consisting of measured sizes and missing values:

$$(NA, NA, m, m, NA, m, m, NA, NA, m, m, m, NA)$$

where m is a measured size and NA a missing value. NAs at the start of the data sequence are study years prior to the individual's first capture, and NAs at the end are study years after the last capture. The goal is to replace the NAs with imputed values by sampling from their conditional distribution given the data.

Filling in final NAs is easy: iterate the growth model with assumed or fitted parameters, starting from the last observation. For example, if the fitted model is $z' \sim Norm(\mu = a + bz, \sigma = (c + dz))$, then starting from final observed value zk at time k we impute

```
zk1 <- rnorm(1,a+b*zk,c+d*zk)
zk2 <- rnorm(1,a+b*zk1,c+d*zk1)
```

and so on out to time S. Initial NAs are harder: we must assume a distribution of size at first census (Worthington et al. 2014). Often this will be unknown, so we will leave initial NAs as missing data.

For the rest, we need a way to impute the leftmost NA in a group between two measured sizes (then we do the next one in the group the same way, using the imputed value for the first as if it were data, and continue until the whole group is imputed). Because growth is Markovian, the distribution of this size depends on the immediately previous size (observed, or already imputed) and on the next observed size, but not on the rest of the data. So given sizes b at time i and B at time $i + T, T > 1$, we need to sample from the conditional distribution at time $i + 1$. Let $G^k(z_k, z)$ denote the conditional density of size z_k after k time steps. The conditional density at time $i + 1$ is then

$$p_1(z|b, B) = \frac{G^{T-1}(B, z)G(z, b)}{G^T(B, b)}. \tag{10.5.1}$$

Exact sampling from this distribution is difficult, but it is easy to sample from a midpoint-rule type discrete approximation. Let $\{z_i\}$ be a set of m mesh points for midpoint rule on the IPMs size range; $\mathbf{G}$ the matrix with entries $\mathbf{G}_{ij} = G(z_j, z_i)$; and $\kappa(z)$ the index value k of the mesh point z_k closest to z. The approximating discrete distribution is

$$\tilde{p}_1(z_i|b, B) = \frac{\mathbf{G}^{T-1}_{\kappa(B),i} \mathbf{G}_{i,\kappa(b)}}{\mathbf{G}^T_{\kappa(B),\kappa(b)}}, i = 1, 2, \cdots, m. \tag{10.5.2}$$

The NA groups will be fairly short unless recaptures are rare, so it will be fairly quick to calculate all the necessary $\mathbf{G}, \mathbf{G}^2, \cdots$ matrices once and for all, for a large m. The numerator values in (10.5.2) are the elementwise product of a row in $\mathbf{G}^{T-1}$ with a column in $\mathbf{G}$, so the vector of probabilities p1 can be calculated by one line of code. Then meshpts[sample(1:m,1,prob=p1)] is an imputed value drawn from the distribution.

Box 10.3: Imputing unobserved sizes in a mark-recapture data set.

The multiple imputation method assumes that the growth model has already been fitted. The discrete approximation in Box 10.3 can also be used to fit any parametric growth model, prior to doing multiple imputation. Using the notation in Box 10.3, the $\mathbf{G}$ matrix is a function of the growth model parameter vector, θ. For a sequence of one individual's time-indexed observed sizes $(z(t_1), z(t_2), \cdots, z(t_n))$, and corresponding mesh point indices $\kappa_i = \kappa(t_i)$, the approximate log-likelihood of the observed size transitions is

$$\sum_{i=1}^{n-1} \log \mathbf{G}_{\kappa_{i+1},\kappa_i}^{(t_{i+1}-t_i)}(\theta) \qquad (10.5.3)$$

The sum of (10.5.3) across all individuals is the overall approximate log-likelihood, which can be maximized to estimate θ. Typically only low powers of $\mathbf{G}$ are needed, so it's feasible to use thousands of mesh points and make the discretization error small enough to ignore. Because the likelihood can be computed despite the missing observations, nothing fancier than maximum likelihood is needed to fit a growth model.

10.5.2 Count data

Going one step further from the ideal, suppose individuals are not even tagged, and the data at each census are just the sizes $z_1, z_2, \cdots$ of a random sample from the population. None of the methods we've presented so far can be used. But fitting is still possible if you are willing to specify parametric demographic models, for example specifying survival as logistic regression on log body mass, and so on. That gives you a kernel $K(z', z, \theta)$ where θ is a parameter vector (slopes and intercepts of regression equations, etc.); for the moment, we assume that the IPM is deterministic.

The idea is simple. A value of θ, and an assumed initial state distribution $n(z, 0)$, together imply all subsequent population states $n(z, t)$. The data at census t are then a sample from the size distributions $p(z, t) = n(z, t)/\int_{\mathbf{Z}} n(y, t)dy$. Parameter values are then chosen to maximize the likelihood of the data (assuming the data at time t are sampled from $p(z, t)$), or to minimize the sum of squared errors between the predicted distributions and histograms of the data.

In the biological modeling literature this is often called *trajectory matching*. Trajectory matching is statistically justified when deviations between the data and model predictions are entirely due to sampling error. There is a large literature on trajectory matching for partial differential equation models of structured populations; a key reference is Banks et al. (1991), and Banks et al. (2011) is a recent application. In some applications the parameter vector θ includes spline coefficients for nonlinear rate equations, giving the models some of the flexibility discussed in Section 10.1.

Ghosh et al. (2012) were the first to suggest and try trajectory matching for IPMs, using a Bayesian approach that allowed the state distribution to vary randomly around the projections of a deterministic IPM. This was computationally feasible only for very simple models, so they also considered a simpler "pseudo-IPM approximation." This calculates $p(z, t+1)$ from the data at time t as if those data were the whole population rather than a sample from it. Similar fitting methods have been used on other kinds of dynamic models (e.g., Dennis et al. 1995; Turchin et al. 2003). To justify these "one-step ahead" methods statistically, we have to pretend that sampling error is zero and model predictions are imperfect only because the dynamics are stochastic.

The pseudo-IPM approximation goes like this (to keep things simple, we omit the random variation around the predicted state distribution). Let $\{z_1$

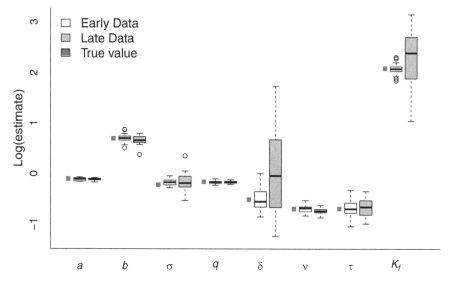

Fig. 10.9 Parameter estimates from 8 years of count data (random sample of size 250) from individual-based simulations of model (10.5.7). Red squares show the true value; white boxplots are the estimates from Early data (years 0 to 7), grey boxplots are the estimates from Late data (years 25–32), 20 replicates of each). Source files: `CountIPMfit.R` and `PlotCountFitEstimates.R`

$(t-1), z_2(t-1), \cdots, z_k(t-1)\}$ be the population sample at time $t-1$. The sample data at time t are assumed to be drawn from the distribution

$$\hat{p}(z', t) = \frac{1}{k} \sum_{j=1}^{k} K(z', z_j(t-1), \theta). \tag{10.5.4}$$

A further approximation by Ghosh et al. (2012) is that the counts of individuals in size ranges $A_1 = [L, L+h), A_2 = [L+h, L+2h), \cdots, A_m = [L+(m-1)h, U]$ at time t are Poisson and independent with mean

$$\lambda_k = \int_{L+(k-1)h}^{L+kh} \hat{p}(z, t)dz. \tag{10.5.5}$$

This is the same as our Poisson approximation for demographic stochasticity, except that bins are independent because the data are a random sample from the population.

It's relatively easy in practice to estimate parameters by maximizing the Poisson one-step-ahead likelihood. In Figure 10.9 we show results from one-step-ahead fitting of a simple IPM based on those used in Ghosh et al. (2012),

$$\text{Survival } s(z) = q \text{ (constant)}$$
$$\text{Growth } z' \sim Norm(b + az, \sigma^2)$$
$$\text{Fecundity } b(z) = \delta z^4/(K_f^4 + z^4) \qquad (10.5.6)$$
$$\text{Offspring } \log z' \sim Norm(\nu, \tau^2)$$

The "data" were generated by individual-based simulations starting from 1000 small individuals at $t = 0$ ($z \sim Norm$ with mean=3, sd=0.8). Parameters were estimated for samples of size 250 in 8 successive years, either years 0 to 7 ("Early" data) or 25 to 32 ("Late" data).

The results are very good for all parameters using the Early data: low bias, and good precision. With the Late data there is very low precision for two parameters, δ and K_f. It's easy to understand why. Once the size distribution stops changing, there's no information about differences in fecundity between small and large individuals. We can still infer average per-capita fecundity, but not which individuals are producing the new recruits. So this is not a flaw in the Ghosh et al. (2012) approach - the problem is that we can't fit a line to just one point, regardless of how clever we are about it. What's important and surprising about Ghosh et al. (2012) is that an IPM can sometimes be fitted successfully, with good estimates of individual-level parameters, without any individual-level data whatsoever.

10.6 Coda

Much of this chapter and indeed much of this book (starting with its title) are about blurring the line between statistical and dynamic models for populations. In the IPM framework, your statistical models for individual demography become the dynamic model for the population. With mark-recapture or count data, the population-level IPM provides the likelihood function for statistical estimation of individual-level demographic parameters, and then for projection or decision-making. The new directions in this chapter are opportunities to blur things further: allowing more accurate statistical models, demographic models based on within-individual physiological and metabolic processes, a wider range of possible assumptions about the dynamics, finite populations, and less ideal data. As models include more and more information about individuals, and computers can handle more complex models and calculations, new analyses will be needed to extract the population, community, and evolutionary implications of what we learn about individual plants and animals, and about the factors that affect their paths through life.

10.7 Appendix: covariance calculation for a size-structured model with post-reproductive census and Poisson distribution of offspring numbers

We consider the following size-structured model with post-reproductive census. Individual i at time t with size z_i survives to time $t+1$ with probability $s(z_i)$. If it survives, (i) its size z' at time $t+1$ is chosen from the distribution $G(z', z_i)$; (i) it produces $Poisson(b(z_i))$ offspring, whose sizes are chosen independently from size distribution $c(z')$. Each individual's fate is independent of the fate of other individuals. The kernel for the corresponding IPM is $K(z', z) = s(z)G(z', z) + s(z)b(z)c(z')$.

We keep the notation from Box 10.1 and additionally define:

- functions $c(S) = \int_S c(z')dz', m(z, S) = b(z)c(S), g(z, S) = \int_S G(z', z)dz'$.
- Bernoulli random σ_i for survival of individual i, $Pr(\sigma_i = 1) = s(z_i)$.
- Bernoulli random α_i for $z_i' \in A$ conditional on survival of i, $Pr(\alpha_i = 1) = g(z_i, A)$.
- random $A_i \sim Poisson(m(z, A))$ for offspring of individual i in A, conditional on survival of i.
- similarly β_i, B_i for subsequent size and offspring in B, conditional on survival of i.

Let n_A, n_B denote the total numbers of individuals in A and B at time $t + 1$. We then have

$$n_A(t + 1) = \sum_i \sigma_i \alpha_i + \sum_i \sigma_i A_i \equiv X_1 + X_2 \qquad (10.7.1)$$

$$n_B(t + 1) = \sum_j \sigma_j \beta_j + \sum_j \sigma_j B_j \equiv Y_1 + Y_2 \qquad (10.7.2)$$

so $Cov[n_A(t + 1), n_B(t + 1)] = Cov[X_1, Y_1] + Cov[X_1, Y_2] + Cov[X_2, Y_1] + Cov(Y_1, Y_2)$.

$Cov[X_1, Y_1]$ is given by equation (10.3.4). We need the following result: if σ is Bernoulli with $s = Pr(\sigma = 1)$ and σ, X, Y are independent random variables with X and Y having finite means $\bar{X}, \bar{Y}$ and finite variances, then

$$Cov[\sigma X, \sigma Y] = s(1 - s)\bar{X}\bar{Y}. \qquad (10.7.3)$$

(Proof: $Cov[\sigma X, \sigma Y] = E[\sigma^2 XY] - E[\sigma X]E[\sigma Y] = s\bar{X}\bar{Y} - s^2\bar{X}\bar{Y}$). Using (10.7.3) and the fact that different individuals have independent fates,

$$Cov[X_1, Y_2] = \sum_{i,j} Cov[\sigma_i \alpha_i, \sigma_j B_j] = \sum_i Cov[\sigma_i \alpha_i, \sigma_i B_i]$$

$$= \sum_i s(z_i)(1 - s(z_i))E[\alpha_i]E[B_i] \qquad (10.7.4)$$

$$= \sum_i s(z_i)(1 - s(z_i))g(z_i, A)m(z_i, B).$$

By symmetry, $Cov[X_2, Y_1] = \sum_i s(z_i)(1 - s(z_i))g(z_i, B)m(z_i, A)$. Calculations nearly identical to (10.7.5) give

$$Cov[X_2, Y_2] = \sum_i s(z_i)(1 - s(z_i))m(z_i, A)m(z_i, B). \qquad (10.7.5)$$

Note that $Cov[X_1, Y_1] < 0$ but the other covariances are positive, reflecting the fact that only surviving individuals have offspring. Combining all terms and doing a bit of algebra, we get

$$Cov[n_A(t+1), n_B(t+1)] =$$
$$\sum_i s_i^{-1}[(1 - s_i)K(A, z_i)K(B, z_i) - \gamma(z_i, A)\gamma(z_i, B)] \quad (10.7.6)$$

The sum in (10.7.6) runs over the individuals at time t. But as in Box 10.1, it can be approximated alá midpoint rule by pretending that all individuals in size interval k have the midpoint size z_k.

References

Abrams P, Matsuda H, Harada Y (1993) Evolutionarily unstable fitness maxima and stable fitness minima of continuous traits. Evol Ecol 7(5):465–487

Adler PB, Dalgleish HJ, Ellner SP (2012) Forecasting plant community impacts of climate variability and change: when do competitive interactions matter? J Ecol 100(2):478–487

Adler PB, Ellner SP, Levine JM (2010) Coexistence of perennial plants: an embarrassment of niches. Ecol Lett 13(8):1019–1029

Allen EJ, Allen LJS, Arciniega A, Greenwood PE (2008) Construction of equivalent stochastic differential equation models. Stoch Anal Appl 26:274–297

Andrew ME, Ustin SL (2010) The effects of temporally variable dispersal and landscape structure on invasive species spread. Ecol Appl 20(3):593–608

Banks HT, Botsford LW, Kappel F, Wang C (1991) Estimation of growth and survival in size-structured cohort data: an application to larval striped bass (*Morone saxatilis*). J Math Biol 30:125–150

Banks HT, Sutton KL, Thompson WC, Bocharov G, Doumic M, Schenkel T, Argilaguet J, Giest S, Peligero C, Meyerhans A (2011) A new model for the estimation of cell proliferation dynamics using CFSE data. J Immunol Methods 373(1–2):143–160

Barfield M, Holt RD, Gomulkiewicz R (2011) Evolution in stage-structured populations. Am Nat 177(4):397–409

Beder JH, Gomulkiewicz R (1998) Computing the selection gradient and evolutionary response of an infinite-dimensional trait. J Math Biol 36:299–319

Bell G, Collins S (2008) Adaptation, extinction and global change. Evol Appl 1(1):3–16

Benton TG, Solan M, Travis JMJ, Sait SM (2007) Microcosm experiments can inform global ecological problems. Trends Ecol Evol 22(10):516–521

Bonner SJ, Morgan BJT, King R (2010) Continuous covariates in mark-recapture-recovery analysis: a comparison of methods. Biometrics 66:1256–1265

Box GEP, Cox DR (1964) An analysis of transformations. J R Stat Soc Ser B Methodol 26(2):211–252

Brunet E, Derrida B, Mueller A, Munier S (2006) Phenomenological theory giving the full statistics of the position of fluctuating pulled fronts. Phys Rev E 73:056126

Bruno JF, Ellner SP, Vu I, Kim K, Harvell CD (2011) Impacts of aspergillosis on sea fan coral demography: modeling a moving target. Ecol Monogr 81(1):123–139

Bullock J, Clarke R (2000) Long distance seed dispersal by wind: measuring and modelling the tail of the curve. Oecologia 124(4):506–521

Bullock JM, Galsworthy SJ, Manzano P, Poschlod P, Eichberg C, Walker K, Wichmann MC (2011) Process-based functions for seed retention on animals: a test of improved descriptions of dispersal using multiple data sets. Oikos 120(8):1201–1208

© Springer International Publishing Switzerland 2016   315
S.P. Ellner et al., *Data-driven Modelling of Structured Populations*,
Lecture Notes on Mathematical Modelling in the Life Sciences,
DOI 10.1007/978-3-319-28893-2

Cagnacci F, Boitani L, Powell RA, Boyce MS (2010) Theme issue "Challenges and opportunities of using GPS-based location data in animal ecology". Philos Trans R Soc B Biol Sci 365(1550):2155

Cam E, Gimenez O, Alpizar-Jara R, Aubry LM, Authier M, Cooch EG, Koons DN, Link WA, Monnat J-Y, Nichols JD, Rotella JJ, Royle JA, Pradel R (2013) Looking for a needle in a haystack: inference about individual fitness components in a heterogeneous population. Oikos 122(5):739–753

Cam E, Link WA, Cooch EG, Monnat JY, Danchin E (2002) Individual covariation in life-history traits: seeing the trees despite the forest. Am Nat 159(1):96–105

Carlo TA, Garcia D, Martinez D, Gleditsch JM, Morales JM (2013) Where do seeds go when they go far? Distance and directionality of avian seed dispersal in heterogeneous landscapes. Ecology 94(2):301–307

Carlo TA, Tewksbury JJ, del Rio CM (2009) A new method to track seed dispersal and recruitment using N-15 isotope enrichment. Ecology 90(12):3516–3525

Caswell H (2000) Prospective and retrospective perturbation analyses: their roles in conservation biology. Ecology 81:619–627

Caswell H (2001) Matrix population models. construction, analysis and interpretation, 2nd edn. Sinauer Associates, Sunderland

Caswell H (2007) Sensitivity analysis of transient population dynamics. Ecol Lett 10(1):1–15

Caswell H (2008) Perturbation analysis of nonlinear matrix population models. Demogr Res 18:59–116

Caswell H (2009) Stage, age and individual stochasticity in demography. Oikos 118:1763–1782

Caswell H (2011a) Beyond R_0: demographic models for variability of lifetime reproductive output. PLoS ONE 6:e20809

Caswell H (2011b) Perturbation analysis of continuous-time absorbing Markov chains. Numer Linear Algebra Appl 18(6, SI):901–917

Caswell H (2012) Matrix models and sensitivity analysis of populations classified by age and stage: a vec-permutation matrix approach. Theor Ecol 5(3):403–417

Caswell H, Takada T, Hunter CM (2004) Sensitivity analysis of equilibrium in density-dependent matrix population models. Ecol Lett 7:380–387

Chambert T, Rotella JJ, Garrott RA (2014) An evolutionary perspective on reproductive individual heterogeneity in a marine vertebrate. J Anim Ecol 83(5):1158–1168

Chambert T, Rotella JJ, Higgs MD, Garrott RA (2013) Individual heterogeneity in reproductive rates and cost of reproduction in a long-lived vertebrate. Ecol Evol 3(7):2047–2060

Charlesworth B (1994) Evolution in age-structured populations. Cambridge University Press, Cambridge

Chesson P (1982) The stabilizing effect of a random environment. J Math Biol 15:1–36

Chesson P (2000) Mechanisms of maintenance of species diversity. Annu Rev Ecol Syst 31:343–366

Chevin L-M (2015) Evolution of adult size depends on genetic variance in growth trajectories: a comment on analyses of evolutionary dynamics using integral projection models. Meth Ecol Evol 6(9):981–986

Childs DZ, Coulson TN, Pemberton JM, Clutton-Brock TH, Rees M (2011) Predicting trait values and measuring selection in complex life histories: reproductive allocation decisions in Soay sheep. Ecol Lett 14(10):985–992

Childs DZ, Rees M, Rose KE, Grubb PJ, Ellner SP (2003) Evolution of complex flowering strategies: an age- and size-structured integral projection model. Proc R Soc Lond Ser B Biol Sci 270(1526):1829–1838

Childs DZ, Rees M, Rose KE, Grubb PJ, Ellner SP (2004) Evolution of size-dependent flowering in a variable environment: construction and analysis of a stochastic integral projection model. Proc R Soc Lond Ser B Biol Sci 271(1537):425–434

Childs DZ, Sheldon BC, Rees M (2016) The evolution of labile traits in sex- and age-structured populations. J Anim Ecol 85:329–342

Choquet R, Reboulet A-M, Pradel R, Gimenez O, Lebreton J-D (2004) M-SURGE: new software specifically designed for multistate capture-recapture models. Anim Biodivers Conserv 27:207–215

Christiansen F (1991) On conditions for evolutionary stability for a continuous character. Am Nat 138(1):37–50

Chu C, Adler PB (2015) Large niche differences emerge at the recruitment stage to stabilize grassland coexistence. Ecol Monogr 85:373–392

Clark J, Fastie C, Hurtt G, Jackson S, Johnson C, King G, Lewis M, Lynch J, Pacala S, Prentice C, Schupp E, Webb T, Wyckoff P (1998) Reid's paradox of rapid plant migration - dispersal theory and interpretation of paleoecological records. Bioscience 48(1):13–24

Clark J, LaDeau S, Ibanez I (2004) Fecundity of trees and the colonization-competition hypothesis. Ecol Monogr 74(3):415–442

Clark J, Silman M, Kern R, Macklin E, HilleRisLambers J (1999) Seed dispersal near and far: patterns across temperate and tropical forests. Ecology 80(5):1475–1494

Clark JS (1998) Why trees migrate so fast: confronting theory with dispersal biology and the paleorecord. Am Nat 152(2):204–224

Clobert J, Baguette M, Benton T, Bullock J (eds) (2012) Dispersal ecology and evolution. Oxford University Press, Oxford

Clutton-Brock T, Pemberton J (2004) Soay sheep. Dynamics and selection in an Island population. Cambridge University Press, Cambridge

Cochran ME, Ellner S (1992) Simple methods for calculating age-specific life history parameters from stage-structured models. Ecol Monogr 62:345–364

Collins S (2013) New model systems for experimental evolution. Evolution 67:1847–1848

Coulson T, Catchpole EA, Albon SD, Morgan BJT, Pemberton JM, Clutton-Brock TH, Crawley MJ, Grenfell BT (2001) Age, sex, density, winter weather, and population crashes in Soay sheep. Science 292(5521):1528–1531

Coulson T, MacNulty D, Stahler D, vonHoldt B, Wayne R, Smith D (2011) Modeling effects of environmental change on wolf population dynamics, trait evolution, and life history. Science 334:1275–1278

Coulson T, Tuljapurkar S, Childs DZ (2010) Using evolutionary demography to link life history theory, quantitative genetics and population ecology. J Anim Ecol 79(6):1226–1240

Cousens RD, Hill J, French K, Bishop ID (2010) Towards better prediction of seed dispersal by animals. Funct Ecol 24(6):1163–1170

Dahlgren JP, Ehrlén J (2011) Incorporating environmental change over succession in an integral projection model of population dynamics of a forest herb. Oikos 120(8):1183–1190

Dahlgren JP, Garcia MB, Ehrlén J (2011) Nonlinear relationships between vital rates and state variables in demographic models. Ecology 92(5):1181–1187

Dalgleish HJ, Koons DN, Hooten MB, Moffet CA, Adler PB (2011) Climate influences the demography of three dominant sagebrush steppe plants. Ecology 92(1):75–85

Datta MS, Korolev KS, Cvijovic I, Dudley C, Gore J (2013) Range expansion promotes cooperation in an experimental microbial metapopulation. Proc Natl Acad Sci USA 110:7354–7359

Davison A, Hinkley D (1997) Bootstrap methods and their application. Cambridge University Press, Cambridge

Dawson AE (2013) Models for forest growth and mortality: linking demography to competition and climate. Ph.D. Thesis, University of Alberta, Department of Mathematical and Statistical Sciences. http://hdl.handle.net/10402/era.31581

Day S, Junge O, Mishaikow K (2004) A rigorous numerical method for the global analysis of infinite dimensional discrete dynamical systems. SIAM J Appl Dyn Syst 3:117–160

Day S, Kalies W (2013) Rigorous computation of the global dynamics of integrodifference equations with smooth nonlinearities. SIAM J Numer Anal 51:2957–2983

de Roos AM (2008) Demographic analysis of continuous-time life-history models. Ecol Lett 11:1–15

de Roos AM, Persson L (2013) Population and community ecology of ontogenetic development. Monographs in population biology, vol. 51. Princeton University Press, Princeton

de Valpine P (2009) Stochastic development in biologically structured population models. Ecology 90(10):2889–2901

Dennis B, Desharnais R, Cushing J, Costantino R (1995) Nonlinear demographic dynamics: mathematical models, statistical methods, and biological experiments. Ecol Monogr 65(3):261–281

Dennis B, Desharnais R, Cushing J, Henson S, Costantino R (2001) Estimating chaos and complex dynamics in an insect population. Ecol Monogr 71(2):277–303

Dercole F, Rinaldi S (2008) Analysis of evolutionary processes: the adaptive dynamics approach and its applications. Princeton University Press, Princeton

Dieckmann U, Heino M, Parvinen K (2006) The adaptive dynamics of function-valued traits. J Theor Biol 241:370–389

Dieckmann U, Law R (1996) The dynamical theory of coevolution: a derivation from stochastic ecological processes. J Math Biol 34(5–6):579–612

Dieckmann U, Law R, Metz JAJ (eds) (2000) The geometry of ecological interactions: simplifying spatial complexity. Cambridge studies in adaptive dynamics, vol. 1. Cambridge University Press, Cambridge

Dunford N, Schwartz JT (1958) Linear operators part i: general theory, Wiley classics library edition. Wiley, New York

Eager E, Rebarber R, Tenhumberg B (2014a) Global asymptotic stability of plant-seed bank models. J Math Biol 69:1–37

Eager E, Rebarber R, Tenhumberg B (2014b) Modeling and analysis of a density-dependent stochastic integral projection model for a disturbance specialist plant and its seed bank. Bull Math Biol 76:1809–1834

Eager EA, Haridas CV, Pilson D, Rebarber R, Tenhumberg B (2013) Disturbance frequency and vertical distribution of seeds affect long-term population dynamics: a mechanistic seed bank model. Am Nat 182(2):180–190

Easterling MR (1998) The integral projection model: theory, analysis and application. Ph.D. Thesis, North Carolina State University

Easterling MR, Ellner SP, Dixon PM (2000) Size-specific sensitivity: applying a new structured population model. Ecology 81(3):694–708

Efron B, Tibshirani RJ (1993) An introduction to the bootstrap. Chapman and Hall/CRC, Boca Raton

Ehrlén J (2000) The dynamics of plant populations: does the history of individuals matter? Ecology 81(6):1675–1684

Ellner S, Hairston NG Jr, (1994) Role of overlapping generations in maintaining genetic variation in a fluctuating environment. Am Nat 143(3):403–417

Ellner SP (2013) Rapid evolution: from genes to communities, and back again? Funct Ecol 27:1087–1099

Ellner SP, Easterling MR (2006) Appendix c: stable population theory for integral projection models. Am Nat 167:410–428. www.jstor.org/stable/full/10.1086/499438

Ellner SP, Guckenheimer J (2006) Dynamic models in biology. Princeton University Press, Princeton

Ellner SP, Rees M (2006) Integral projection models for species with complex demography. Am Nat 167(3):410–428

Ellner SP, Rees M (2007) Stochastic stable population growth in integral projection models. J Math Biol 54:227–256

Ellner SP, Schreiber S (2012) Temporally variable dispersal and demography can accelerate the spread of invading species. Theor Popul Biol 82:283–298

Engen S, Lande R, Sæther B-E (2013) A quantitative genetic model of r- and K-selection in a fluctuating population. Am Nat 181(6):725–736

Engen S, Lande R, Sæther B-E, Festa-Bianchet M (2007) Using reproductive value to estimate key parameters in density-independent age-structured populations. J Theor Biol 244:308–317

Erhardt V (2009) corcounts: generate correlated count random variables. R package version 1.4.

Eshel I (1983) Evolutionary and continuous stability. J Theor Biol 103(1):99–111

Eshel I, Motro U (1981) Kin selection and strong evolutionary stability of mutual help. Theor Popul Biol 19(3):420–433

Ewens WJ (1979) Mathematical population genetics. Biomathematics, vol. 9. Springer, New York

Ewens WJ (2004) Mathematical population genetics 1: Theoretical introduction. Biomathematics, vol. 9. Springer, New York

Falconer DS, Mackay TF (1996) Introduction to quantitative genetics, 4th edn. Pearson Education Limited, Harlow

Feichtinger G (1971) Stochastische modelle demographischer prozesse. Lecture notes in operations research and mathematical systems, vol. 44. Springer, Berlin

Feichtinger G (1973) Markovian models for some demographic processes. Stat Hefte 14:310–334

Fieberg J, Ellner S (2001) Stochastic matrix models for conservation and management: a comparative review of methods. Ecol Lett 4:244–266

Fox J, Weisberg HS (2011) An R companion to applied regression. Sage Publications, Los Angeles

Furstenburg H, Kesten H (1960) Products of random matrices. Ann Math Stat 31:457–469

Geritz SAH, Kisdi E, Meszena G, Metz JAJ (1998) Evolutionarily singular strategies and the adaptive growth and branching of the evolutionary tree. Evol Ecol 12(1):35–57

Geritz SAH, Metz JAJ, Kisdi E, Meszena G (1997) Dynamics of adaptation and evolutionary branching. Phys Rev Lett 78(10):2024–2027

Geritz SAH, van der Meijden E, Metz JAJ (1999) Evolutionary dynamics of seed size and seedling competitive ability. Theor Popul Biol 55(3):324–343

Ghosh S, Gelfand AE, Clark JS (2012) Inference for size demography from point pattern data using integral projection models. J Agric Biol Environ Stat 17(4):641–677

Goodman LA (1969) Analysis of population growth when birth and death rates depend upon several factors. Biometrics 25(4):659–681

Gratten J, Pilkington JG, Brown EA, Clutton-Brock TH, Pemberton JM, Slate J (2012) Selection and microevolution of coat pattern are cryptic in a wild population of sheep. Mol Ecol 21(12):2977–2990

Gremer JR, Venable DL (2014) Bet hedging in desert winter annual plants: optimal germination strategies in a variable environment. Ecol Lett 17(3):380–387

Grubb P (1986) Problems posed by sparse and patchily distributed species in species-rich communities. In: Diamond J, Case T (eds) Community ecology. Harper and Row, New York, pp 207–225

Hadfield JD (2010) MCMC methods for multi-response generalized linear mixed models: the MCMCglmm R package. J Stat Softw 33(2):1–22

Hairston NJ, Ellner S, Kearns C (1993) Overlapping generations: the storage effect and the maintenance of biotic diversity. In: Rhodes OE, Chesser RK, Smith MH (eds) Population dynamics in ecological space and time. University of Chicago Press, Chicago, pp 109–145

Hanski I, Saccheri I (2006) Molecular-level variation affects population growth in a butterfly metapopulation. PLoS Biol 4(5):719–726

Hardin DP, Takáč P, Webb GF (1988) Asymptotic properties of a continuous-space discrete-time population model in a random environment. J Math Biol 26:361–374

Haridas C, Tuljapurkar S (2005) Elasticities in variable environments: properties and implications. Am Nat 166:481–495

Harris TE (1963) The theory of branching processes. Die Grundlehren der Mathematischen Wissenschaften, vol. 119. Springer, Berlin

Haymes KL, Fox GA (2012) Variation among individuals in cone production in Pinus palustris (Pinaceae). Am J Bot 99(4):640–645

Hegland SJ, Jongejans E, Rydgren K (2010) Investigating the interaction between ungulate grazing and resource effects on Vaccinium myrtillus populations with integral projection models. Oecologia 163(3):695–706

Heiss F, Winschel V (2008) Likelihood approximation by numerical integration on sparse grids. J Econ 144(1):62–80

Hesse E, Rees M, Müeller-Schäerer H (2008) Life-history variation in contrasting habitats: flowering decisions in a clonal perennial herb (Veratrum album). Am Nat 172(5):E196–E213

Hirsch M, Smith H (2005) Monotone dynamical systems. In: Canada A, Drabek P, Fonda A (eds) Handbook of differential equations, ordinary differential equations, vol. 2. Elsevier, New York, pp 239–357

Hodges JL, Le Cam L (1960) The poisson approximation to the poisson binomial distribution. Ann Math Stat 31(3):737–740

Horvitz C, Schemske DW, Caswell H (1997) The relative "importance" of life-history stages to population growth: prospective and retrospective analyses. In: Tuljapurkar S, Caswell H (eds) Structured population models in marine, terrestrial, and freshwater systems. Chapman and Hall, New York, pp 247–271

Hunter C, Caswell H, Runge M, Regehr E, Amstrup S, Stirling I (2010) Climate change threatens polar bear populations: a stochastic demographic analysis. Ecology 91:2883–2897

Hutchinson RA, Viers JH, Quinn JF (2007) Perennial pepperweed inventory at the cosumnes river preserve. A technical report to the california bay-delta authority ecosystem restoration program. University of California, Davis

Isaacson E, Keller HB (1966) Analysis of numerical methods. Wiley, New York

Iwasa Y, Pomiankowski A, Nee S (1991) The evolution of costly mate preferences. II. The "handicap" principle. Evolution 45(6):1431–1442

Jacquemyn H, Brys R, Jongejans E (2010) Size-dependent flowering and costs of reproduction affect population dynamics in a tuberous perennial woodland orchid. J Ecol 98(5):1204–1215

Johnston SE, Gratten J, Berenos C, Pilkington JG, Clutton-Brock TH, Pemberton JM, Slate J (2013) Life history trade-offs at a single locus maintain sexually selected genetic variation. Nature 502(7469):93–95

Jones FA, Muller-Landau HC (2008) Measuring long-distance seed dispersal in complex natural environments: an evaluation and integration of classical and genetic methods. J Ecol 96(4):642–652

Jongejans E, Shea K, Skarpaas O, Kelly D, Ellner SP (2011) Importance of individual and environmental variation for invasive species spread: a spatial integral projection model. Ecology 92(1):86–97

Jörgens K (1982) Linear integral operators (translated by G. Roach). Pitman, Boston

Kachi N (1983) Population dynamics and life history strategy of Oenothera erythrosepala in a sand dune system. Ph.D Thesis, University of Tokyo

Kachi N, Hirose T (1983) Bolting induction in Oenothera erythrosepala borbas in relation to rosette size, vernalization, and photoperiod. Oecologia 60(1):6–9

Kachi N, Hirose T (1985) Population dynamics of Oenothera glazioviana in a sanddune system with special reference to the adaptive significance of size-dependent reproduction. J Ecol 73:887–901

Kalisz S, McPeek M (1992) Demography of an age-structured annual: resampled projection matrices, elasticity analyses, and seed bank effects. Ecology 73(3):1082–1093

Karlin S, Taylor HM (1981) A second course in stochastic processes. Academic, San Diego

Katul G, Porporato A, Nathan R, Siqueira M, Soons M, Poggi D, Horn H, Levin S (2005) Mechanistic analytical models for long-distance seed dispersal by wind. Am Nat 166:366–381

Kemeny JG, Snell JL (1960) Finite Markov chains. D. Van Nostrand Company, New York

King R, Brooks S, Morgan B, Coulson T (2006) Factors influencing Soay sheep survival: a Bayesian analysis. Biometrics 62(1):211–220

Kingsolver JG, Gomulkiewicz R, Carter PA (2001) Variation, selection and evolution of function-valued traits. Genetica 112:87–104

Kirkpatrick M, Heckman N (1989) A quantitative genetic model for growth, shape, reaction norms, and other infinite-dimensional characters. J Math Biol 27(4):429–450

Kooijman S (2010) Dynamic energy budget theory for metabolic organisation, 3rd edn. Cambridge University Press, Cambridge

Kot M (2003) Do invading organisms do the wave? Can Appl Math Q 10:139–170

Kot M, Lewis MA, van den Driessche P (1996) Dispersal data and the spread of invading organisms. Ecology 77:2027–2042

Kot M, Schaffer W (1986) Discrete time growth-dispersal models. Math Biosci 80(1):109–136

Krasnosel'skij M, Lifshits JA, Sobolev A (1989) Positive linear systems: the method of positive operators. Heldermann, Berlin

Kuss P, Rees M, Ægisdóttir HH, Ellner SP, Stöcklin J (2008) Evolutionary demography of long-lived monocarpic perennials: a time-lagged integral projection model. J Ecol 96(4):821–832

Lande R (1982) A quantitative genetic theory of life history evolution. Ecology 63(3):607–615

Lande R (2007) Expected relative fitness and the adaptive topography of fluctuating selection. Evolution 61(8):1835–1846

Lande R, Engen S, Sæther B-E (2003) Stochastic population dynamics in ecology and conservation. Oxford series in ecology and evolution. Oxford University Press, Oxford

Lande R, Engen S, Sæther B-E (2009) An evolutionary maximum principle for density-dependent population dynamics in a fluctuating environment. Philos Trans R Soc B Biol Sci 364(1523):1511–1518

Lange K, Holmes W (1981) Stochastic stable population growth. J Appl Probab 18:325–344

Langrock R, King R (2013) Maximum likelihood analysis of mark-recapture-recovery models in the presence of continuous covariates. Ann Appl Stat 7:1709–1732

Lehoucq RB, Sorensen DC, Yang C (1998) ARPACK users' guide: solution of large-scale eigenvalue problems with implicitly restarted Arnoldi methods. Society for Industrial and Applied Mathematics, Philadelphia

Leininger SP, Foin TC (2009) *Lepidium latifolium* reproductive potential and seed dispersal along salinity and moisture gradients. Biol Invasions 11(10):2351–2365

Levin S, Muller-Landau H (2000) The evolution of dispersal and seed size in plant communities. Evol Ecol Res 2:409–435

Levin SA (1974) Dispersion and population interactions. Am Nat 108:207–228

Li C-K, Schneider H (2002) Applications of Perron-Frobenius theory to population dynamics. J Math Biol 44:450–462

Liaw A, Wiener M (2002) Classification and regression by randomForest. R News 2(3):18–22

Lui R (1982) A nonlinear integral operator arising from a model in population genetics. I. Monotone initial data. SIAM J Math Anal 13:913–937

Lutscher F, Nisbet RM, Pachepsky E (2010) Population persistence in the face of advection. Theor Ecol 3:271–284

Luxemburg WAJ, Zaanen A (1963) Compactness of integral operators in Banach function spaces. Math Ann 149:150–180

Lynch M, Walsh B (1998) Genetics and analysis of quantitative traits. Sinauer Associates, Sunderland

McCauley E, Nisbet RM, de Roos AM, Murdoch WW, Gurney WSC (1996) Structured population models of herbivorous zooplankton. Ecol Monogr 66(4):479–501

McClintock BT, King R, Thomas L, Matthiopoulos J, McConnell BJ, Morales JM (2012) A general discrete-time modeling framework for animal movement using multistate random walks. Ecol Monogr 82(3):335–349

Merow C, Dahlgren JP, Metcalf CJE, Childs DZ, Evans ME, Jongejans E, Record S, Rees M, Salguero-Gómez R, McMahon SM (2014) Advancing population ecology with integral projection models: a practical guide. Meth Ecol Evol 5(2):99–110

Metcalf CJE, Ellner SP, Childs DZ, Salguero-Gómez R, Merow C, McMahon SM, Jongejans E, Rees M (2015) Statistical modelling of annual variation for inference on stochastic population dynamics using Integral Projection Models. Meth Ecol Evol 6:1007–1017

Metcalf CJE, Horvitz CC, Tuljapurkar S, Clark DA (2009a) A time to grow and a time to die: a new way to analyze the dynamics of size, light, age, and death of tropical trees. Ecology 90:2766–2778

Metcalf CJE, McMahon SM, Salguero-Gómez R, Jongejans E (2013) IPMpack: an R package for integral projection models. Meth Ecol Evol 4(2):195–200

Metcalf CJE, Rees M, Buckley YM, Sheppard AW (2009b) Seed predators and the evolutionarily stable flowering strategy in the invasive plant, *Carduus nutans.* Evol Ecol 23(6):893–906

Metcalf CJE, Rose KE, Childs DZ, Sheppard AW, Grubb PJ, Rees M (2008) Evolution of flowering decisions in a stochastic, density-dependent environment. Proc Natl Acad Sci USA 105(30):10466–10470

Metcalf CJE, Rose KE, Rees M (2003) Evolutionary demography of monocarpic perennials. Trends Ecol Evol 18(9):471–480

Metcalf CJE, Stephens DA, Rees M, Louda SM, Keeler KH (2009c) Using Bayesian inference to understand the allocation of resources between sexual and asexual reproduction. J R Stat Soc Ser C Appl Stat 58:143–170

Metz JAJ, Diekmann O (1986) The dynamics of physiologically structured populations. Springer, Berlin/New York

Metz JAJ, Mollison D, van den Bosch F (2000) The dynamics of invasion waves. In: Dieckmann U, Law R, Metz JAJ (eds) The geometry of ecological interactions: simplifying spatial complexity, chap. 23. Cambridge University Press, Cambridge, pp 482–512

Metz JAJ, Mylius SD, Diekmann O (2008) When does evolution optimize? Evol Ecol Res10(5):629–654

Metz JAJ, Nisbet RM, Geritz SAH (1992) How should we define "fitness" for general ecological scenarios? Trends Ecol Evol 7(6):198–202

Miller TEX, Louda SM, Rose KA, Eckberg JO (2009) Impacts of insect herbivory on cactus population dynamics: experimental demography across an environmental gradient. Ecol Monogr 79(1):155–172

Miller TEX, Williams JL, Jongejans E, Brys R, Jacquemyn H (2012) Evolutionary demography of iteroparous plants: incorporating non-lethal costs of reproduction into integral projection models. Proc R Soc Lond Ser B Biol Sci 279(1739):2831–2840

Mollison D (1972) The spatial propagation of simple epidemics. In: Proceedings of the 6th Berkeley symposium on mathematical statistics and probability, vol. 3, pp 579–614

Morales J, Haydon D, Frair J, Holsinger K, Fryxell J (2004) Extracting more out of relocation data: building movement models as mixtures of random walks. Ecology 85(9):2436–2445

Morales JM, Carlo TA (2006) The effects of plant distribution and frugivore density on the scale and shape of dispersal kernels. Ecology 87(6):1489–1496

Mueller S, Scealy JL, Welsh AH (2013) Model selection in linear mixed models. Stat Sci 28(2):135–167

Mylius SD, Diekmann O (1995) On evolutionarily stable life histories, optimization and the need to be specific about density dependence. Oikos 74(2):218–224

Neubert MG, Caswell H (2000) Demography and dispersal: calculation and sensitivity analysis of invasion speed for structured populations. Ecology 81:1613–1628

Neubert MG, Kot M, Lewis MA (1995) Dispersal and pattern formation in a discrete-time predator-prey models. Theor Popul Biol 48:7–43

Nichols J, Sauer J, Pollock KH, Hestbeck JB (1992) Estimating transition probabilities for stage-based population projection matrices using capture-recapture data. Ecology 73:306–312

Nicole F, Dahlgren JP, Vivat A, Till-Bottraud I, Ehrlen J (2011) Interdependent effects of habitat quality and climate on population growth of an endangered plant. J Ecol 99(5):1211–1218

Nisbet RM, Gurney WSC, Murdoch WW, McCauley E (1989) Structured population models - a tool for linking effects at individual and population level. Biol J Linn Soc 37(1–2):79–99

Nisbet RM, Muller EB, Lika K, Kooijman SALM (2000) From molecules to ecosystems through dynamic energy budget models. J Anim Ecol 69:913–926

Ovaskainen O, Meerson B (2010) Stochastic models of population extinction. Trends Ecol Evol 25(11):643–652

Ozgul A, Childs DZ, Oli MK, Armitage KB, Blumstein DT, Olson LE, Tuljapurkar S, Coulson T (2010) Coupled dynamics of body mass and population growth in response to environmental change. Nature 466(7305):482–485

Ozgul A, Coulson T, Reynolds A, Cameron TC, Benton TG (2012) Population responses to perturbations: the importance of trait-based analysis illustrated through a microcosm experiment. Am Nat179(5):582–594

Pacala SW, Canham CD, Saponara J, Silander JA, Kobe RK, Ribbens E (1996) Forest models defined by field measurements: estimation, error analysis and dynamics. Ecol Monogr 66(1):1–43

Parvinen K, Dieckmann U, Heino M (2006) Function-valued adaptive dynamics and the calculus of variations. J Math Biol 52(1):1–26

Parvinen K, Heino M, Dieckmann U (2013) Function-valued adaptive dynamics and optimal control theory. J Math Biol 67(3):509–533

Pfister CA (1998) Patterns of variance in stage-structured populations: evolutionary predictions and ecological implications. Proc Natl Acad Sci USA 95(1):213–218

Pfister CA, Stevens FR (2002) The genesis of size variability in plants and animals. Ecology 83(1):59–72

Pfister CA, Stevens FR (2003) Individual variation and environmental stochasticity: implications for matrix model predictions. Ecology 84(2):496–510

Pinheiro JC, Bates DM (2000) Mixed–effects models in S and S-PLUS. Springer, New York

Plummer M (2003) JAGS: a program for analysis of Bayesian graphical models using Gibbs sampling. In: Hornik K, Leisch F, Zeileis A (eds) Proceedings of the 3rd international workshop on distributed statistical computing (DSC), Vienna, 20–22 March 2003

Pons J, Pausas J (2007) Acorn dispersal estimated by radio-tracking. Oecologia 153(4):903–911

Price G (1970) Selection and covariance. Nature 227:520–521

Ramsay J, Hooker G, Graves S (2009) Functional data analysis with R and Matlab. Springer, New York

Ramsay JO, Silverman BW (2005) Functional data analysis. Springer series in statistics, 2nd edn. Springer, New York

Rebarber R, Tenhumberg B, Townley S (2012) Global asymptotic stability of density dependent integral population projection models. Theor Popul Biol 81:81–87

Rees M, Childs DZ, Metcalf CJE, Rose KE, Sheppard AW, Grubb PJ (2006) Seed dormancy and delayed flowering in monocarpic plants: selective interactions in a stochastic environments. Am Nat 168(2):E53–E71

Rees M, Childs DZ, Rose KE, Grubb PJ (2004) Evolution of size-dependent flowering in a variable environment: partitioning the effects of fluctuating selection. Proc R Soc Ser B Biol Sci 271(1538):471–475

Rees M, Ellner SP (2009) Integral projection models for populations in temporally varying environments. Ecol Monogr 79:575–594

Rees M, Ellner SP (2016) Evolving integral projection models: evolutionary demography meets eco-evolutionary dynamics. Methods in Ecology and Evolution 7:157–170

Rees M, Osborne CP, Woodward FI, Hulme SP, Turnbull LA, Taylor SH (2010) Partitioning the components of relative growth rate: how important is plant size variation? Am Nat 176(6):E152–E161

Rees M, Rose KE (2002) Evolution of flowering strategies in *Oenothera glazioviana*: an integral projection model approach. Proc R Soc Ser B Biol Sci 269(1499):1509–1515

Rees M, Sheppard AW, Briese D, Mangel M (1999) Evolution of size-dependent flowering in *Onopordum illyricum*: a quantitative assessment of the role of stochastic selection pressures. Am Nat 154(6):628–651

Richards SA, Whittingham MJ, Stephens PA (2011) Model selection and model averaging in behavioural ecology: the utility of the IT-AIC framework. Behav Ecol Sociobiol 65:77–89

Robertson A (1966) A mathematical model of the culling process in dairy cattle. Anim Prod 8:95–108

Roff DA (2002) Life history evolution. Sinauer Associates, Sunderland

Rose KE, Louda SM, Rees M (2005) Demographic and evolutionary impacts of native and invasive insect herbivores on *Cirsium canescens*. Ecology 86(2):453–465

Rose KE, Rees M, Grubb PJ (2002) Evolution in the real world: stochastic variation and the determinants of fitness in *Carlina vulgaris*. Evolution 56(7):1416–1430

Rueffler C, Metz JAJ (2013) Necessary and sufficient conditions for R_0 to be a sum of contributions of fertility loops. J Math Biol 66:1099–1022

Salguero-Gomez R, Siewert W, Casper BB, Tielboerger K (2012) A demographic approach to study effects of climate change in desert plants. Philos Trans R Soc B Biol Sci 367(1606, SI):3100–3114

Santure AW, de Cauwer I, Robinson MR, Poissant J, Sheldon BC, Slate J (2013) Genomic dissection of variation in clutch size and egg mass in a wild great tit (*Parus major*) population. Mol Ecol 22(15):3949–3962

Sauer J, Slade N (1987) Size-based demography of vertebrates. Ann Rev Ecol Syst 18:71–90

Schreiber SJ, Ross N (2016) Individual-based integral projection models: the role of size-structure on extinction risk and establishment success. Methods in Ecology and Evolution (in press)

Simmonds EG, Coulson T (2015) Analysis of phenotypic change in relation to climatic drivers in a population of Soay sheep *Ovis aries*. Oikos 124:543–552

Skarpaas O, Shea K (2007) Dispersal patterns, dispersal mechanisms, and invasion wave speeds for invasive thistles. Am Nat 170(3):421–430

Skellam JG (1951) Random dispersal in theoretical populations. Biometrika 38 (1-2):196–218

Slatkin M(1973) Gene flow and selection in a cline. Genetics 75(4):733–756

Smallegange IM, Coulson T (2013) Towards a general, population-level understanding of eco-evolutionary change. Trends Ecol Evol 28(3):143–148

Smith HL, Thieme HR (2013) Persistence and global stability for a class of discrete time structured population models. Discrete Contin Dyn Syst A 33:4627–4646

Snyder RE (2003) How demographic stochasticity can slow biological invasions. Ecology 84:1333–1339

Stearns SC (1992) The evolution of life histories. Oxford University Press, London

Steiner UK, Tuljapurkar S (2012) Neutral theory for life histories and individual variability in fitness components. Proc Natl Acad Sci USA 109(12):4684–4689

Steiner UK, Tuljapurkar S, Coulson T, Horvitz C (2012) Trading stages: life expectancies in structured populations. Exp Gerontol 47(10):773–781

Stevens MHH (2009) A primer of ecology with R. Use R! Springer, New York

Stinchcombe JR, Function-valued Traits Working Group, and Kirkpatrick, M. (2012). Genetics and evolution of function-valued traits: understanding environmentally responsive phenotypes. Trends Ecol Evol 27:637–647

Takada T, Nakajima H (1992) An analysis of life history evolution in terms of the density-dependent Lefkovitch matrix model. Math Biosci 112(1):155–176

Takada T, Nakajima H (1998) Theorems on the invasion process in stage-structured populations with density-dependent dynamics. J Math Biol 36(5):497–514

R Core Team (2015) R: a language and environment for statistical computing. R Foundation for Statistical Computing, Vienna

Teller BJ, Adler PB, Edwards CB, Hooker G, Ellner SP (2016) Linking demography with drivers: climate and competition. Methods in Ecology and Evolution 7:171–183

Thieme HR (2009) Spectral bound and reproduction number for infinite-dimensional population structure and time heterogeneity. SIAM J Appl Math 70(1):188–211

Travis JMJ, Harris CM, Park KJ, Bullock JM (2011) Improving prediction and management of range expansions by combining analytical and individual-based modelling approaches. Meth Ecol Evol 2(5):477–488

Tuljapurkar S (1990) Population dynamics in variable environments. Springer, New York

Tuljapurkar S, Haridas CV (2006) Temporal autocorrelation and stochastic population growth. Ecol Lett 9:327–337

Tuljapurkar S, Horvitz C (2006) From stage to age in variable environments: life expectancy and survivorship. Ecology 87(6):1497–1509

Tuljapurkar S, Horvitz C, Pascarella J (2003) The many growth rates and elasticities of populations in random environments. Am Nat 162(4):489–502

Tuljapurkar S, Horvitz C, Pascarella J (2004) Correction. Am Nat 164(6):821–823

Tuljapurkar S, Steiner UK, Orzack SH (2009) Dynamic heterogeneity in life histories. Ecol Lett 12(1):93–106

Turchin P (1998) Quantitative analysis of movement: measuring and modeling population redistribution in plants and animals. Sinauer Associates, Sunderland

Turchin P, Wood SN, Ellner SP, Kendall BE, Murdoch WW, Fischlin A, Casas J, McCauley E, Briggs CJ (2003) Dynamical effects of plant quality and parasitism on population cycles of larch budmoth. Ecology 84(5):1207–1214

van Noordwijk A, de Jong G (1986) Acquisition and allocation of resources: their influence on variation in life-history tactics. Am Nat 128(1):137–142

Viers JH, Hutchinson RA, Quinn JF (2008) Modeling perennial pepperweed populations at the cosumnes river preserve. A Technical Report to the California Bay-Delta Authority Ecosystem Restoration Program. University of California, Davis

Vindenes Y, Engen S, Sæther B-E (2011) Integral projection models for finite populations in a stochastic environment. Ecology 92(5):1146–1156

Vindenes Y, Langangen Ø (2015) Individual heterogeneity in life histories and eco-evolutionary dynamics. Ecol Lett 18(5):417–432

Vindenes Y, Sæther B-E, Engen S (2012) Effects of demographic structure on key properties of stochastic density-independent population dynamics. Theor Popul Biol 82(4):253–263

Weinberger HF (1982) Long-time behavior of a class of biological models. SIAM J Math Anal 13:353–396

Weiner J, Martinez S, MüllerSchärer H, Stoll P, Schmid B (1997) How important are environmental maternal effects in plants? A study with *Centaurea maculosa*. J Ecol 85(2):133–142

Wesselingh RA, Klinkhamer PGL, de Jong TJ, Boorman LA (1997) Threshold size for flowering in different habitats: effects of size-dependent growth and survival. Ecology 78(7):2118–2132

Williams B, Nichols JD, Conroy MJ (2002) Analysis and management of animal populations. Academic, San Diego

Williams JL (2009) Flowering life-history strategies differ between the native and introduced ranges of a monocarpic perennial. Am Nat 174(5):660–672

Williams JL, Miller TEX, Ellner SP (2012) Avoiding unintentional eviction from integral projection models. Ecology 93:2008–2014

Wood SN (2011) Fast stable restricted maximum likelihood and marginal likelihood estimation of semiparametric generalized linear models. J R Stat Soc (B) 73(1):3–36

Worthington H, King R, Buckland ST (2014) Analysing mark-recapture-recovery data in the presence of missing covariate data via multiple imputation. J Agric Biol Environ Stat 20(1):28–46

Zabreyko P, Koshelev AI, Krasnosel'skii MA, Mikhlin SG, Rakovshchik LS, Stet'senko VY (1975) Integral equations - a reference text. Noordhoff International Publishing, Leyden

Zachmann L, Moffet C, Adler P (2010) Mapped quadrats in sagebrush steppe: long-term data for analyzing demographic rates and plant-plant interactions. Ecology 91:3427

Zadoks J-C (2000) Foci, small and large: a specific class of biological invasion. In: Dieckmann U, Law R, Metz JAJ (eds) The geometry of ecological interactions: simplifying spatial complexity, chap. 23. Cambridge University Press, Cambridge, pp 482–512

Zuidema PA, Jongejans E, Chien PD, During HJ, Schieving F (2010) Integral projection models for trees: a new parameterization method and a validation of model output. J Ecol 98(2):345–355

Index

© Springer International Publishing Switzerland 2016

327

S.P. Ellner et al., *Data-driven Modelling of Structured Populations*,
Lecture Notes on Mathematical Modelling in the Life Sciences,
DOI 10.1007/978-3-319-28893-2

Lightning Source UK Ltd.
Milton Keynes UK
UKOW06f1232210616

276765UK00002B/26/P